# LONDON MATHEMATICAL SOCIETY LECTURE NOTE SERIES

Managing Editor: Professor N.J. Hitchin,
*Mathematical Institute, University of Oxford, 24–29 St. Giles, Oxford OX1 3LB, UK*

All the titles listed below can be obtained from good booksellers or from Cambridge University Press. For a complete series listing visit
http://publishing.cambridge.org/stm/mathematics/lmsn/

London Mathematical Society Lecture Note Series 326

# Modular Representations of Finite Groups of Lie Type

## JAMES E. HUMPHREYS
*University of Massachusetts, Amherst*

# CAMBRIDGE
## UNIVERSITY PRESS

University Printing House, Cambridge CB2 8BS, United Kingdom

Cambridge University Press is part of the University of Cambridge.

It furthers the University's mission by disseminating knowledge in the pursuit of education, learning and research at the highest international levels of excellence.

www.cambridge.org
Information on this title: www.cambridge.org/9780521674546

© Cambridge University Press 2005

First published 2006

A catalogue record for this publication is available from the British Library

ISBN 978-0-521-67454-6 Paperback

# Contents

# Preface

"I confess I like to take account of possibilities. Don't you know mathematics are my hobby? Did you ever study algebra? I always have an eye on the unknown quantity."

Henry James, *The Story of a Year*

Ich predige die Mathematik ...... Die Beschäftigung mit der Mathematik, sage ich, ist das beste Mittel gegen die Kupidität. Staatsanwalt Paravant, der stark angefochten war, hat sich drauf geworfen, er hat es jetzt mit der Quadratur des Kreises und spürt große Erleichterung.

Thomas Mann, *Der Zauberberg*

– Ah! c'etait impossible, les cours duraient parfois fort tard.
– Même après 2 heures du matin? demandait le baron.
– Des fois.
– Mais l'algèbre s'apprend aussi facilement dans un livre.
– Même plus facilement, car je ne comprends pas grand'chose aux cours.
– Alors? D'ailleurs l'algèbre ne peut te servir à rien.
– J'aime bien cela. Ca dissipe ma neurasthénie.

Marcel Proust, *Sodome et Gomorrhe*

Whatever its therapeutic value may be, group theory offers plenty of diversions and challenges. In particular, the study of finite groups of Lie type and their representations (ordinary or modular) leads to deep questions, many of which are still unsolved.

A more accurate but cumbersome title for this book would be: "A guide to the modular representation theory of finite groups of Lie type in the defining characteristic $p$". As a subtheme, the relationship between ordinary and modular representations is explored, in the context of Deligne–Lusztig characters. However, I have stopped short of treating the "cross-characteristic" case, where representations are studied over fields of prime characteristic $\ell$ different from $p$ (see the recent monograph by Cabanes and Enguehard [**78**]). In that case the questions and methods are quite different, being less connected with algebraic group representations and more intertwined with Deligne–Lusztig theory.

The approach here emphasizes the remarkably close parallels with the representation theories of the ambient simple algebraic group and its Frobenius kernels. Roughly speaking, the key representations of our finite group over a field of $p^r$ elements closely resemble those of the $r$th Frobenius kernel (the case $r = 1$ corresponding to the Lie algebra of the algebraic group).

The treatment here is necessarily far from self-contained. It relies heavily on the theory exposed in Jantzen's book *Representations of Algebraic Groups*, whose second edition is referred to here as [**RAGS**]. However, the main prerequisites for study of the finite groups are concentrated in just the core chapters of Jantzen's Part II. (These remain largely unchanged from the first edition, though the coverage of Lusztig's Conjecture is significantly expanded in the new edition.)

One of my goals has been to make the subject more accessible to those working in neighboring parts of group theory, number theory, and topology. With this in mind, I have adopted some basic notational conventions at the outset in 1.8 and then tried to keep the chapters relatively independent. When it seems most useful I have given proofs in detail, but in the later chapters I have opted for informal exposition accompanied by examples and precise references.

The literature of this subject extends in many directions and often deals with special cases not yet possible to treat in a general framework. Even the large reference list I have assembled here is incomplete, though I have tried to include the primary sources and to attribute results correctly.

It is common in the theory of groups of Lie type to find "generic" patterns that are independent of $p$ when $p$ is sufficiently large. Partly for this reason, I have confined the treatment at first to Chevalley groups and their twisted analogues in types A, D, $E_6$. A concluding chapter summarizes what can be said about the groups of Suzuki and Ree, which are defined only in characteristic 2 or 3.

Here is a quick overview of the main topics discussed:

- Standard characterizations of the finite groups of Lie type
- Simple modules, from the classical results of Curtis and Steinberg to Lusztig's Conjecture
- Blocks, projective modules, and Cartan invariants
- Extensions of simple modules, or more generally Loewy series of projectives

- Cohomology of the finite groups relative to algebraic group cohomology, leading to the study of support varieties
- Decomposition behavior of ordinary characters (and Deligne–Lusztig characters) modulo $p$
- Special features of groups of type $G_2$, finite general linear groups, and the groups of Suzuki and Ree

Notation requires special attention, since there are parallel developments involving algebraic groups and finite groups. Various compromises are made in the literature, none entirely satisfactory. My own compromise is to use boldface letters such as $\mathbf{G}$ for algebraic groups (whereas on a chalkboard one might underline or use script letters) and roman letters such as $G$ for their finite counterparts. Typically $G = \mathbf{G}^F$, the fixed points under a Frobenius map $F$.

I am grateful to many colleagues who have provided valuable feedback on earlier drafts of the book. Among these must be singled out Jens Carsten Jantzen, whose contributions to the subject have been of fundamental importance.

# CHAPTER 1

# Finite Groups of Lie Type

We begin with a brief review of the standard ways in which finite groups of Lie type are classified, constructed, and described. One complication is the multiplicity of approaches to describing this family of groups, which leads in turn to differing notational conventions in the literature. Our viewpoint will be mainly that of algebraic groups over finite fields (1.1), reformulated in terms of Frobenius maps (1.3). But occasional use will be made of the convenient axiomatic approach afforded by $BN$-pairs: see 1.7 and Chapter 7 below.

Even within the framework of algebraic groups, there is more than one way to organize the finite groups. Steinberg's unified description of the groups as fixed points of endomorphisms of algebraic groups is undoubtedly the most elegant and useful. However, in our treatment of modular representations it will be convenient to keep the groups of Ree and Suzuki (defined only in characteristic 2 or 3) largely separate from the other groups: see Chapter 20. These groups arise less directly from the ambient algebraic groups and of course do not exhibit any "generic" behavior for large primes $p$ as other groups of Lie type do.

To conclude this introductory chapter we establish in 1.8 some standard notation.

## 1.1. Algebraic Groups over Finite Fields

The finite groups of Lie type are close relatives of the groups $\mathbf{G}(k)$ of rational points of algebraic groups defined over a finite field $k$. Here $\mathbf{G}$ is an affine variety with group operations given by regular functions, identified with its points over an algebraically closed field $K$. When the subfield $k$ is perfect, $\mathbf{G}$ is *defined over* $k$ precisely when it is the common set of zeros of a family of polynomials with coefficients in $k$. A standard example is the special linear group $\mathrm{SL}(n, K)$, which is defined over the prime field in $K$.

Here we recall some essential facts, referring to several textbooks for details: Borel [**55**, §16], Humphreys [**220**, §34–35], Springer [**385**]. Using a basic theorem of Lang (see 1.4 below), one shows without too much difficulty:

THEOREM. *Let $k$ be a finite field having $q = p^r$ elements and let $\mathbf{G}$ be any connected algebraic group defined over $k$. Then:*

(a) $\mathbf{G}$ *is* **quasisplit**, *meaning it has a Borel subgroup defined over $k$. Moreover, all such Borel subgroups are conjugate under $\mathbf{G}(k)$.*

(b) $\mathbf{G}$ *has a maximal torus $\mathbf{T}$ defined over $k$, which lies in a Borel subgroup $\mathbf{B}$ defined over $k$. Here $\mathbf{B} = \mathbf{T} \ltimes \mathbf{U}$, with $\mathbf{U}$ the unipotent radical of $\mathbf{B}$.*

(c) *Given an* **isogeny** $\varphi$ *(an epimorphism with finite kernel) from $\mathbf{G}$ onto another connected algebraic group $\mathbf{H}$ over $k$, the finite groups $\mathbf{G}(k)$ and*

$\mathbf{H}(k)$ *have the same order even though the map* $\mathbf{G}(k) \to \mathbf{H}(k)$ *induced by* $\varphi$ *need not be surjective.*

While the proof of part (c) involves some algebraic geometry, the coincidence of orders can be verified directly in special cases. Consider for example the isogeny $SL(n, K) \to PGL(n, K)$ obtained by restricting the natural map $GL(n, K) \to GL(n, K)/K^{\times} = PGL(n, K)$ (where $K^{\times}$ is identified with scalar matrices). Taking $G = GL(n, k)$, note that $\det : G \to k^{\times}$ is surjective and has kernel $SL(n, k)$, while the surjective map $G \to G/k^{\times} = PGL(n, k)$ has kernel $k^{\times}$. Thus $|SL(n, k)| = |PGL(n, k)|$.

Our main concern is with connected groups, though some nonconnected ones (such as normalizers of maximal tori or centralizers of arbitrary elements in reductive groups) also occur naturally. While any finite linear group over a finite field may be regarded as an algebraic group, the Borel–Tits structure theory essentially relies on connectedness assumptions.

We take as background the standard theory exposed in the texts mentioned above (and used heavily by Jantzen in [**RAGS**]). This focuses on a connected reductive group $\mathbf{G}$ such as $GL(n, K)$, whose derived group is connected and semisimple. In turn, such a semisimple group decomposes as the almost-direct product of simple algebraic groups (having no proper connected normal algebraic subgroups). Chevalley's classification of these simple groups shows that they fall into essentially the same families over $K$ as over $\mathbb{C}$.

Each simple algebraic group has a Lie type A–G (indexed by the **rank** $\ell = \dim \mathbf{T}$) and corresponding **root system** $\Phi$. But within each type there may be several distinct groups. There is always a **simply connected** group $\mathbf{G}$ and an **adjoint** group isomorphic to $\mathbf{G}$ modulo its finite center. There may also be intermediate groups: quotients of $\mathbf{G}$ by central subgroups. The simply connected group is equal to its derived group, while its center is naturally isomorphic to the quotient of the weight lattice $X$ by the root lattice $\mathbb{Z}\Phi$.

## 1.2. Classification Over Finite Fields

The classification of the groups $\mathbf{G}(k)$ when $\mathbf{G}$ is a simple algebraic group defined over a finite field $k$ begins with the fact (based on Lang's Theorem) that $\mathbf{G}$ is quasisiplit over $k$. If moreover $\mathbf{G}$ has a $k$-split maximal torus (isomorphic over $k$ to a product of copies of the multiplicative group), then $\mathbf{G}$ is called **split** over $k$, otherwise **nonsplit**. Split groups exist for all Lie types, but nonsplit groups only for types $A_{\ell}$ (with $\ell > 1$), $D_{\ell}$ (with $\ell \geq 4$), and $E_6$. This classification was worked out independently by Hertzig, Steinberg, and Tits, mainly in the framework of Galois cohomology.

While there are strikingly close parallels in the structure and representations of split and nonsplit groups of the same type over $\mathbb{F}_q$, the split groups are usually much easier to work with. For example, the important subgroups of $\mathbf{G}$ defined over $\mathbb{F}_q$ play a similar role in the structure of $\mathbf{G}(k)$: maximal tori, Borel subgroups, root groups, etc. But for a nonsplit group there are added complications in the description of tori and root groups. It is common in the literature to find separate treatments of split and nonsplit cases.

A further complication is that within most Lie types there are several distinct finite groups, as in the algebraic group classification: for example, $SL(n, q), PGL(n, q)$, and $PSL(n, q)$ are all of type $A_{n-1}$. When $\mathbf{G}$ is simply connected as in the case of $SL(n, K)$, one usually calls the finite group $\mathbf{G}(k)$ **universal**. (We remark that Schmid [**358**] proposes a broader concept of "universal" group for arbitrary root systems.)

To show that finite groups of all possible Lie types actually exist, it is efficient to follow Chevalley's uniform method (recalled in 1.5 below). For classical types A–D, one can instead make identifications with known classical matrix groups over $k$. A careful review of the concrete descriptions of classical groups over finite fields can be found in the Atlas [**117**, Chap. 2].

The reader should be cautioned that varying notational schemes are found in the literature. For example, the notation $SU(n, q)$ is used here for the subgroup of $GL(n, q^2)$ which is often denoted more literally as $SU(n, q^2)$. Our choice emphasizes the close parallel between $SL(n, q)$ and $SU(n, q)$, seen in their order formulas and representation theories. More functorial notation such as $SL_n(\mathbb{F}_q)$ or $SL_n(q)$ is also widely used.

In dealing with modular representations in the defining characteristic, it is convenient to work in the setting of simply connected simple algebraic groups. Here the main unsolved problems about characters and dimensions can be formulated efficiently in the language of weights. To make the transition to closely related groups such as quotients by central subgroups and their derived groups, one has to use standard representation-theoretic techniques unrelated to Lie theory. Usually this is routine for the problems we study here. But in some situations the transition can be delicate, as seen for example in the work of Lusztig and others on ordinary characters. (It is instructive to study the organization of character tables for families of related groups in the Atlas [**117**].)

Sometimes it is more natural to study arbitrary semisimple groups, or reductive groups such as $GL(n, K)$ which are not semisimple. Though we typically formulate results only for simple algebraic groups, the reader should be aware of a few complications. For example, the algebraic group $PGL(n, K) = GL(n, K)/K^\times$ is isomorphic to the simple adjoint group of type $A_{n-1}$, which over an algebraically closed field is the same as $PSL(n, K)$. However, over a finite subfield there is usually a difference between the group $PGL(n, k)$ and its subgroup $PSL(n, k)$, the latter typically being the derived group of the former and having index equal to the order of the group of $n$th roots of 1 in $k^\times$.

## 1.3. Frobenius Maps

The algebraic group approach does not directly yield the groups of Suzuki and Ree. To unify the description of all finite groups of Lie type, Steinberg [**395**] studied an arbitrary algebraic group endomorphism $\sigma : \mathbf{G} \to \mathbf{G}$ whose group of fixed points $\mathbf{G}_\sigma$ is finite. Here $\mathbf{G}$ is defined and split over $\mathbb{F}_q$. The most basic example is the **standard Frobenius map** relative to $q = p^r$: if $\mathbf{G}$ is given explicitly as a matrix group, this map just raises each matrix entry to the $q$th power. (The map can be described intrinsically in terms of the algebra of regular functions on $\mathbf{G}$.) The resulting finite group of fixed points coincides with the group of rational points $\mathbf{G}(\mathbb{F}_q)$.

More complicated endomorphisms are obtained by composing the standard Frobenius map relative to $q$ with a nontrivial graph automorphism $\pi$ arising from a symmetry of the Dynkin diagram of $\mathbf{G}$. (These maps commute.) The only simple groups with a nontrivial graph automorphism are those of types $A_\ell$ (with $\ell > 1$),

$D_\ell$ (for $\ell \geq 4$), and $E_6$. Type $D_4$ is unusual in having such automorphisms of both order 2 and order 3. The group of fixed points is isomorphic to the group of rational points over $\mathbb{F}_q$ of a quasisplit but nonsplit group of the same type as $\mathbf{G}$. For example, one gets $SU(3, q)$ rather than $SL(3, q)$ from the ambient algebraic group $SL(3, K)$. Note that the square or cube of the endomorphism is the standard Frobenius map relative to $q$.

In each of these cases the endomorphism $\mathbf{G} \to \mathbf{G}$ characterizes the $\mathbb{F}_q$-structure of $\mathbf{G}$. Following current usage, we denote the map by $F$ and call it a **Frobenius map** relative to $p^r$ (and $\pi$ if a nontrivial graph automorphism is involved). Write $G := \mathbf{G}^F$ for the finite group of fixed points of $F$.

For groups with root systems of types $B_2, F_4, G_2$, more exotic endomorphisms of $\mathbf{G}$ can be constructed, yielding the groups of Suzuki in type $B_2$ and Ree in types $F_4$ and $G_2$. (See 5.3 and Chapter 20 for further details.) One combines selective $q$th powers with graph symmetries interchanging long and short root subgroups. More precisely, let $q = p^{2r+1}$ be an odd power of $p = 2$ for types $B_2$ and $F_4$ (resp. $p = 3$ for type $G_2$), and assume $\mathbf{G}$ is defined over $\mathbb{F}_q$ (hence split). Then let $\alpha \mapsto \overline{\alpha}$ interchange short and long roots as a graph automorphism would do with lengths ignored. Map an element $x_\alpha(c)$ of a root subgroup $\mathbf{U}_\alpha$ to $x_{\overline{\alpha}}(c')$, where $c' := c^{p^r}$ if $\alpha$ is long and $c' := c^{p^{r+1}}$ if $\alpha$ is short. For the chosen $p$ this defines an endomorphism of $\mathbf{G}$ whose square is just the standard Frobenius map corresponding to $q$. (Some authors call this type of endomorphism a "Frobenius map", but we do not.)

The upshot of Steinberg's analysis is that these are the only possible endomorphisms of $\mathbf{G}$ having a finite fixed point subgroup. The language of Frobenius maps is used systematically (but with some variation in the definition) in most current work. See for example Cabanes–Enguehard [**78**], Carter [**86**, 14.1], Digne–Michel [**134**, Chap. 3], Geck [**184**], Gorenstein–Lyons–Solomon [**190**, 2.1].

## 1.4. Lang Maps

The basic theorem of Lang mentioned earlier can be reformulated for a connected algebraic group $\mathbf{G}$ defined over $\mathbb{F}_q$ in the more general form suggested by Steinberg [**395**, §10]. Starting with a Frobenius map $F : \mathbf{G} \to \mathbf{G}$ relative to $q$, the associated **Lang map** $L : \mathbf{G} \to \mathbf{G}$ is defined by $L(g) = F(g)g^{-1}$. Then the theorem states that $L$ is *surjective*. Here $F$ may be standard or may (in case $\mathbf{G}$ is reductive) involve also a graph automorphism.

In the literature there are several variants of the definition of $L$ using different left–right conventions, for example, $g \mapsto g^{-1}F(g)$ or $g \mapsto gF(g^{-1})$. This has no effect on the surjectivity or its applications.

The proof of Lang's Theorem uses the fact that the fibers of $L$ are the orbits of $G = \mathbf{G}^F$ acting by right translation, together with an easy calculation showing that the differential at each point is bijective. See for example Springer [**385**, 4.4.17]. The argument translates also into the language of quotients: The quotient of the affine variety $\mathbf{G}$ by the right translation action of $G$ is isomorphic to $\mathbf{G}$ itself, with $L$ as the quotient map. In particular, the $G$-invariant regular functions on $\mathbf{G}$ are of the form $L^*f = f \circ L$ for $f \in K[\mathbf{G}]$. (These statements apply equally well to the restricted Lang map on a closed connected subgroup $\mathbf{H}$ of $\mathbf{G}$.)

## 1.5. Chevalley Groups and Twisted Groups

To construct all possible finite groups of Lie type in a uniform way, one follows the lead of Chevalley's 1955 Tôhoku paper. He showed how to obtain the simple groups of split type by a process of reduction modulo $p$. This starts with the choice of a good $\mathbb{Z}$-basis for any simple Lie algebra over $\mathbb{C}$, whose adjoint operators raised to powers and divided by corresponding factorials still leave the basis invariant. The operators corresponding to root vectors act nilpotently, so the usual exponential power series is just a polynomial operator leaving the Chevalley basis invariant. It makes sense to reduce modulo a prime $p$ (no division by $p$ being required), leading to a matrix group over $\mathbb{F}_p$ generated by "exponentials". Extension of scalars gives the desired groups over arbitrary finite fields.

With a few very small exceptions, these groups are simple and are the derived groups of the various $\mathbf{G}(\mathbb{F}_q)$ obtained from split adjoint groups (for example $\mathrm{PSL}(n, q) \leq \mathrm{PGL}(n, q)$). The exceptions are the solvable groups $\mathrm{A}_1(2)$ and $\mathrm{A}_1(3)$; the group $\mathrm{B}_2(2)$ of order 720, which is isomorphic to the symmetric group $S_6$; and the group $\mathrm{G}_2(2)$ of order 12096, which has a simple subgroup of index 2 isomorphic to $^2\mathrm{A}_2(3)$.

In his 1967–68 Yale lectures [**394**], Steinberg replaced the adjoint representation by an arbitrary irreducible representation over $\mathbb{C}$ and generalized Chevalley's procedure to obtain groups of all isogeny types as well as the various kinds of twisted groups. (See also Carter's book [**86**].)

Unless other specified, we always work with a simply connected group $\mathbf{G}$, defined and split over $\mathbb{F}_p$. The finite group $G = \mathbf{G}^F$ of fixed points under the standard Frobenius map relative to $q$ is called a (universal) **Chevalley group**, while the term **twisted group** refers to the fixed points under a Frobenius map involving a nontrivial graph automorphism. But note that some authors include all of these groups under the rubric "Chevalley group". And it is common to regard the groups of Suzuki and Ree as types of twisted groups, but we find it more convenient to keep these separate.

Table 1 summarizes our labelling by Lie type, together with the order of the corresponding universal group. Chevalley groups are listed first, followed by twisted groups and the groups of Suzuki and Ree.

As in the case of classical groups, notational conventions for Lie types vary in the literature: see for example Gorenstein–Lyons–Solomon [**188**, pp. 8–9] and the Atlas of Finite Groups [**117**]. In the case of the twisted groups of types $\mathrm{A}, \mathrm{D}, \mathrm{E}_6$, some sources use $q^2$ rather than $q$ in the labels. And in the case of the groups of Suzuki and Ree, $q^2$ may be replaced by $q$ in the labels as well as the order formulas. This issue is discussed carefully in the Atlas [**117**, 3.2], where alternate notations (such as $\mathrm{Sz}(q)$ for $^2\mathrm{B}_2(q^2)$) are also described. We prefer notation which shows the analogy between orders of the twisted and nontwisted groups.

## 1.6. Example: SL(3, q) and SU(3, q)

The similarities and differences between split and nonsplit groups of the same type show up clearly in the simplest situation, groups of types $\mathrm{A}_2(q)$ and $^2\mathrm{A}_2(q)$. With our notational conventions, the orders of the two groups are given by similar polynomials of degree $8 = \dim \mathrm{SL}(3, K)$:

$$|\mathrm{SL}(3, q)| = q^3(q^2 - 1)(q^3 - 1) \text{ and } |\mathrm{SU}(3, q)| = q^3(q^2 + 1)(q^3 - 1).$$

| Lie type | Order of universal group |
|---|---|
| $A_\ell(q), \ell \geq 1$ | $q^{\binom{\ell+1}{2}} \prod_{i=2}^{\ell+1}(q^i - 1)$ |
| $B_\ell(q), \ell \geq 2$ | $q^{\ell^2} \prod_{i=1}^{\ell}(q^{2i} - 1)$ |
| $C_\ell(q), \ell \geq 2$ | $q^{\ell^2} \prod_{i=1}^{\ell}(q^{2i} - 1)$ |
| $D_\ell(q), \ell \geq 4$ | $q^{\ell(\ell-1)}(q^\ell - 1) \prod_{i=1}^{\ell-1}(q^{2i} - 1)$ |
| $E_6(q)$ | $q^{36}(q^2 - 1)(q^5 - 1)(q^6 - 1)(q^8 - 1)(q^9 - 1)(q^{12} - 1)$ |
| $E_7(q)$ | $q^{63}(q^2 - 1)(q^6 - 1)(q^8 - 1)(q^{10} - 1)(q^{12} - 1)(q^{14} - 1)(q^{18} - 1)$ |
| $E_8(q)$ | $q^{120}(q^2 - 1)(q^8 - 1)(q^{12} - 1)(q^{14} - 1)(q^{18} - 1)(q^{20} - 1)(q^{24} - 1)(q^{30} -$ |
| $F_4(q)$ | $q^{24}(q^2 - 1)(q^6 - 1)(q^8 - 1)(q^{12} - 1)$ |
| $G_2(q)$ | $q^6(q^2 - 1)(q^6 - 1)$ |
| $^2A_\ell(q), \ell \geq 2$ | $q^{\binom{\ell+1}{2}} \prod_{i=2}^{\ell+1}(q^i - (-1)^i)$ |
| $^2D_\ell(q), \ell \geq 4$ | $q^{\ell(\ell-1)}(q^\ell + 1) \prod_{i=1}^{\ell-1}(q^{2i} - 1)$ |
| $^3D_4(q)$ | $q^{12}(q^2 - 1)(q^6 - 1)(q^8 + q^4 + 1)$ |
| $^2E_6(q)$ | $q^{36}(q^2 - 1)(q^5 + 1)(q^6 - 1)(q^8 - 1)(q^9 + 1)(q^{12} - 1)$ |
| $^2B_2(q^2), q^2 = 2^{2r+1}$ | $q^4(q^2 - 1)(q^4 + 1)$ |
| $^2F_4(q^2), q^2 = 2^{2r+1}$ | $q^{24}(q^2 - 1)(q^6 + 1)(q^8 - 1)(q^{12} + 1)$ |
| $^2G_2(q^2), q^2 = 3^{2r+1}$ | $q^6(q^2 - 1)(q^6 + 1)$ |

TABLE 1. Universal groups of Lie type

Consider how the finite torus $\mathbf{T}^F$ looks in each case. In $SL(3, q)$ we just get a direct product of two copies of $\mathbb{F}_q^\times$, one for each simple root $\alpha, \beta$. But in $SU(3, q) \subset SL(3, q^2)$ the situation is quite different. Here the graph automorphism interchanges $\alpha$ and $\beta$, so a typical diagonal matrix fixed by $F$ has eigenvalues $c, c^q, 1/c^{q+1}$ as $c$ runs over the cyclic group $\mathbb{F}_{q^2}^\times$ of order $q^2 - 1$. Thus $\mathbf{T}^F$ is cyclic.

There is also a sharp contrast in the structure of the upper triangular unipotent subgroup $\mathbf{U}^F$ in the two cases. This subgroup of $SL(3, q)$ (a Sylow $p$-subgroup) has order $q^3$, with structure like that of $\mathbf{U}$ as a product of three root groups. In $\mathrm{SU}(3, q)$ the order of $\mathbf{U}^F$ is the same, but the group structure changes: a unipotent matrix

$$
\begin{pmatrix}
1 & a & c \\
0 & 1 & b \\
0 & 0 & 1
\end{pmatrix}
$$

with entries in $\mathbb{F}_{q^2}$ is fixed by $F$ precisely when $b = a^q$ and $c + c^q + a^{q+1} = 0$. Here there is a single positive "root group".

Differences like these propagate through all pairs of split and nonsplit groups based on the same root system. See Carter [**86**, Chap. 13] or Steinberg [**394**, §11] for full details about the structure of twisted groups (along with those of Suzuki and Ree).

## 1.7. Groups With a *BN*-Pair

In work leading to the classification of finite simple groups, it has often been useful to describe groups of Lie type in a uniform axiomatic setting independent of algebraic groups: see Gorenstein–Lyons–Solomon [**189**, §30]. For this purpose the most efficient formalism is that of *BN*-**pairs** (called **Tits systems** by Bourbaki in [**56**, Chap. IV]), as introduced by Tits in the aftermath of Chevalley's 1955 Tôhoku paper. This captures the essence of the Bruhat decomposition as it appears in various settings, without explicit introduction of the root system.

Given a group $G$ with subgroups $B$ and $N$, the data defining a *BN*-pair consists formally of a quadruple $(G, B, N, S)$ subject to the following requirements:

- $G$ is generated by its subgroups $B$ and $N$.
- $T := B \cap N$ is a normal subgroup of $N$.
- $W := N/T$ is generated by a set $S$ of involutions.
- For $s \in S$ and $w \in W$, $sBw \subseteq BwB \cup BswB$.
- For $s \in S$, $sBs \neq B$.

Here we adopt the usual convention of writing expressions like $sBw$ and $BwB$ when the choice of a representative in $N$ of an element of $W$ makes no difference. (Note that the letter $H$ is often used in the literature for the group we call $T$.)

The axioms lead quickly to the **Bruhat decomposition**: $G$ is the disjoint union (indexed by $W$) of the double cosets $BwB$. As a further consequence of the axioms, $W$ is seen to be a **Coxeter group** with distinguished set of generators $S$. The cardinality of $S$ is then called the **rank** of the *BN*-pair.

A minor adjustment can be made without significant loss of generality, to insure that the *BN*-pair is "saturated": $T = \bigcap_{w \in W} wBw^{-1}$. By the axioms, $T$ is always included in the right side. If the inclusion were proper, we could simply replace $T$ by the right side and enlarge $N$ accordingly. (We always assume saturation.)

With our previous notational conventions, the group $G = \mathbf{G}^F$ has a natural *BN*-pair structure: When $G$ is a Chevalley group, take $B = \mathbf{B}^F$ for a Borel subgroup $\mathbf{B}$ of $\mathbf{G}$ corresponding to $\Phi^+$ and $N = \mathbf{N}^F$ for $\mathbf{N} = N_{\mathbf{G}}(\mathbf{T})$, the normalizer of an $\mathbb{F}_q$-split maximal torus $\mathbf{T}$ lying in $\mathbf{B}$. But when $G$ is a twisted group, the *BN*-pair structure encodes the "relative" structure of a nonsplit algebraic group over $\mathbb{F}_q$: the Weyl group of the *BN*-pair is the subgroup of the usual Weyl group fixed by the induced action of $F$, while the rank of the *BN*-pair is the "relative" rank of the algebraic group measuring the dimension of a maximal $\mathbb{F}_q$-split torus. For example, one views $\mathrm{SU}(3, q)$ as having a *BN*-pair of rank 1.

Which finite groups—especially simple groups—have a $BN$-pair? A group with a $BN$-pair of rank 1 is the same thing as a doubly transitive permutation group: $B$ can be interpreted as the isotropy group of a point, while $N$ stabilizes a set consisting of this point and one other (and $T$ is the subgroup fixing both). Thus it is unrealistic to expect a complete list of finite groups with such a $BN$-pair. But surprisingly enough, Tits [420] was able to show (by geometric methods) that all finite simple groups with a $BN$-pair of rank $\geq 3$ are of Lie type.

A more serious restriction is needed in rank 2 to reflect the special internal structure of a group of Lie type: one needs a (saturated) **split $BN$-pair of characteristic** $p$. For this, add to the axioms the assumption that $B$ has a normal $p$-subgroup $U$ complementary to $T$, while $T$ is abelian and has order relatively prime to $p$. Using an impressive array of group-theoretic methods, it has been shown that all simple groups with a split $BN$-pair of characteristic $p$ are of Lie type (or degenerate versions thereof): see Hering–Kantor–Seitz [204], Kantor–Seitz [266], Fong–Seitz [175].

In order to bypass this rather sophisticated literature, one can impose the further condition that the $BN$-pair is **strongly split** as defined by Cabanes–Enguehard [78, §2] and Genet [185] (see 8.7 below). This yields Levi decompositions for parabolic subgroups as well as a version of the Chevalley commutation formulas. (See also Tinberg [418].)

There are numerous treatments in the literature of groups with a $BN$-pair, directed toward different applications. See for example Bourbaki [56, Chap. IV], Cabanes–Enguehard [78, Part I], Carter [86, 8.2–8.3] and [89, Chap. 2], Curtis [121, 123] and [122, §3], Curtis–Reiner [127, §65, §69], Gorenstein–Lyons–Solomon [189, §30], Humphreys [220, §29].

## 1.8. Notational Conventions

Unless otherwise specified, $\mathbf{G}$ will denote a simple, simply connected algebraic group defined and split over $\mathbb{F}_p$, identified with its group $\mathbf{G}(K)$ of rational points over an algebraically closed field $K$ of characteristic $p > 0$. Its root system is $\Phi$, with a simple system $\Delta$. Usually we fix a Borel subgroup $\mathbf{B}$ (corresponding to a set of positive roots $\Phi^+$) and a maximal torus $\mathbf{T} \subset \mathbf{B}$, both split over $\mathbb{F}_p$. The rank of $\mathbf{G}$ is $\ell = \dim \mathbf{T} = |\Delta|$.

Thus $\mathbf{G} \cong \mathrm{SL}(\ell+1, K), \mathrm{Spin}(2\ell+1, K), \mathrm{Sp}(2\ell, K), \mathrm{Spin}(2\ell, K)$, or one of the five exceptional groups of types $\mathrm{E}_6, \mathrm{E}_7, \mathrm{E}_8, \mathrm{F}_4, \mathrm{G}_2$.

If $F : \mathbf{G} \to \mathbf{G}$ is a Frobenius map relative to $q$ (and possibly a nontrivial graph automorphism $\pi$), its fixed point subgroup $G := \mathbf{G}^F$ is a universal Chevalley group or a twisted group of type $\mathrm{A}_\ell$ ($\ell > 1$), $\mathrm{D}_\ell$, or $\mathrm{E}_6$ over $\mathbb{F}_q$. (Suzuki and Ree groups are discussed separately in Chapter 20.)

Additional notation will be introduced as we go along. (See the list of frequently used notation at the end of the book.)

# CHAPTER 2

# Simple Modules

To study the representations over $K$ of an arbitrary finite group $G$, one usually concentrates first on those which are realized within the group algebra $KG$. The main examples are simple modules and projective modules.

When $G$ is a finite group of Lie type, there are two natural approaches to the study of simple $KG$-modules:

- Describe them intrinsically in the setting of groups with split $BN$-pairs.
- Describe them as restrictions of simple modules for the ambient algebraic group $\mathbf{G}$.

While the second approach is less direct, it has yielded (so far) much more detailed information than the first approach and will therefore be our main focus here. We defer until Chapter 7 the more self-contained development due to Curtis and Richen, based on $BN$-pairs.

Even though it is possible to classify the simple $KG$-modules in a coherent way from the algebraic group viewpoint, we still do not know in most cases their dimensions or (Brauer) characters. Modulo knowledge of the formal characters of simple $\mathbf{G}$-modules (still incomplete in most cases), which we call **standard character data** for $\mathbf{G}$, it is often possible to derive further results about the category of finite dimensional $KG$-modules: projectives, extensions, etc. This is usually the approach we follow, motivated by Lusztig's Conjecture for $\mathbf{G}$ (see 3.11 below).

After a detailed review of simple modules for the algebraic groups, following [**RAGS**], we turn to the finite groups. Besides the cited papers, we can point to useful surveys in the Atlas [**117**, Chap. 2] and Gorenstein–Lyons–Solomon [**190**, 2.8]. For unexplained notation see 1.8.

## 2.1. Representations and Formal Characters

For the study of finite groups of Lie type we normally find it most convenient to work with a simple algebraic group. But the treatment of representations in [**RAGS**] allows $\mathbf{G}$ to be semisimple—or even reductive, which is needed for inductive purposes when passing to Levi subgroups.

In this and the following section, $\mathbf{G}$ can be any connected semisimple group over $K$, with a fixed Borel subgroup $\mathbf{B} = \mathbf{TU}$ corresponding to a choice of simple system $\Delta$ and positive roots $\Phi^+$. For brevity we always denote by $X$ the character group $X(\mathbf{T})$ (with the convention that the group law is written additively) and by $X^+$ the subset of dominant weights relative to $\Phi^+$. Then $X$ is partially ordered by $\mu \leq \lambda$ iff $\lambda - \mu$ is a sum (possibly 0) of positive roots. We call this the **natural ordering** of $X$.

Denote by $W$ the **Weyl group** $N_{\mathbf{G}}(\mathbf{T})/\mathbf{T}$ of $\mathbf{G}$ relative to the maximal torus $\mathbf{T}$. So $W$ is generated by "simple" reflections (relative to roots in $\Delta$) and has a corresponding length function $\ell(w)$. There is a unique longest element $w_\circ$ in $W$, which sends $\Phi^+$ to $-\Phi^+$; its length is $m := |\Phi^+|$.

The dual root system is denoted by $\Phi^\vee$. When $\Phi$ is irreducible, $\Phi^\vee$ has the same type as $\Phi$ except that types $B_\ell$ and $C_\ell$ are dual to each other. The natural pairing $\langle x, \alpha^\vee \rangle$ is defined for all $x \in \mathbb{R} \otimes_{\mathbb{Z}} X$.

Representations of $\mathbf{G}$ are always assumed to be *rational* and *finite dimensional*, unless otherwise specified. Typically we use the equivalent language of $\mathbf{G}$-modules. It is a basic fact that Jordan decomposition in $\mathbf{G}$ is preserved under homomorphisms of algebraic groups. This implies that any $\mathbf{G}$-module $M$ is a direct sum of its **weight spaces** $M_\lambda$ $(\lambda \in X)$ relative to $\mathbf{T}$:

$$M_\lambda := \{x \in M \mid t \cdot x = \lambda(t)x \text{ for all } t \in \mathbf{T}\}.$$

In turn, $M$ has a **formal character** $\operatorname{ch} M$ in the group ring $\mathbb{Z}[X]$ of $X$. To view $X$ as a multiplicative group in this context, we introduce canonical basis elements $e(\lambda)$ of the free abelian group $\mathbb{Z}[X]$ indexed by $\lambda \in X$. These are multiplied by the rule $e(\lambda)e(\mu) = e(\lambda + \mu)$. Now the formal character is defined by

$$\operatorname{ch} M := \sum_{\lambda \in X} \dim M_\lambda \, e(\lambda).$$

Finite dimensionality of $M$ implies that $\dim M_\lambda = \dim M_{w\lambda}$ for all $w \in W$, so $\operatorname{ch} M$ lies in the subring of $W$-invariants $\mathcal{X} := \mathbb{Z}[X]^W$. We call this the **formal character ring** of $\mathbf{G}$ relative to $\mathbf{T}$. As a result, $\operatorname{ch} M$ can be rewritten as a $\mathbb{Z}^+$-linear combination of various orbit sums $s(\mu) := \sum_{w \in W^\mu} e(w\mu)$, with $\mu \in X^+$ and $W^\mu$ a set of coset representatives of $W$ modulo the isotropy group $W_\mu$ of $\mu$.

For the study of representations it is usually most convenient to assume that $\mathbf{G}$ is *simply connected*. This translates into the assumption that $X$ is the full weight lattice of the abstract root system $\Phi$. Other semisimple groups with the same root system are obtained by factoring out subgroups of the finite group $Z(\mathbf{G})$. Then it is not difficult to sort out which representations of $\mathbf{G}$ induce representations of the quotient group: it is just a question of which weights lie in the character group of the corresponding quotient of $\mathbf{T}$ (a sublattice of $X$).

## 2.2. Simple Modules for Algebraic Groups

As shown by Chevalley in the late 1950s, the highest weight classification of irreducible representations for a semisimple algebraic group over an algebraically closed field is essentially characteristic-free. (See [**RAGS**, II.2], [**220**, §31].)

THEOREM. *Let $\mathbf{G}$ be a semisimple algebraic group over $K$. Fix notation as in 2.1, and let $\mathbf{B}^- = \mathbf{T}\mathbf{U}^-$ be the Borel subgroup of $\mathbf{G}$ opposite to $\mathbf{B}$. Then:*

(a) *Every simple (rational) $\mathbf{G}$-module $M$ has a unique highest weight $\lambda \in X^+$ in the natural partial ordering of $X$. Whenever $M_\mu \neq 0$, we have $\mu \leq \lambda$.*

(b) *The weight space $M_\lambda$ is one-dimensional, spanned by a $\mathbf{U}$-invariant vector $v^+$, called a **maximal vector** or **highest weight vector**.*

(c) *$M$ is spanned by the vectors $u \cdot v^+$, for $u \in \mathbf{U}^-$.*

(d) *Two simple modules with the same highest weight $\lambda$ are isomorphic, so $M$ may be denoted unambiguously by $L(\lambda)$.*

(e) *For every $\lambda \in X^+$ there exists a simple $\mathbf{G}$-module of highest weight $\lambda$.*

All of this is developed systematically in [**RAGS**, II.2]. The characterization of simple modules by highest weights depends essentially on the fact that **T** acts by commuting semisimple operators, coupled with the fact that **U** acts as a family of unipotent operators and then fixes at least one nonzero vector. Standard conjugacy theorems for maximal tori and Borel subgroups of **G** make it permissible to work with fixed subgroups of these types.

The existence assertion (e) will be recalled in 2.3 below.

The theorem stops far short of answering all the obvious questions about the nature of the modules $L(\lambda)$, and indeed the answers are not yet fully known: see Chapter 3 below. For example, what is the formal character $\operatorname{ch} L(\lambda)$ and what is $\dim L(\lambda)$? Thanks to the Weyl group invariance of weight space dimensions, $\operatorname{ch} L(\lambda)$ can also be written as a $\mathbb{Z}^+$-linear combination of the orbit sums $s(\mu)$ for $\mu \in X^+$, with the orbit of $\lambda$ having coefficient 1:

$$\operatorname{ch} L(\lambda) = \sum_{\mu \in X^+} m_\mu s(\mu) \text{ with } m_\mu \in \mathbb{Z}^+ \text{ and } m_\lambda = 1.$$

The weights which actually occur in $L(\lambda)$ all lie between $\lambda$ and $w_\circ \lambda$ in the natural ordering.

A few observations can be made even without developing the theory further:

EXAMPLES. (1) $L(0)$ is just the trivial 1-dimensional **G**-module.

(2) Regarding the dual vector space $L(\lambda)^*$ as a **G**-module in the standard way by $(g \cdot f)(v) := f(g^{-1} \cdot v)$, this module is also simple. In fact it is isomorphic to $L(\lambda^*)$, where $\lambda^* := -w_\circ \lambda$. We refer to it as the **dual** or **contragredient**. It often happens that simple modules are self-dual: when $\Phi$ is irreducible, $w_\circ = -1$ if $\Phi$ is of type $A_1, B_\ell, C_\ell, D_\ell$ for $\ell$ even, $E_7, E_8, F_4, G_2$ (see for example [**215**, Ex. 13.5] or [**234**, Ex. 3.7]).

(3) If the Dynkin diagram of **G** admits a nontrivial graph automorphism $\pi$, one can "twist" the **G**-module $L(\lambda)$ by composing the representation **G** $\to \operatorname{GL}(L(\lambda))$ with $\pi^{-1}$. The resulting module $L(\lambda)^\pi$ is again simple and has as highest weight $\pi(\lambda)$, where $\pi$ induces on **T** and hence on $X(\mathbf{T})$ a natural automorphism. For example, in type $A_\ell$ with $\ell > 1$, $\pi$ interchanges simple roots $\alpha_i$ and $\alpha_{\ell-i+1}$, while $\pi$ has a similar effect on fundamental dominant weights.

(4) Given a maximal vector $v^+$ in $L(\lambda)$, the stabilizer of $Kv^+$ in **G** includes **B** and is therefore one of the **standard parabolic subgroups** generated by **B** together with a family of root subgroups $\mathbf{U}_{-\alpha}$ for $\alpha$ running over a subset of the set $\Delta$ of simple roots.

## 2.3. Construction of Modules

There is no known way to construct explicitly all simple modules $L(\lambda)$ for an arbitrary semisimple algebraic group—nor is it likely that a way will be found. (However, some specific modules can be realized naturally in exterior or symmetric powers: see for example Chapter 19.) The situation here is somewhat parallel to that encountered in the study of infinite dimensional highest weight modules for a semisimple Lie algebra over $\mathbb{C}$, but in fact is appreciably more complicated. This parallel appears most clearly in the Weyl module approach examined in Chapter 3 below, which involves reduction modulo $p$.

Chevalley's existence theorem is intrinsic to characteristic $p$, but reveals very little detail about the simple modules. The approach is somewhat akin to that used in finite group theory, where simple $KG$-modules show up as submodules of the left (or right) regular representation. For an algebraic group the role of the group algebra is played by the algebra $K[\mathbf{G}]$ of regular functions, which is infinite dimensional but locally finite dimensional. Here $\mathbf{G}$ acts naturally by right translations: $(g \cdot f)(x) = f(xg)$ for $g, x \in \mathbf{G}$.

By starting with a hypothetical $L(\lambda)$ (with $\lambda \in X^+$) and a maximal vector $v^+$ one can identify a suitable regular function which generates a simple submodule of this type in $K[\mathbf{G}]$. This is explained in [**RAGS**, II.2] (and in somewhat different ways in [**394**, Thm. 40] and [**220**, 31.4]). Jantzen adopts the geometric framework of sheaf cohomology on the flag variety $\mathbf{G}/\mathbf{B}^-$, the opposite Borel subgroup being used to avoid sign problems with weights. (So the reader must be cautious about his notation for Borel subgroups.) In this language, the 0th cohomology group of a line bundle determined by a dominant weight $\lambda$ is realizable as a subspace of the function algebra. This space $H^0(\lambda)$ is finite dimensional (since the flag variety is projective) and carries a natural $\mathbf{G}$-action. Concretely,

$$H^0(\lambda) := \{f \in K[\mathbf{G}] \mid f(gb) = \lambda(b)^{-1}f(g) \text{ for all } g \in \mathbf{G}, b \in \mathbf{B}^-\}$$

This construction can also be interpreted as "induction" (in the sense of algebraic groups) of a 1-dimensional $\mathbf{B}^-$-module to $\mathbf{G}$. The $\mathbf{G}$-module $H^0(\lambda)$ has a unique simple submodule, the sought-for $L(\lambda)$.

Later in the development, Kempf's Vanishing Theorem shows that $H^0(\lambda)$ is the only non-vanishing cohomology group of the line bundle associated with $\lambda \in X^+$. Then a general principle about base change from characteristic 0 shows that the formal character of $H^0(\lambda)$ is given by Weyl's classical formula. As a result, the weight space dimensions in $L(\lambda)$ are bounded by those in the classical theory. (This comparison reappears in a dual form in the study of Weyl modules.)

## 2.4. Contravariant Forms

One very useful tool in the study of $\mathbf{G}$-modules is a bilinear form associated to certain maps of $\mathbf{G}$-modules. (This plays an important role in the proof of the restriction theorem below.) The ideas here emerged in somewhat different ways in early work of Steinberg [**394**, pp. 228–229] and Wong [**436**, **437**], after which Jantzen developed them further in a series of papers. Here we summarize the construction of the form, following [**RAGS**, II.8.17].

The starting point is an involutive anti-automorphism $\tau$ of $\mathbf{G}$ which fixes the maximal torus $\mathbf{T}$ pointwise while interchanging the root groups $\mathbf{U}_\alpha$ and $\mathbf{U}_{-\alpha}$ for all $\alpha \in \Phi$. The existence of such a map $\tau$ is a byproduct of the general treatment of isomorphisms of reductive groups: see [**RAGS**, II.1]. When $\mathbf{G} = \mathrm{SL}(n, K)$, the usual transpose map is a natural choice for $\tau$.

Next one uses $\tau$ to define a functor on the category of rational $\mathbf{G}$-modules, taking a module $M$ to a module $^\tau M$ (see [**RAGS**, II2.12]). Here $^\tau M$ is the dual vector space $M^*$ with a twisted $\mathbf{G}$-action given by $g \cdot f = f \circ \tau(g)$. Since $\tau$ fixes $\mathbf{T}$ pointwise, the formal characters of $M$ and $^\tau M$ are the same. In particular, since a simple module $L(\lambda)$ is determined within isomorphism by its formal character, we have $^\tau L(\lambda) \cong L(\lambda)$.

To construct a bilinear form, start with an arbitrary $\mathbf{G}$-module homomorphism $\varphi : M \to {}^\tau M$. Since $^\tau M$ is the dual space $M^*$, it makes sense to define a bilinear form on $M$ by the recipe $(v, v') = \varphi(v)(v')$. The key property of such a form is then easy to check:

$$(g \cdot v, v') = (v, \tau(g) \cdot v') \text{ for all } g \in \mathbf{G} \text{ and } v, v' \in M.$$

Any bilinear form with this property is called a **contravariant form** on $M$. From contravariance one gets immediately:

$$(M_\lambda, M_\mu) = 0 \text{ for distinct weights } \lambda, \mu \in X.$$

One also sees readily that the radical of the form is a **G**-submodule, which must be proper if $\varphi \neq 0$.

Later (in Chapter 4) we will see a computational role for contravariant forms on other modules, but for the moment we consider only simple modules:

PROPOSITION. *For all* $\lambda \in X^+$, *the simple* **G**-*module* $L(\lambda)$ *has a nondegenerate contravariant form. Any such form is symmetric, and two such forms are proportional.*

The proof goes as follows. Start with any module isomorphism

$$\varphi : L(\lambda) \to {}^\tau L(\lambda) \cong L(\lambda)$$

(essentially unique by Schur's Lemma). As noted above, the radical of the associated contravariant form is a proper **G**-submodule, so the form must be nondegenerate. Again by Schur's Lemma, any two such forms must be proportional. But the fact that $\tau$ has order 2 implies that the form $\langle v, v' \rangle := (v', v)$ is also contravariant, whence $\langle v, v' \rangle = c(v', v)$ and $c^2 = 1$. This forces the form to be either symmetric or alternating. Since weight spaces for distinct weights are orthogonal while $\dim L(\lambda)_\lambda = 1$, the form is nonzero on this weight space and hence cannot be alternating.

## 2.5. Representations of Frobenius Kernels

In characteristic $p$ the Lie algebra $\mathfrak{g}$ of **G** has many finite dimensional representations which fail to "integrate" to representations of **G**. In the other direction, most simple **G**-modules are no longer simple as $\mathfrak{g}$-modules. To pass from **G**-modules to $\mathfrak{g}$-modules, one takes differentials of the coordinate functions involved; it is sometimes more convenient, however, to regard this as a kind of restriction to a normal subgroup.

The language of group schemes, developed in [**RAGS**, Part I], makes this idea precise. The standard Frobenius map $F : \mathbf{G} \to \mathbf{G}$ associated with $p^r$ has a scheme-theoretic kernel $\mathbf{G}_r$, called the $r$th **Frobenius kernel**. (Note that the same kernel results from composing $F$ with a graph automorphism of **G**.) The algebra of regular functions defining $\mathbf{G}_r$ may be identified naturally with the dual of the Hopf algebra $\mathrm{Dist}(\mathbf{G}_r)$ of distributions on $\mathbf{G}_r$, also known as a **hyperalgebra**, whose dimension is $p^{r \dim \mathfrak{g}}$. For brevity we refer to $\mathrm{Dist}(\mathbf{G}_r)$-modules simply as $\mathbf{G}_r$-modules.

Recall that $\mathfrak{g}$ has an extra operation $X \mapsto X^{[p]}$ which makes it a **restricted** Lie algebra. In particular, the $p$th power of $\mathrm{ad}\, X$ equals $\mathrm{ad}\, X^{[p]}$. More generally, a representation $\varphi$ of $\mathfrak{g}$ is called *restricted* if the $p$th power $\varphi(X)$ for each $X \in \mathfrak{g}$ coincides with the operator $\varphi(X^{[p]})$. Equivalently, $\varphi$ is a representation of the **restricted enveloping algebra** (written as $u(\mathfrak{g})$ or $U_0(\mathfrak{g})$): the quotient of the universal enveloping algebra $U(\mathfrak{g})$ obtained by identifying an associative $p$th power $X^p$ ($X \in \mathfrak{g}$) with the element $X^{[p]}$. The definitions show that $u(\mathfrak{g})$ is isomorphic to

$\mathrm{Dist}(\mathbf{G}_1)$. Thus the rational representations of $\mathbf{G}_1$ are essentially the same as the restricted representations of $\mathfrak{g}$.

Our basic philosophy is that the rational representations of $\mathbf{G}_r$ of greatest interest parallel closely the representations of the finite Chevalley (or twisted) group $G = \mathbf{G}^F$ when $F$ is a Frobenius map relative to $p^r$ (involving possibly a graph automorphism $\pi$). Notice that the dimension of the Hopf algebra $KG$ is a polynomial in $p$ whose highest degree term is the same as the hyperalgebra dimension.

The following theorem generalizes from the Lie algebra case the basic observation of Curtis [118]. We say that a dominant weight $\lambda$ is $p^r$-**restricted** if $0 \le \langle \lambda, \alpha \rangle < p^r$ for all simple roots $\alpha$, and denote the set of these weights (relative to a fixed maximal torus $\mathbf{T}$) by $X_r$.

THEOREM. *Let $X_r$ be the set of $p^r$-restricted weights in $X^+$. Then the simple $\mathbf{G}$-modules $L(\lambda)$ with $\lambda \in X_r$ remain simple and nonisomorphic on restriction to the $r$th Frobenius kernel $\mathbf{G}_r$. Every simple $\mathbf{G}_r$-module is isomorphic to one of these.*

The proof is developed in Jantzen [**RAGS**, II.3]. The main point is to see that the condition placed on $\lambda$ implies there cannot be any extra "highest weight vectors" in $L(\lambda)$ for $\mathbf{G}_r$: vectors stable under the Frobenius kernel $\mathbf{B}_r$ but not proportional to a given maximal vector of weight $\lambda$. Such vectors would occur in proper nonzero $\mathbf{G}_r$-submodules. One complication is that weights in $X$ which are congruent modulo $p^r$ look the same for $\mathbf{T}_r$. This undermines the role of the usual partial ordering of weights. Even so, the fact that $\mathbf{G}_r$ is a *normal* subgroup scheme makes the full action of $\mathbf{T}$ available by working with the "mixed" group scheme $\mathbf{G}_r\mathbf{T}$. (This is where restricting $\mathbf{G}$-modules to finite subgroups becomes more subtle.)

The representation theory of the group schemes $\mathbf{G}_r$ and $\mathbf{G}_r\mathbf{T}$ will be explored further in Chapter 10 when we study projective $KG$-modules.

## 2.6. Invariants in the Function Algebra

When we discuss below the way in which some of the modules $L(\lambda)$ restrict to the finite group $G = \mathbf{G}^F$, it will be essential to reformulate the construction of simple modules inside $K[\mathbf{G}]$ outlined in 2.3. We mainly follow Jantzen [**257**, App. 1], apart from the switching of positive and negative root groups.

To start with, the density of the big cell in the Bruhat decomposition shows that the restriction map from functions on $\mathbf{G}$ to functions on $\mathbf{U}$ is *injective* on the copy of $L(\lambda)$ embedded in $K[\mathbf{G}]$ via the inclusion $H^0(\lambda) \hookrightarrow K[\mathbf{G}]$ (see 2.3). Moreover, this restriction map preserves the right translation action of $\mathbf{U}$. Taken by itself, the algebra $K[\mathbf{U}]$ has a natural grading by the *negative* part of the root lattice, coming from the (inverse!) conjugation action of $\mathbf{T}$ on $\mathbf{U}$:

$$(t \cdot f)(u) := f(t^{-1}ut).$$

The constant functions, denoted $K$, are the only $\mathbf{U}$-invariants and comprise the 0th graded part.

Now the subspace $M$ of $K[\mathbf{U}]$ corresponding to the embedded copy of $L(\lambda)$ is $\mathbf{T}$-stable, but a maximal vector of weight $\lambda$ in $L(\lambda)$ is fixed by $\mathbf{U}$ and thus acquires "weight" 0 in this grading. A similar downward shift occurs for all weight spaces in $L(\lambda)$. Though it is not strictly necessary here, one could make the $\mathbf{B}$-action on $L(\lambda)$ compatible with the $\mathbf{B}$-action on $M$ by shifting the $\mathbf{T}$-grading of $K[\mathbf{U}]$ by $\lambda$ (i.e., tensoring the representation of $\mathbf{T}$ by this character). With or without this adjustment, the action of $\mathbf{U}$ on the copy of $L(\lambda)$ in $K[\mathbf{U}]$ has the usual effect of taking a given graded element of "weight" $\mu$ to itself plus others of "weights" obtained from $\mu$ by subtracting positive roots.

In this framework, the essence of Theorem 2.5 can be restated:

THEOREM. *For* $\lambda \in X_r$, *embed* $L(\lambda)$ *as a* $\mathbf{B}$-*submodule* $M$ *of* $K[\mathbf{U}]$ *as above. Then the space of invariants* $M^{\mathbf{U}_r}$ *under right translation by the* $r$th *Frobenius kernel of* $\mathbf{U}$ *just consists of the constants* $K$.

Our goal in this chapter is to use the theorem as a steppingstone to the parallel case of restriction of modules from $\mathbf{G}$ to $G$. To this end we have to give a more precise description of the full ring of invariants in $K[\mathbf{U}]$ under $\mathbf{U}_r$ in terms of the Frobenius map $F$. This requires a notational distinction in case $F$ involves a nontrivial graph automorphism $\pi$, viewed as an automorphism of $\mathbf{G}$ which leaves $\mathbf{T}$ invariant. Write $\lambda \mapsto \widetilde{\lambda}$ for the induced action of $\pi$ on $X$.

PROPOSITION. *Let* $F$ *be a Frobenius map relative to* $q = p^r$ *and possibly a graph automorphism* $\pi$ *of* $\mathbf{G}$. *Restricting* $F$ *to* $\mathbf{U}$, *its (injective) comorphism* $F^*$ *sending* $f$ *to* $f \circ F$ *satisfies:*.

(a) *The algebra of invariants* $K[\mathbf{U}]^{\mathbf{U}_r}$ *is precisely the image of* $F^*$.
(b) *If* $f \in K[\mathbf{U}]$ *has "weight"* $\mu$ *for the* $\mathbf{T}$-*action, then* $f \circ F$ *has "weight"* $q\mu$ *when* $\pi = 1$ *or* $q\widetilde{\mu}$ *when* $\pi \neq 1$.

The point in (a) is that the Frobenius endomorphism on $\mathbf{U}$ is a quotient map in the sense of group schemes, realizing the quotient of $\mathbf{U}$ by $\mathbf{U}_r$. (Since $\pi$ is an automorphism of algebraic groups, it has no effect on the quotient construction.)

## 2.7. Steinberg's Tensor Product Theorem

In characteristic 0 the analogues of the simple $\mathbf{G}$-modules $L(\lambda)$ are usually unrelated to each other, but in characteristic $p$ the standard Frobenius maps relative to powers of $p$ yield strong interactions among simple modules. As a byproduct, the determination of characters and dimensions is reduced in principle to the study of the $p^\ell$ restricted weights in $X_1$. This basic theorem was proved by Steinberg [**393**], inspired by the earlier treatment of the more elementary case SL$(2, K)$ by Brauer–Nesbitt [**61**].

Starting with the standard Frobenius map $F$ relative to $p$, any (rational) $\mathbf{G}$-module $M$ can be "twisted" by the $r$th power of $F$ (for $r \geq 1$) via the composition $\mathbf{G} \to \mathbf{G} \to \mathrm{GL}(M)$ of $F^r$ with the given representation. Write $M^{[r]}$ for this $r$th **Frobenius twist**.

THEOREM. *Let* $\lambda \in X^+$.

(a) *The* $\mathbf{G}$-*module* $L(p^r \lambda)$ *is isomorphic to the Frobenius twist* $L(\lambda)^{[r]}$.
(b) *If* $\lambda \in X_r$, *write its* $p$-*adic expansion* $\lambda = \lambda_0 + p\lambda_1 + p^2\lambda_2 + \cdots + p^{r-1}\lambda_{r-1}$ *with all* $\lambda_i \in X_1$. *Then the* $\mathbf{G}$-*module* $L(\lambda)$ *is isomorphic to*

$$L(\lambda_0) \otimes L(p\lambda_1) \otimes \cdots \otimes L(p^{r-1}\lambda_{r-1}).$$

*Thus*

$$L(\lambda) \cong L(\lambda_0) \otimes L(\lambda_1)^{[1]} \otimes \cdots \otimes L(\lambda_{r-1})^{[r-1]}.$$

We shall refer to this as the **Tensor Product Theorem**. Steinberg's original proof is intertwined with that of the restriction theorem for finite groups discussed below, using previous study by Curtis of the Lie algebra representations. Jantzen [**RAGS**, II.3.17] follows Cline–Parshall–Scott [**115**] in giving a more elegant conceptual proof within the framework of the algebraic group and its Frobenius kernels.

## 2.8. Example: SL(2, K)

Here and in later chapters the special case $\mathbf{G} = \mathrm{SL}(2, K)$ illustrates some of the basic ideas involved in the general case, albeit often in an oversimplified form. Since the diagonal subgroup $\mathbf{T}$ of $\mathbf{G}$ has dimension 1, it is usually convenient to identify the character group $X$ of $\mathbf{T}$ with the additive group $\mathbb{Z}$. The dominant weights correspond to $\mathbb{Z}^+$, with 1 being the fundamental weight and 2 the positive root. The Weyl group has order 2.

As mentioned already, Brauer–Nesbitt [**61**, §30] constructed the simple modules with $p$-restricted highest weights and proved the tensor product theorem in this case. From the general theory one gets only a rough picture of $L(\lambda)$, but their direct construction makes everything explicit. The natural action of $\mathbf{G}$ on 2-dimensional space already yields the simple module $L(1)$, while the symmetric powers of this module yield further modules with weight multiplicities equal to 1. Concretely, $\mathbf{G}$ is acting here on the $d+1$-dimensional space of homogeneous polynomials of degree $d$ in two variables. The weights are $d, d-2, d-4, \ldots, -(d-2), -d$. Straightforward computation shows that for $0 \leq d < p$, these modules are actually simple. Then the tensor product theorem takes over to produce all other $L(\lambda)$.

It is easy to see directly how the center of $\mathrm{SL}(2, K)$ is acting: Say $p \neq 2$. The negative of the identity matrix acts as the identity precisely when the highest weight is even. So the corresponding modules $L(0), L(2), \ldots$ yield all of the simple modules for $\mathrm{PSL}(n, K) = \mathrm{PGL}(n, K)$.

## 2.9. Brauer Theory

When $G$ is an arbitrary finite group, of order divisible by $p$, the study of $KG$-modules usually begins with the simple modules. A basic question is how many nonisomorphic ones there are over $K$ (or a smaller splitting field). Here and later we refer the reader to the standard texts, notably Curtis and Reiner [**125, 126**], for the basic theory. Recall that an element of $G$ is called $p$-**regular** if its order is not divisible by $p$. Similarly, a conjugacy class is called $p$-regular if the common order of its elements is not divisible by $p$.

THEOREM. *If $G$ is an arbitrary finite group, the number of nonisomorphic KG-modules is equal to the number of p-regular conjugacy classes in $G$.*

The same result holds when $K$ is replaced by an arbitrary splitting field of characteristic $p$. Proofs can be found in [**125**, §83B] and [**126**, 17.11].

## 2.10. Counting Semisimple Classes

Consider now the finite group $G = \mathbf{G}^F$. Here there is a definitive result:

THEOREM. *Let $G = \mathbf{G}^F$ over $\mathbb{F}_q$, with $\mathbf{G}$ simply connected of rank $\ell$. Then the number of p-regular (= semisimple) classes of $G$ is $q^\ell$.*

This is due to Steinberg [**393**] (see also [**395**, 14.8]). Other accounts are given by Carter [**89**, (3.7)] and Humphreys [**237**, (8.7)].

For an algebraic group **G** which is not simply connected, it requires extra work to count the number of $p$-regular classes in a finite subgroup $\mathbf{G}^F$ of Lie type. (Examples in the two Atlases [**117, 243**] are instructive.)

## 2.11. Restriction to Finite Subgroups

Following work by Curtis [**120**] on representations of Lie algebras and Chevalley groups over prime fields (in the framework of Chevalley's 1955 Tôhoku paper), the following fundamental Restriction Theorem was first proved in full generality by Steinberg [**393**]:

THEOREM. *Let* $G = \mathbf{G}^F$ *over a field of* $p^r$ *elements.*

(a) *For each* $\lambda \in X_r$, *the rational* **G**-*module* $L(\lambda)$ *remains irreducible on restriction to* $G$.

(b) *If* $\lambda \in X_r$, *all* $U$-*invariants in* $L(\lambda)$ *are multiples of a maximal vector* $v^+$ *of weight* $\lambda$.

(c) *The simple* $KG$-*modules* $L(\lambda)$ *with* $\lambda \in X_r$ *are pairwise nonisomorphic and exhaust the* $q^\ell$ *isomorphism classes of simple* $KG$-*modules.*

With some extra work the theorem can also be adapted to the Suzuki and Ree groups: see Chapter 20 below.

The proof of the Restriction Theorem involves three distinct steps:

(1) Determine the number of nonisomorphic simple $KG$-modules, i.e., the number of $p$-regular classes in $G$.

(2) Prove that this many simple modules for **G** remain simple on restriction to $G$.

(3) Prove that no two of the restricted modules are isomorphic.

The first step has already been discussed in 2.10. The other steps will be carried out in the following sections. Here we just add a few related comments:

REMARKS. (1) It is noteworthy that twisted and untwisted groups of the same Lie type over a field of $p^r$ elements have precisely the same simple modules. This is not at all obvious *a priori*. More subtle resemblances of this sort will be found when we discuss projective modules and cohomology.

(2) When $\lambda$ is dominant but not $p^r$-restricted, the restriction of $L(\lambda)$ to $G$ becomes an ordinary tensor product of a simple module with another module: this follows immediately from Steinberg's Tensor Product Theorem.

(3) When passing from a universal group like $\mathrm{SL}(n, q)$ to a quotient by a central subgroup like $\mathrm{PSL}(n, q)$, it takes further work to sort out which of the $L(\lambda)$ with $\lambda \in X_r$ yield linear (not just projective) representations of the various quotients of $G$ by central subgroups.

(4) One might hope to prove the above theorem by studying directly how $L(\lambda)$ restricts to $G$, while ignoring the Frobenius kernels (or Lie algebra). But the need to keep track of weight spaces makes such a direct approach problematic.

## 2.12. Proof of Irreducibility

Here we follow the proof in Jantzen [**257**, App. 1], based on ideas of Kempf [**270**]. It is written down for Chevalley groups, but needs only small adaptations to twisted groups. The approach is somewhat simpler and more conceptual than Steinberg's, exploiting fully the parallel with representations of Frobenius kernels.

Recall from 2.6 that when $\pi$ is nontrivial, we write $\widetilde{\mu}$ for the image of a weight $\mu$ under the associated automorphism of $X$.

Given a weight $\lambda \in X_r$, we have to show that the simple $\mathbf{G}$-module $L(\lambda)$ remains simple on restriction to $G$. We know that $L(\lambda)^{\mathbf{U}} = L(\lambda)_\lambda$. Set $U := \mathbf{U}^F$. We claim it is enough to prove that

$$(1) \qquad\qquad L(\lambda)^U = L(\lambda)_\lambda.$$

Indeed, assuming (1), let $v^+$ span $L(\lambda)_\lambda$. Being a $p$-group, $U$ has a nonzero fixed vector in any nonzero $U$-submodule of $L(\lambda)$. So (1) implies that $v^+$ (hence $KGv^+$) lies in every simple submodule of $L(\lambda)$. It follows that $KGv^+$ is the unique simple $KG$-submodule of $L(\lambda)$. To show that it is all of $L(\lambda)$, we appeal to the existence of a nondegenerate contravariant form (see 2.4) on $L(\lambda)$ with $(v^+, v^+) \neq 0$. The orthogonal complement of the $KG$-submodule $KGv^+$ is again a $KG$-submodule, but doesn't contain $v^+$ and is therefore 0. Thus $KGv^+ = L(\lambda)$.

It remains to prove (1). For this we shall exploit the parallel result 2.6 for the $r$th Frobenius kernel $\mathbf{U}_r$ of $\mathbf{U}$, based on the assumption that $\lambda \in X_r$:

$$(2) \qquad\qquad L(\lambda)^{\mathbf{U}_r} = L(\lambda)_\lambda.$$

In order to make effective use of this fact we invoke the embedding of $L(\lambda)$ into the regular functions $K[\mathbf{U}]$ (see 2.6). This embedding preserves the weight spaces under $\mathbf{T}$, but with a downward shift by $\lambda$. Now we can make a further reduction: to obtain (1) from (2) it will suffice to prove that for any finite dimensional $\mathbf{B}$-submodule $M$ of $K[\mathbf{U}]$ (such as the embedded $L(\lambda)$),

$$(3) \qquad\qquad M \cap K[\mathbf{U}]^{\mathbf{U}_r} = K \text{ implies } M \cap K[\mathbf{U}]^V = K.$$

To prove this we need a more explicit description of the ring $K[\mathbf{U}]^U$, similar to that given for the $\mathbf{U}_r$-invariants in 2.6. Here the role of the Frobenius map $F$ is played by the *Lang map* $L : \mathbf{U} \to \mathbf{U}$, where $L(u) := F(u)u^{-1}$ (see 1.4). This is a surjective map, whose comorphism $L^*$ sends $f$ to $f \circ L$. Recall that $L$ is a quotient map, which implies that the ring of invariants can be described as

$$K[\mathbf{U}]^U = L^*K[\mathbf{U}] = \{f \circ L \mid f \in K[\mathbf{U}]\}.$$

Now we can use the $\mathbf{T}$-action on the given $\mathbf{B}$-module $M$ to finish the proof. The "weights" of $\mathbf{T}$ on $K[\mathbf{U}]$ are nonnegative $\mathbb{Z}$-linear combinations of negative roots. The key observation is that for any $f \in K[\mathbf{U}]$ of "weight" $\mu$, $f \circ L$ is the sum of $f \circ F$ (which has weight $q\mu$ or $q\widetilde{\mu}$ depending on whether $\pi = 1$ or not) and various weight vectors of strictly higher weight.

Returning to the implication in (3), we are given $0 \neq f \in K[\mathbf{U}]$ such that $f \circ L \in M$. Write $f = \sum_\mu f_\mu$ and $f \circ L = \sum_\mu g_\mu$ as sums of "weight" vectors. Since $f \circ L$ lies in the $\mathbf{T}$-module $M$, all $g_\mu \in M$. Let $\nu$ be *minimal* for $f_\nu \neq 0$. From the above discussion we see $f_\nu \circ F = g_{q\nu}$ (or $g_{q\widetilde{\nu}}$). In particular, $f_\nu \circ F$ lies in both $M$ and $F^*K[\mathbf{U}]$, which is by assumption $K$. This means $q\nu = 0$ (or $q\widetilde{\nu} = 0$), hence $\nu = 0$. By minimality of $\nu$, we conclude $f = f_0 \in K$. $\square$

### 2.13. Proof of Distinctness: Chevalley Groups

It remains to show that for distinct $\lambda$ and $\mu$ in $X_r$, the $G$-modules $L(\lambda)$ and $L(\mu)$ are not isomorphic. The problem here is that the restrictions of $\lambda$ and $\mu$ to $T$

sometimes coincide. This is clear already when $G = \mathrm{SL}(2, p)$: identifying weights as usual with integers, the weights $0$ and $p - 1$ both restrict to the trivial character of the diagonal subgroup of $G$.

Whenever $G$ is a Chevalley group, the finite torus $T$ is a direct product of $\ell$ copies of $k^\times$ (with $q = p^r$ and $k = \mathbb{F}_q$). This abelian group has just $(q - 1)^\ell$ distinct homomorphisms $T \to k^\times$. If we order the fundamental weights as $\varpi_1, \ldots, \varpi_\ell$, then it is clear that the $q^{\ell-1}$ weights $\lambda = \sum a_i \varpi_i$ with $0 \leq a_i \leq q - 2$ yield distinct homomorphisms $T \to k^\times$. But if we allow some $a_i$ to equal $q - 1$, the restriction of the weight to $T$ cannot be distinguished from the case when the coordinate is 0. In this way the algebraic group classification by highest weight breaks down.

In such ambiguous situations it is easy to distinguish the cases $a_i = 0$ and $a_i = q - 1$ by noting that in the first case the corresponding root group $U_{-\alpha_i}$ fixes $v^+$, whereas in the second case it does not. Thus when $\lambda$ and $\mu$ are distinct weights which agree on $T$, the different actions of $U^-$ on $v^+$ force the restrictions of $L(\lambda)$ and $L(\mu)$ to be nonisomorphic.

In the algebraic group setting (2.2) the subgroup of $\mathbf{G}$ stabilizing $Kv^+$ is a standard parabolic subgroup $\mathbf{P}$ determined by a subset of the simple roots: $U_{-\alpha} \subset \mathbf{P}$ precisely when $\langle \lambda, \alpha^\vee \rangle = 0$. Since $\mathbf{G}$ is split, $P = \mathbf{P}^F$ is the stabilizer of $Kv^+$ in $G$ and is determined by the same root data. At the same time we know from 2.12 that $v^+$ is the only $U$-invariant vector in $L(\lambda)$ (up to scalars). Thus $L(\lambda)$ as a $KG$-module intrinsically determines the parabolic subgroup $P \subset G$ stabilizing the line through $v^+$.

Now suppose that $L(\lambda)$ is isomorphic to $L(\mu)$ as a $KG$-module, for distinct weights $\lambda$ and $\mu$. A module isomorphism must make $U$-invariant vectors ($=$ maximal vectors) correspond, so $\lambda$ and $\mu$ must agree on $T$. This means that for some simple $\alpha$, the integers $\langle \lambda, \alpha^\vee \rangle$ and $\langle \mu, \alpha^\vee \rangle$ are distinct but congruent modulo $q - 1$. Since these integers lie between 0 and $q - 1$, one must be 0 and the other must be $q - 1$. This forces $U_{-\alpha}$ to fix the first maximal vector but not the second, implying that the parabolic stabilizers in $G$ differ. This contradiction completes the proof in the split case.

## 2.14. Proof of Distinctness: Twisted Groups

When $G$ is a twisted group, a further refinement is needed, since the finite torus and unipotent group become more complicated to describe. The prototype is $\mathrm{SU}(3, q)$, where two simple roots $\alpha$ and $\widetilde{\alpha}$ are involved. The group $T = \mathbf{T}^F$ can be identified with the set of ordered pairs $(c, c^q)$, where $c$ runs over the multiplicative group of $\mathbb{F}_{q^2}$. Applying a $q$-restricted weight $\lambda = m\varpi_1 + n\varpi_2$ then yields $c^{m+nq}$. For $0 \leq m, n \leq q - 1$, these powers of $c$ are distinct (and determine $m, n$ uniquely). But when $m = n = 0$ and $m = n = q - 1$, the powers of $c$ coincide. These two cases are distinguished by the fact that action of the associated "root group" involving both $U_{-\alpha}$ and $U_{-\widetilde{\alpha}}$ on a highest weight vector of $L(\lambda)$ is trivial in the first case but nontrivial in the second. The argument is similar for any twisted group. $\square$

This line of argument yields a nice formulation, which carries over to the setting of split $BN$-pairs considered below:

COROLLARY. *The $q^\ell$ nonisomorphic simple $KG$-modules $L(\lambda)$ (with $\lambda \in X_r$) are in natural bijection with the pairs $(\chi, P)$, where $\chi : T \to K^\times$ is a homomorphism, $P$ is a standard parabolic subgroup of $G$, and $\chi$ is trivial on the torus subgroup belonging to the $\pi$-orbit of $\alpha$ precisely when the corresponding "root subgroup" lies in $P$.*

## 2.15. Action of a Sylow $p$-Subgroup

For use in Chapter 9 we add to the previous development a refinement which also occurs naturally in the *a priori* approach of Curtis–Richen [**122**, Thm. 4.2(b)].

PROPOSITION. *If $\lambda \in X_r$ and $v^+$ is a maximal vector in $L(\lambda)$, then the $KU^-$-submodule generated by $v^+$ is all of $L(\lambda)$.*

PROOF. It is simplest at first to assume that $G$ is a Chevalley group, but the formalism of split $BN$-pairs permits the same argument to be used in all cases (including the groups of Suzuki and Ree). Then the notations below have to be re-interpreted.

Abbreviate $U^-$ by $V$. Similarly, when $\alpha > 0$ write $V_\alpha$ for the root group $U_{-\alpha}$ obtained by conjugating $U_\alpha$ by a representative in $N$ of the reflection $s_\alpha$.

We need to show that $g \cdot v^+ \in KV \cdot v^+$ for all $g \in G$. Thanks to the Bruhat decomposition, it is enough to show that all $w \cdot v^+$ lie in $KV \cdot v^+$, where $w$ runs over representatives of the Weyl group in $N$.

From the rank 1 case we see immediately when $s = s_\alpha$ is a simple reflection that $s \cdot v^+ \in KV_\alpha \cdot v^+$, using the assumption that $\lambda \in X_r$.

For an arbitrary $w \in W$, write $w = s_1 \cdots s_t$ in reduced form, and write $U_i$ and $V_i$ respectively for the root groups $U_{\alpha_i}$ and $V_{\alpha_i}$. Because the expression is reduced, the split $BN$-pair axioms yield readily the fact that

$$s_1 s_2 \cdots s_{k-1} V_k \subseteq V \text{ for each } k \leq t.$$

This permits an easy inductive argument, based on the previous paragraph. The idea is clear from the case $t = 3$:

$$
\begin{aligned}
s_1 s_2 s_3 \cdot v^+ \;\; &\in \;\; K\, s_1 s_2 V_3 \cdot v^+ \\
&= \;\; K\, s_1\, {}^{s_2}V_3\, s_2 \cdot v^+ \\
&\subseteq \;\; K\, s_1\, {}^{s_2}V_3\, V_2 \cdot v^+ \\
&= \;\; K\, {}^{s_1 s_2}V_3\, {}^{s_1}V_2\, s_1 \cdot v^+ \\
&\subseteq \;\; K\, {}^{s_1 s_2}V_3\, {}^{s_1}V_2\, V_1 \cdot v^+ \\
&\subseteq \;\; KV \cdot v^+
\end{aligned}
$$

$\square$

CHAPTER 3

# Weyl Modules and Lusztig's Conjecture

As we saw in Chapter 2, the study of simple $KG$-modules for a finite group $G$ of Lie type (in the defining characteristic) can be reduced to the study of simple $\mathbf{G}$-modules $L(\lambda)$ with $p$-restricted highest weights. In the algebraic group setting, a number of ideas and techniques have been introduced in the hope of getting more complete information about the formal characters of these modules, based on the study of Weyl modules.

First we summarize the main results, most of which are developed systematically in [**RAGS**, II.6–II.8] (though in a different logical order):

- Weyl modules (3.1)
- Premet's comparison of weights of simple modules and Weyl modules when the highest weight is $p$-restricted (3.2)
- relationship with cohomology of line bundles on the flag variety (3.3)
- affine Weyl group (3.4–3.5)
- Linkage Principle and translation functors (3.6)
- Steinberg modules (3.7)
- contravariant form on a Weyl module (3.8)
- Jantzen filtrations and sum formula (3.9)

If $p$ is not too small (say $p \geq h$, the Coxeter number), Lusztig's 1979 Conjecture (3.11) offers the best hope for a uniform theoretical explanation of the characters in a setting essentially independent of $p$. At this writing it remains unproved in general. But it has been proved indirectly by Andersen–Jantzen–Soergel [**18**] for "sufficiently large" $p$ (though without a definite estimate of how large $p$ should be).

More direct computational methods, which are effective for very small $p$ and low ranks, will be discussed in the next chapter. Both there and in the current chapter, the finite group $G$ remains largely in the background as a motivating ingredient.

### 3.1. Weyl Modules

In 2.3 we outlined the existence proof for simple $\mathbf{G}$-modules with a given highest weight $\lambda \in X^+$. This involves a theoretical realization of $L(\lambda)$ as a submodule of the algebra of regular functions, but reveals little about the formal character or dimension.

Another approach to constructing simple $\mathbf{G}$-modules is much less intrinsic to characteristic $p$ but leads to a helpful formulation of the relationship between their (still unknown) formal characters and the familiar Weyl character formula. This approach by reduction modulo $p$ from a $\mathbb{Z}$-form of the corresponding $\mathbf{G}_{\mathbb{C}}$-module also suggests effective algorithms for computation in low ranks (see Chapter 4 below).

One begins with a simple module $V(\lambda)_{\mathbb{C}}$ for a corresponding semisimple Lie algebra $\mathfrak{g}_{\mathbb{C}}$ over $\mathbb{C}$; these are also indexed by dominant weights. As explained in Steinberg [394] or Humphreys [215, Chap. VII], there is a special $\mathbb{Z}$-form $V(\lambda)_{\mathbb{Z}} \subset V(\lambda)_{\mathbb{C}}$ which permits reduction modulo $p$ in such a way that $\mathbf{G}$ then acts naturally on the resulting **Weyl module** $V(\lambda) := K \otimes V(\lambda)_{\mathbb{Z}}$. Although $V(\lambda)$ usually fails to be simple, it has a unique simple quotient $L(\lambda)$. Other composition factors $L(\mu)$ all satisfy $\mu < \lambda$, encouraging an inductive approach to the determination of $\operatorname{ch} L(\lambda)$. In particular, $V(\lambda) = L(\lambda)$ for weights $\lambda$ such as 0 which are minimal relative to the partial ordering of $X^+$.

The advantage of working with Weyl modules in characteristic $p$ is that their formal characters are unchanged in the process of reducing modulo $p$ and are therefore given by Weyl's character formula. For brevity we write $\chi(\lambda) := \operatorname{ch} V(\lambda)$ and $\chi_p(\lambda) := \operatorname{ch} L(\lambda)$ for $\lambda \in X^+$. Both of these expressions lie in the ring $\mathbb{Z}[X]^W$ of $W$-invariants in the group ring $\mathbb{Z}[X]$ of $X$. Moreover, the partial order on $X^+$ allows us to relate the sets of formal characters in two equivalent ways:

$$(1) \qquad \chi(\lambda) = \sum_{\mu \leq \lambda} a_\mu \, \chi_p(\mu), \text{ with } a_\mu \in \mathbb{Z}^+ \text{ and } a_\lambda = 1.$$

$$(2) \qquad \chi_p(\lambda) = \sum_{\mu \leq \lambda} b_\mu \chi(\mu), \text{ with } b_\mu \in \mathbb{Z} \text{ and } b_\lambda = 1.$$

Eventually the unknown coefficients in (2) become the focus of Lusztig's Conjecture, for suitably limited $p$ and $\lambda$, but only after enough theory has been developed to refine this way of writing the equation. The main goal of this chapter is to outline this theory carefully, with reference mainly to the extensive treatment in [**RAGS**]. But first we look at a more special question.

## 3.2. Restricted Highest Weights

Comparison of simple modules with Weyl modules has been the key to most theoretical progress. While the Tensor Product Theorem 2.7 would allow us to focus exclusively on those $L(\lambda)$ having $p$-restricted highest weights $\lambda$, this does not work so well in practice due to the fact that $V(\lambda)$ usually has dominant subweights which are not $p$-restricted. Such weights may well be the highest weights of composition factors. On the other hand, it is possible to limit the study of Weyl modules to weights below some reasonable upper bound in the weight lattice. This is the way Lusztig's Conjecture is formulated.

Before outlining the framework leading to the conjecture, we make some further observations about weights. Consider the set of all weights of a $\mathbf{G}$-module $M$, ignoring their multiplicities:

$$\Pi(M) := \{\mu \in X \mid M_\mu \neq 0\}$$

It is well known that $\Pi(V(\lambda))$ is as large as it could possibly be: it contains all $\mu \in X^+$ satisfying $\mu \leq \lambda$, together with their $W$-conjugates. (See Bourbaki [**57**, VIII §7, Prop. 5] or Humphreys [**215**, 21.3].)

Evidently $\Pi(L(\lambda)) \subset \Pi(V(\lambda))$ for all $\lambda \in X^+$, though the multiplicities usually differ. Leaving aside the question of precise multiplicities, Premet [**343**] is able to show that in almost all cases a simple $\mathbf{G}$-module $L(\lambda)$ with $p$-restricted highest weight $\lambda$ has precisely the same set of weights as the Weyl module $V(\lambda)$. (This confirms an earlier conjecture of I.D. Suprunenko which she had checked in some special cases.)

THEOREM. *Assume $p > 2$ if $\mathbf{G}$ is of type $B_\ell, C_\ell$, or $F_4$ and $p > 3$ if $\mathbf{G}$ is of type $G_2$. Then for all $p$-restricted weights $\lambda \in X^+$, we have $\Pi(V(\lambda)) = \Pi(L(\lambda))$.*

The strategy of Premet's proof is somewhat complicated and requires detailed study of weights and roots for each type of root system. This is probably unavoidable, since the excluded cases really involve counterexamples to the conclusion of the theorem: one or more weights occurring in $V(\lambda)$ may no longer occur in $L(\lambda)$.

The first such example occurs for $B_2 = C_2$, where the natural 5-dimensional representation of the orthogonal group in characteristic 2 fails to be irreducible and collapses to the natural 4-dimensional representation of the symplectic group. The 0 weight thus disappears. With the standard labelling of the Dynkin diagrams for $B_\ell$ and $C_\ell$, there are similar phenomena for $p = 2$ and arbitrary $\ell$. The natural $2\ell + 1$-dimensional representation of $B_\ell$ (with highest weight $\varpi_1$) loses its 0 weight space to yield a $2\ell$-dimensional irreducible representation, matching the dimension of the natural representation of $C_\ell$ having this highest weight. The simple module of highest weight $\varpi_\ell$ for $C_\ell$ is similarly obtained from the Weyl module by discarding a weight space and has dimension equal to that of the simple Weyl module for $B_\ell$ having this highest weight. (The special isogenies in characteristic 2 between these groups guarantee such a coincidence of dimensions, as explained in 5.4 below.)

Computational methods discussed in Chapter 4 also show how the theorem breaks down for exceptional groups with two root lengths. For the group of type $G_2$, with the first simple root short, $\dim V(\varpi_1) = 7$ but $L(\varpi_1)$ loses the 0 weight space when $p = 2$. When $p = 3$ the adjoint module $V(\varpi_2)$ of dimension 14 loses the $W$-orbit of weights through $\varpi_1$ as well as half of its 0 weight space, resulting in $\dim L(\varpi_2) = 7$. For $F_4$ and $p = 2$, Table 2 in 4.4 shows a similar loss of weight spaces (for the orbit of $\varpi_4$) when $V(\varpi_1)$ is the natural 52-dimensional module.

These exceptions make it clear that one cannot expect to prove Premet's Theorem in a straightforward conceptual way. Instead one has to look closely at the action of root groups on weight vectors. What is the source of the difficulty? Given a maximal vector $v^+$ in $V(\lambda)$, the assumption that $\lambda \in X_1$ means that $m = \langle \lambda, \alpha^\vee \rangle < p$ for all *simple* roots $\alpha$. So the theory for $SL(2, K)$ shows that all weights in the string from $\lambda$ to $\lambda - m\alpha$ occur in $L(\lambda)$. But for a *nonsimple* root $\alpha$, it may well happen that $m \geq p$, making the analysis much more difficult. This shows up in the examples just mentioned.

## 3.3. Cohomology of Line Bundles

Weyl modules can also be interpreted naturally in the more geometric context of sheaf cohomology of line bundles on the flag variety $\mathcal{B}$ of $\mathbf{G}$: see [**RAGS**, II.2.13]. Here the line bundles come from characters $\lambda \in X$ and may be denoted $\mathcal{L}(\lambda)$. To retain the role of dominant weights relative to $\mathbf{B}$, Jantzen defines $\mathcal{B}$ to be $\mathbf{G}/\mathbf{B}^-$, where $\mathbf{B}^-$ is the Borel subgroup opposite to $\mathbf{B}$ (corresponding to $\Phi^-$). (But this convention is not universal in the literature.)

For brevity denote by $H^i(\lambda)$ the cohomology group $H^i(\mathcal{B}, \mathcal{L}(\lambda))$ for any $\lambda \in X$ and $i \geq 0$. The space $H^0(\lambda)$ of global sections carries a natural structure of finite dimensional $\mathbf{G}$-module, as do the higher cohomology groups.

When $\lambda$ is *dominant*, $H^0(\lambda)$ is nonzero and can be interpreted as the $\mathbf{G}$-module "induced" from the 1-dimensional $\mathbf{B}^-$-module defined by $\lambda$. Moreover, it has a unique simple submodule, which is isomorphic to $L(\lambda)$. This notion of induction has many features reminiscent of induction for finite groups, such as Frobenius reciprocity. But there are some unexpected outcomes, due essentially to the fact that $\mathcal{B}$ is a projective variety. For example, inducing the trivial 1-dimensional $\mathbf{B}^-$-module to $\mathbf{G}$ still yields the trivial module: $H^0(0) = K$.

It is well-known that $H^i(\lambda) = 0$ for all $i > m = \dim \mathcal{B}$. Here $m$ is the number of positive roots, which coincides with the order of the longest element $w_\circ$ of the Weyl group. As in characteristic 0, Serre duality [**RAGS**, II.4.2] shows that $H^i(\lambda)$ is dual to $H^{m-i}(-\lambda - 2\rho)$ as a $\mathbf{G}$-module. The weight $\rho$ is defined to be the half-sum of positive roots (equal to the sum of fundamental weights in $X$). It satisfies $\langle \rho, \alpha^\vee \rangle = 1$ for all $\alpha \in \Delta$.

Suppose $\lambda \in X^+$. In [**RAGS**, II.2.13(1)] Jantzen actually *defines* $V(\lambda)$ to be the dual of $H^0(-w_\circ\lambda)$ (or by Serre duality, $H^m$ for a related weight). This is consistent with the fact that $L(\lambda)$ is the unique simple quotient of $V(\lambda)$, whereas the unique simple submodule of $H^0(-w_\circ\lambda)$ is $L(-w_\circ\lambda) = L(\lambda^*) = L(\lambda)^*$.

To reconcile the two notions of Weyl module, one needs a much deeper result in characteristic $p$, *Kempf's Vanishing Theorem* [**RAGS**, II.4.5]. This states simply that for $\lambda \in X^+$, all $H^i(\lambda) = 0$ when $i > 0$. (While this resembles Kodaira vanishing in characteristic 0, it turns out that for nondominant $\lambda$ there are sometimes multiple nonvanishing cohomology groups.) When combined with the fact that $\mathcal{B}$ has a natural $\mathbb{Z}$-form and that Euler characteristic remains constant under base change, Kempf's theorem shows that the formal character $\operatorname{ch} H^0(\lambda)$ is the same as that of $V(\lambda)_{\mathbb{C}}$ for dominant $\lambda$. This is just the classical Weyl character. On the other hand, Frobenius reciprocity allows one to embed our earlier module $V(\lambda)$ into the Serre dual of $H^0(-w_\circ\lambda)$. Dimension comparison shows that equality must hold.

A further consequence is the characterization of Weyl modules as universal highest weight modules, analogous to (infinite dimensional) Verma modules for a semisimple Lie algebra over $\mathbb{C}$ (13.6).

THEOREM. *In the category of finite dimensional rational $\mathbf{G}$-modules, the Weyl module $V(\lambda)$, $\lambda \in X^+$, plays the role of universal highest weight module: Every $\mathbf{G}$-module generated by a maximal vector of weight $\lambda$ is isomorphic to a quotient of $V(\lambda)$.*

REMARK. Donkin has suggested an alternate notational scheme, which substitutes $\Delta(\lambda)$ for $V(\lambda)$ and $\nabla(\lambda)$ for $H^0(\lambda)$. This emphasizes the dual role of the two $\mathbf{G}$-modules and lends itself to related axiomatic treatments of "highest weight categories" in the spirit of Cline–Parshall–Scott.

## 3.4. The Affine Weyl Group

Associated to any irreducible root system $\Phi$ with weight lattice $X$ and associated Weyl group $W$ is an **affine Weyl group** $W_a$. Historically this infinite discrete reflection group was found to be a useful auxiliary tool in the study of compact Lie groups. Later it appeared more intrinsically as the "Weyl group" for a suitable $BN$-pair structure on a $p$-adic Lie group, then as the "Weyl group" of an affine Lie algebra. Following Bourbaki [**56**, Chap. 5] (see also [**234**, Chap. 4]) one can define $W_a$ to be the semidirect product of the Weyl group $W$, acting in the euclidean space $E := X \otimes_{\mathbb{Z}} \mathbb{R}$ as a reflection group, with the group of affine translations induced by the *coroot* lattice:

$$W_a := W \ltimes \mathbb{Z}\Phi^\vee,$$

where $W$ acts as usual on the lattice. As an abstract group, $W_a$ is a Coxeter group, generated by a set of simple reflections in $W$ together with one additional affine reflection. As a Coxeter group, $W_a$ has a length function $\ell(w)$ and a natural partial ordering $w \leq w'$: the **Chevalley–Bruhat ordering**.

In characteristic $p$, the affine Weyl group plays a somewhat mysterious role in the representation theory of **G** and its Lie algebra $\mathfrak{g}$. This was first articulated by Verma [**426**] around 1971. Here the description involves a subtle shift, replacing the coroot lattice by the root lattice $\mathbb{Z}\Phi$. Thus a group **G** of type $B_\ell$ requires an affine Weyl group of type $C_\ell$ in Bourbaki's language (and vice versa). The prime $p$ also plays an essential role here, multiplying all translations by a factor of $p$. This modification in the geometric description of the affine Weyl group leads to the alternate notation used in [**RAGS**, II.6]: $W_p = W \ltimes p\mathbb{Z}\Phi$. But $W_p$ is still abstractly isomorphic as a Coxeter group to the corresponding affine Weyl group (using the root lattice). Its extra generator involves an affine reflection relative to the highest short root.

A further complication in the study of **G**-modules, implicit in the Tensor Product Theorem, is the eventual need to work also with subgroups $W_{p^r}$ of $W_p$. These groups are also abstractly isomorphic to the same underlying Coxeter group.

For a thorough treatment of affine Weyl groups in the characteristic $p$ setting, see [**RAGS**, II.6].

## 3.5. Alcoves

The action of the affine Weyl group on the affine space $E$ is well studied in the texts cited above. One starts with the reflecting hyperplanes for the affine reflections $s_{\alpha,cp}$ in $W_p$ (described below in 3.9). The connected components of the space obtained by removing these hyperplanes from $E$ are called **alcoves**. The closure of any alcove is then a fundamental domain for the action of $W_p$. For type $A_2$, some of the alcove geometry is pictured in 3.10 below.

Since the special weight $\rho$ plays a distinguished role in representation theory, we modify the geometry slightly by taking $-\rho$ (rather than 0) to be the origin. This shift already shows up in classical characteristic 0 results such as the Weyl character formula. Formally, we define the **dot action** of $W_p$ on $E$ by

$$w \cdot x = w(x + \rho) - \rho \text{ for } w \in W_p, x \in E.$$

Weights in a single $W_p$-orbit under this action are said to be **linked**. It is convenient to use as fundamental domain for $W_p$ relative to the dot action the **lowest alcove**

$$C_\circ := \{x \in E \mid 0 < \langle x + \rho, \alpha_0^\vee \rangle < p\},$$

where $\alpha_0$ denotes the highest *short* root (thus $\alpha_0^\vee$ is the highest root of $\Phi^\vee$). Those weights $\lambda \in X$ which lie in the interior of some alcove are called $p$-**regular**. In order for such weights to exist, it is necessary and sufficient that $p > \langle \rho, \alpha_0^\vee \rangle = h - 1$. Here $h$ is the **Coxeter number** of $W$.

When we study finite groups of Lie type, the weights in $X_r$ always play a special role. If $r = 1$, these weights lie in the closures of $|W|/f$ of the $p$-alcoves; here $f$ is the index of $\mathbb{Z}\Phi$ in $X$, usually called the **index of connection** for $\Phi$. For arbitrary $r$ the picture becomes hierarchical: $X_r$ involves the closures of $|W|/f$ of the $p^r$-alcoves, each of which is further subdivided into $p$-alcoves.

## 3.6. Linkage and Translation

Initially the only information we have about the potential composition factors $L(\mu)$ of a Weyl module $V(\lambda)$ is the inequality $\mu \leq \lambda$ in $X^+$. For an arbitrary indecomposable $\mathbf{G}$-module $M$, an easy necessary condition on the weights occurring in $M$ is that they all lie in the same coset of the root lattice in $X$: this follows from the fact that $\mathbf{G}$ is generated by its root subgroups $\mathbf{U}_\alpha$. A much more precise constraint can be formulated in terms of the action of $W_p$. Following work by Humphreys [213], Verma [426] proposed the following general principle, which was eventually proved by Andersen for all $p$. We call it the **Linkage Principle**:

THEOREM. *Let $M$ be an indecomposable (rational) $\mathbf{G}$-module. If $L(\lambda)$ and $L(\mu)$ both occur as composition factors of $M$, then $\mu = w \cdot \lambda$ for some $w \in W_p$.*

Andersen's proof, in the wider setting of all sheaf cohomology groups, leads to a more precise formulation in case $M$ is a highest weight module with highest weight $\lambda$: here $\lambda$ and $\mu$ must be *strongly linked*, meaning that the linkage is carried out by a sequence of affine reflections starting at $\mu$ and taking each successive weight to a higher one in the partial ordering. This just expresses the underlying role of the Chevalley–Bruhat ordering of $W_p$.

For an easy application of the Linkage Principle, let $C_\circ$ be the lowest alcove for $W_p$. If a dominant weight $\lambda$ lies in $\overline{C_\circ}$, then $V(\lambda) = L(\lambda)$ just as in the special case $\lambda = 0$. Indeed, no lower linked weight can be dominant.

Jantzen added another crucial ingredient: **translation functors**. The basic idea is that whenever $\lambda$ and $\mu$ both lie inside a single dominant $p$-alcove $C$, the composition factor multiplicities $[V(\lambda) : L(w \cdot \lambda)]$ and $[V(\mu) : L(w \cdot \mu)]$ are the same. More precisely, if $\lambda$ lies inside $C$ and $\mu$ lies in the "upper closure" $\widehat{C}$ of $C$, then the composition factor multiplicities of $V(\lambda)$ naturally predict those of $V(\mu)$ (but with degeneracies).

The results just outlined lead in 3.11 below to a reformulation of the basic problem which focuses strongly on $W_p$—but only under the condition that $p$-regular weights exist. For this one needs $p \geq h$. This condition allows one to focus just on a single $W_p$-orbit such as the orbit of the 0 weight.

## 3.7. Steinberg Modules

Before going further with the general theory we look at another important application of the Linkage Principle. Usually a Weyl module $V(\lambda)$ is much larger than its simple quotient $L(\lambda)$. But interesting exceptions occur when $\lambda = (p^r - 1)\rho$ for $r = 1, 2, \ldots$. Note that $L((p^r - 1)\rho)$ is *self-dual* (2.2), since $\rho^* = -w_\circ\rho = -(-\rho) = \rho$.

THEOREM. *Let $\rho \in X^+$ be the sum of the fundamental dominant weights.*

(a) *The Weyl module $V((p-1)\rho)$ coincides with its simple quotient $L((p-1)\rho)$.*
(b) $\dim L((p-1)\rho) = p^m$, *where $m = |\Phi^+|$.*
(c) *All other $L(\lambda)$ with $\lambda \in X_1$ have dimension strictly smaller than $p^m$.*
(d) *In general, $V((p^r-1)\rho) = L((p^r-1)\rho)$ and has dimension $p^{rm}$, while all other $L(\lambda)$ with $\lambda \in X_r$ have strictly smaller dimensions.*

PROOF. For (a), apply the Linkage Principle, which shows that no other restricted weight is linked to $(p-1)\rho$.

To prove (b) and (c), recall Weyl's dimension formula (see [**57**, VIII, §9, no. 2] or [**215**, 24.3]):

$$\dim V(\lambda) = \frac{\prod_{\beta > 0} \langle \lambda + \rho, \beta^\vee \rangle}{\prod_{\beta > 0} \langle \rho, \beta^\vee \rangle}.$$

Write $\lambda \in X_1$ as $\lambda = \sum_{\alpha \in \Delta} m_\alpha \varpi_\alpha$, where $\varpi_\alpha$ is the fundamental weight corresponding to $\alpha$. Here $m_\alpha \le p - 1$. For each fixed $\beta > 0$, we get

$$\langle \lambda, \beta^\vee \rangle = \sum_{\alpha \in \Delta} m_\alpha \langle \varpi_\alpha, \beta^\vee \rangle \le (p-1) \sum_{\alpha \in \Delta} \langle \varpi_\alpha, \beta^\vee \rangle = (p-1)\langle \rho, \beta^\vee \rangle,$$

with equality just when all $m_\alpha = p - 1$. Thus the $\beta$ contribution to Weyl's formula can be estimated by

$$\frac{\langle \lambda + \rho, \beta^\vee \rangle}{\langle \rho, \beta^\vee \rangle} \le p,$$

with equality for all $\beta$ just when all $m_\alpha = p - 1$.

(d) This follows immediately from the Tensor Product Theorem (2.7) and the dimension comparison in (b), (c). □

For brevity write $\mathrm{St} := L((p-1)\rho)$, or more generally, $\mathrm{St}_r := L((p^r-1)\rho)$. When $G = \mathbf{G}^F$ over a field of $p^r$ elements, $\mathrm{St}_r$ is referred to simply as the **Steinberg module** for $G$. It turns out to play a pivotal role in the later theory, being a projective $KG$-module as well as simple: see 9.3 below. (Similarly, the restriction of $\mathrm{St}_r$ to the $r$th Frobenius kernel $\mathbf{G}_r$ is a projective module for the corresponding finite dimensional hyperalgebra $\mathrm{Dist}(\mathbf{G}_r)$ introduced in 2.5.)

## 3.8. Contravariant Form on a Weyl Module

A useful tool in studying Weyl modules is a *contravariant form* defined initially in characteristic 0. Recall the discussion in 2.4 and the references there to early work of Wong, Steinberg, and Jantzen. Rather than introduce a form on $L(\lambda)$ directly via an anti-involution on $\mathbf{G}$, one can start with a similarly behaved anti-involution $\tau$ on the universal enveloping algebra of $\mathfrak{g}_\mathbb{C}$. This map interchanges positive and negative root vectors in $\mathfrak{g}_\mathbb{C}$ while fixing a Cartan subalgebra pointwise. Moreover, $\tau$ preserves the Kostant $\mathbb{Z}$-form of the enveloping algebra. This leads much as in 2.4 to a symmetric bilinear form on $V(\lambda)_\mathbb{Z}$. It can be normalized so that $(v^+, v^+) = 1$ for a chosen maximal vector $v^+$. After reduction modulo $p$ the form becomes contravariant in the earlier sense relative to the action of $\mathbf{G}$:

$$(g \cdot v, v') = (v, \tau(g) \cdot v') \text{ for all } g \in \mathbf{G} \text{ and } v, v' \in V(\lambda).$$

As before, distinct weight spaces are orthogonal. But the form on $V(\lambda)$ is typically degenerate. More precisely:

PROPOSITION. *The contravariant form on a Weyl module $V(\lambda)$ with $\lambda \in X^+$ has as radical the unique maximal submodule, thus inducing a nondegenerate contravariant form on the simple quotient $L(\lambda)$.*

PROOF. The contravariance shows that the radical of the form is a submodule, proper because $(v^+, v^+) = 1$. On the other hand, if $M$ is a proper submodule of $V(\lambda)$ and $v \in M$, then all $\tau(g) \cdot v$ $(g \in \mathbf{G})$ lie in the sum of weight spaces not involving $\lambda$. Thus $(g \cdot v^+, v) = (v^+, \tau(g) \cdot v) = 0$, because distinct weight spaces are orthogonal. This shows that $V(\lambda)$ is orthogonal to $v$, forcing $M$ to be included in the radical. □

While recovering the form on $L(\lambda)$ introduced in 2.4, this construction opens the way to working more systematically over $\mathbb{Z}$. This leads in particular to the *Jantzen filtration* [**RAGS**, II.8] of $V(\lambda)$, which has both computational and theoretical interest.

### 3.9. Jantzen Filtration and Sum Formula

Working at first with the special $\mathbb{Z}$-form $V(\lambda)_{\mathbb{Z}}$, Jantzen introduced a filtration in $V(\lambda)$ for arbitrary $p$ and arbitrary $\lambda \in X^+$:

$$V(\lambda) = V(\lambda)^0 \supset V(\lambda)^1 \supset V(\lambda)^2 \supset \cdots,$$

with $V(\lambda)/V(\lambda)^1 \cong L(\lambda)$ and $V(\lambda)^n = 0$ for large enough $n$. This is now called the **Jantzen filtration**. In most cases it is expected to coincide with the natural radical filtration of $V(\lambda)$. What is distinctive about his construction is the resulting **Sum Formula**, expressing $\sum_{i>0} \operatorname{ch} V(\lambda)^i$ as a computable $\mathbb{Z}$-linear combination of Weyl characters.

To explain the formula, we need a little more notation. Write $s_{\alpha,cp}$ for the affine reflection relative to the hyperplane in $E$ obtained by translating the standard reflecting hyperplane (through $-\rho$) for $\alpha \in \Phi^+$ by $(cp\alpha)/2$. For an arbitrary $\mu \in X$, define a formal character expression $\chi(\lambda)$ as follows. If $\langle \mu + \rho, \alpha^\vee \rangle = 0$ for some root $\alpha$, set $\chi(\mu) = 0$. Otherwise there is a unique $w \in W$ for which $\lambda = w \cdot \mu$ lies in $X^+$ and has a Weyl character $\chi(\lambda)$. Then set

$$\chi(\mu) := \det(w) \operatorname{ch}(w \cdot \mu).$$

Finally, write $\nu_p(n) = r$ if $r$ is the precise power of $p$ which divides a nonzero integer $n$. Now the Sum Formula says, for arbitrary $p$ and arbitrary $\lambda \in X^+$:

$$\sum_{i>0} \operatorname{ch} V(\lambda)^i = \sum_{\alpha>0} \quad \sum_{0<cp<\langle\lambda+\rho,\alpha^\vee\rangle} \nu_p(cp)\chi(s_{\alpha,cp} \cdot \lambda).$$

Jantzen originally constructed his filtration by looking at the precise powers of $p$ dividing certain values of the contravariant form on $V(\lambda)_{\mathbb{Z}}$. As a result he found that the form on $V(\lambda)$ induces a nondegenerate form on each quotient in the filtration. But some restrictions on $p$ were needed in the proof of the Sum Formula. Later Andersen was able to define the filtration and prove the Sum Formula in general, by working in the sheaf cohomology setting with images of various canonical homomorphisms. (See [**RAGS**, II.8].)

In various papers Jantzen used this formula along with ad hoc arguments to compute all formal characters (for $\lambda \in X_1$ or not too large) in the low-rank cases $A_2, B_2, G_2, A_3$ and to obtain partial results for types $B_3, C_3, A_4$. The results for $A_2$ and $B_2$ recover or complete early calculations by Braden [**58**] done by more laborious methods. The method is quite simple for $A_2$, where there are only two alcoves in the restricted region and $V(\lambda) = L(\lambda)$ for $\lambda$ in the lowest alcove (or its upper wall). When $\lambda$ is in the upper alcove, with reflected weight $s \cdot \lambda = \mu$ in the lower alcove, the Sum Formula shows quickly that $V(\lambda)$ has just the two composition factors $L(\lambda)$ and $L(\mu)$. For discussion of the harder case $A_3$, see [**RAGS**, 8.20]. Use of the Sum Formula rapidly gets more complicated, as the recursion begins to involve numerous cancellations; eventually multiple composition factors create too much ambiguity to go further. (Many of the ideas are exposed in papers by Jantzen [**244**]–[**250**].)

## 3.10. Generic Behavior of Weyl Modules

We have outlined briefly the main tools developed by 1980 for the study of composition factor multiplicities in Weyl modules, in tandem with at least some of the submodule structure. At this point the overall picture may still strike the reader as hopelessly complicated. But the work of Jantzen and others does lead to a fairly good understanding of the "generic" behavior of Weyl modules. (See especially [**247**].) Here is a rough sketch.

First of all, one should assume $p \geq h$ in order to make any uniform statements. Linkage and translation arguments show that it is usually sufficient to focus on the $W_p$-orbit of a single $p$-regular weight in $C_o$ such as 0. More subtle is the fact that when $\lambda$ belongs to the lowest $p^2$-alcove and is not "too close" to any wall of the dominant Weyl chamber, the composition factors of $V(\lambda)$ lie in a generic pattern. In particular, the total number of composition factors as well as their multiplicities are independent of $\lambda$ (and conjecturally also independent of $p$, as confirmed by Jantzen in low rank cases). For example, the total number is $2, 9, 20, 119, 104$ for types $A_1, A_2, B_2, G_2, A_3$ respectively. Multiplicities begin to occur when one reaches $G_2$ and $A_3$. In each case the generic pattern of alcoves in which linked weights with positive multiplicity lie depends just on the "type" of alcove to which $\lambda$ belongs. The number of types is the number of restricted alcoves: $|W|/f$, as in 3.5. Figure 1 shows the two generic patterns for type $A_2$, which illustrate how far $\lambda$ must be from Weyl chamber walls.

FIGURE 1. Generic patterns for type $A_2$

Predictable but much more complicated generic patterns then occur for larger weights, as the power of $p$ grows. For example, when $\mathbf{G} = \mathrm{SL}(3, K)$ the number of composition factors of a generic Weyl module will be a power of 9, with multiplicities still 1: for example, 81 when $\lambda$ is in sufficiently general position in the lowest $p^3$-alcove. Here one has a configuration of 9 $p^2$-alcoves resembling one of the initial generic patterns, each $p^2$-alcove in turn involving a generic pattern of 9 $p$-alcoves. Besides predicting such patterns, Jantzen's formal character methods allow one to pass systematically from the generic case to more degenerate cases. The generic patterns recur indirectly for Frobenius kernels: see Chapter 10 below.

Somewhat miraculously, the study of reduction modulo $p$ of ordinary characters of finite groups of Lie type leads to very similar generic patterns of composition factors: see Chapters 16–17 below.

## 3.11. Lusztig's Conjecture

When $p \geq h$, Jantzen's work allows one to recast the earlier character comparison 3.1(2) in a way which emphasizes $W_p$ and its Chevalley–Bruhat ordering: when the weight $w \cdot 0$ is dominant, $\chi_p(w \cdot 0)$ is a $\mathbb{Z}$-linear combination of various Weyl characters $\chi(w' \cdot 0)$ with $w' \cdot 0 \in X^+$ and $w' \leq w$. But the unknown integer coefficients remain mysterious. Moreover, some upper bound on $w \cdot 0$ seems required in order to get a reasonable formula depending just on $W_p$ and not on subgroups of type $W_{p^r}$.

Pursuing a close analogy with the Kazhdan–Lusztig Conjecture in characteristic 0 (in the framework of 13.6 below), Lusztig [310] proposed a characteristic $p$ version in which the affine Weyl group $W_p$ replaces the ordinary Weyl group. Jantzen [RAGS, II.8.20, II.C] provides a thorough treatment of the conjecture and its consequences. For more concise expositions, see Andersen [16], Donkin [143]. The reader should be aware that there are somewhat different-looking formulations of the conjecture in the literature. (Also, there is a misstatement in Jantzen's first edition II.7.20(3): one needs to replace $ww_0$ and $w'w_0$ by $w_0w$ and $w_0w'$.)

To make a reasonable conjecture, we have to assume $\lambda$ is not "too large" relative to $p$. Consider *dominant* $p$-regular weights of the form $\lambda = -w \cdot 0$ (the sign being Lusztig's convention). Lusztig imposes a condition on $w$ known as the **Jantzen condition**:

$$\langle -w\rho, \alpha_0^\vee \rangle \leq p(p - h + 2)$$

This means roughly—but not precisely—that $\lambda$ should lie in the lowest $p^2$-alcove (see Boe [53]).

CONJECTURE (Lusztig). *Assume that $p \geq h$ and that $w \in W_p$ satisfies the Jantzen condition. If $-w \cdot 0$ is dominant, then*

$$(1) \qquad \chi_p(-w \cdot 0) = \sum_{y \leq w} (-1)^{\ell(w)-\ell(y)} P_{y,w}(1)\chi(-y \cdot 0),$$

*with the sum taken over those $y \in W_p$ for which $-y \cdot 0$ is dominant.*

Notice here that one can also write $-w \cdot 0 = -w\rho - \rho = ww_0 \cdot 0$.

Building on Andersen's further study of Kazhdan–Lusztig polynomials for the affine Weyl group, Kato [267] proved that Lusztig's Conjecture is compatible with the Tensor Product Theorem. This led him to expect that the conjecture would be valid for all $p$-restricted weights even when some of these fail to satisfy the Jantzen condition above. (This can happen for small $p$ even when $p \geq h$.)

## 3.12. Evidence for the Conjecture

The history of the Kazhdan–Lusztig Conjecture in characteristic 0 suggests that there is no elementary way to understand the connection between our multiplicity problem and the formalism of Kazhdan–Lusztig polynomials. Even so, there are various reasons to have confidence in the truth of Lusztig's Conjecture in characteristic $p$, coming from different directions:

- Heuristic evidence comes from the closely parallel Kazhdan–Lusztig Conjecture, published in 1979 and soon afterwards proved independently by Beilinson–Bernstein and Brylinski–Kashiwara.
- Even before Lusztig formulated his conjecture, Jantzen's combination of methods in [244]–[250] led to explicit computations in a number of low rank cases (with a few small primes excluded): $A_2, A_3, B_2, G_2$. These results were later confirmed to agree with Lusztig's Conjecture.
- Buch and Lauritzen [65] have worked out computer algorithms which led in April 1995 to a confirmation of the Lusztig–Kato conjecture for type $A_4$ and $p = 5$. This agrees with an independent calculation by Scott and his students, described in [363], who also verify the conjecture for $p = 7$.
- Some older computer calculations as well as direct study of special weights, to be discussed in the next chapter, produce explicit results for various small primes $p$. When $p \geq h$ they agree with Lusztig's Conjecture.

The most persuasive support for Lusztig's Conjecture comes from the work of Andersen–Jantzen–Soergel [18]. They develop a sophisticated indirect comparison with quantum groups at a $p$th root of unity, ultimately showing that for a fixed simple type and for "large enough" $p$, Lusztig's Conjecture is true. However, their method gives no definite estimate on how large $p$ must be.

This paper shows that the quantum group category furnishes a successful model in characteristic 0 for the combinatorics of Lusztig's Conjecture. The reduction modulo $p$ of a simple module for the quantum group with suitably bounded highest weight $\lambda$ involves the corresponding simple module $L(\lambda)$ for $\mathbf{G}$ as a quotient. Equality conjecturally holds for $p$ not too small: this is in fact equivalent to Lusztig's Conjecture. (But equality does not hold for some small $p$.) In any case, the comparison with quantum groups already imposes a stronger upper bound for weight multiplicities of $L(\lambda)$ than given by comparison with $V(\lambda)$: the simple modules for the quantum group themselves occur as quotients of modules having the same formal characters as Weyl modules.

More recently, Parshall–Scott [332, Part II] have formulated new equivalent conditions involving Ext under which the Lusztig Conjecture holds (or fails to hold), with special attention paid to type $A_\ell$.

A final remark should be made about the small primes missing from Lusztig's Conjecture, for which no definite conjecture has yet been formulated. One problem here is that the study of Weyl modules with $p$-restricted highest weights may actually require use of not just $W_p$, but also $W_{p^2}, \ldots$ (This becomes more apparent when projective modules are brought into the picture.) So the character formulas may become much more intricate as a result.

# CHAPTER 4

# Computation of Weight Multiplicities

In principle the classical results of Curtis and Steinberg (Chapter 2) reduce the study of simple $KG$-modules for a finite group $G$ of Lie type in the defining characteristic to the study of simple $\mathbf{G}$-modules $L(\lambda)$ with $p$-restricted highest weights. But the comparison with Weyl modules in Chapter 3 indicates that it is important in practice to allow more general highest weights.

For the relevant weights and primes, Lusztig's Conjecture promises what is probably the best possible conceptual understanding of the characters of simple modules for $\mathbf{G}$. It explains for example the sense in which their behavior is controlled by the affine Weyl group, independent of $p$. But from the combinatorial viewpoint, the recursive calculations needed to determine $\operatorname{ch} L(\lambda)$ explicitly remain daunting.

Since the early 1970s some effective computational approaches have been developed, based on the possibility of working with Weyl modules over $\mathbb{Z}$: see 4.2 and 4.4. Here too there is a recursive flavor, similar to that involved in Lusztig's approach to character formulas, but there is no special requirement on $p$. As a result, many explicit results for relatively small highest weights have been obtained with the help of computers. In this chapter we explain the basic methods involved and survey what has actually been done. One byproduct of the computational work is a good understanding of "small" representations for each of the simple types (4.7).

In the case of classical groups, simple modules can also be studied in special cases by working directly with constructions based on the standard representation (sometimes in tandem with Weyl modules): see 4.5.

Related computations in small cases lead to explicit decompositions of tensor products of simple modules, which will be discussed in more detail in Chapter 6. As in Chapter 3, the finite groups play only a motivating role here.

Two caveats: (1) It is almost impossible to guarantee the correctness of all published calculations, though in some cases independent work has produced the same results. (2) Different numbering systems for vertices of Dynkin diagrams are used in the literature, leading to possible confusion. Our convention is always to follow Bourbaki [56].

## 4.1. Weight Spaces in Verma Modules

The problem of computing weight multiplicities for $L(\lambda)$ is formally parallel to the classical problem for semisimple Lie algebras over $\mathbb{C}$. So we begin in that setting (which is explained in more detail in 13.6 below).

Take $\mathbf{G}_\mathbb{C}$ to be a simply connected algebraic group of the same type as $\mathbf{G}$, with Lie algebra $\mathfrak{g}_\mathbb{C}$ and universal enveloping algebra $U = U(\mathfrak{g}_\mathbb{C})$. Fix a Cartan subalgebra $\mathfrak{h}_\mathbb{C}$ and a choice of positive roots $\Phi^+$ and simple roots $\alpha_1, \ldots, \alpha_\ell$, along with a Chevalley basis in $\mathfrak{g}_\mathbb{C}$: $H_1, \ldots, H_\ell$ and root vectors $X_\alpha$ and $Y_\alpha$ for $\alpha > 0$. Let $\mathfrak{g}_\mathbb{Z}$

be the $\mathbb{Z}$-span of this basis; it is stable under the Kostant $\mathbb{Z}$-form $U_\mathbb{Z}$ of $U$, which is generated by 1 along with all $H_i(H_i - 1) \cdots (H_i - n + 1)/n!$ and all $X_\alpha^n/n!$ and $Y_\alpha^n/n!$. Then a triangular decomposition $U = U^- U^0 U^+$ relative to a fixed ordering of the positive roots adapts to $U_\mathbb{Z}$. For example, $U_\mathbb{Z}^+$ has as $\mathbb{Z}$-basis all products $X_{\alpha_1}^{a_1}/(a_1)! \ldots X_{\alpha_m}^{a_m}/(a_m)!$.

For each $\lambda \in \mathfrak{h}^*$ there is a **Verma module** $M = M(\lambda)_\mathbb{C}$, having a unique maximal submodule $M'$ and unique simple quotient $V = V(\lambda)_\mathbb{C}$ (more usually denoted by $L(\lambda)_\mathbb{C}$ in the literature). This quotient is finite dimensional precisely when $\lambda \in X^+$, the only case which interests us here. In any case $M$ is generated by a maximal vector $v^+$ (unique up to scalars) of weight $\lambda$. In fact, the linear map $U^- \to M$ sending $u \mapsto u \cdot v^+$ is an isomorphism. So the only nonzero weight spaces of a Verma module belong to weights $\mu \le \lambda$ in the usual partial ordering, while the dimension $d$ of such a weight space is equal to the value $\mathcal{P}(\lambda - \mu)$ of the **Kostant partition function**: the number of ways to write $\lambda - \mu$ as a sum of positive roots. All of this carries over to the $\mathbb{Z}$-forms under the action of $U_\mathbb{Z}$.

## 4.2. Weight Spaces in Characteristic $p$

During the 1970s a computational approach to weight space multiplicities in characteristic $p$ was developed by Burgoyne and Williamson. This is independent of the more theoretical considerations discussed in Chapter 3, but is limited in terms of practical application. Here we outline the steps involved in the proof of their main result. (See Burgoyne [66] and Burgoyne–Williamson [67]. There are also some related computations in the thesis of Burgoyne's student Zaslawsky [460].)

THEOREM. *For any $\lambda \in X^+$ and any $\mu \in X$, there is an effective algorithm which computes the weight multiplicity* $\dim L(\lambda)_\mu$.

Fix $\lambda \in X^+$. To compute the (classical) dimension of the $\mu$-weight space $\dim V_\mu$, one has efficient methods available, notably Freudenthal's recursive formula. But there is an alternative algorithm over $\mathbb{Z}$. In the notation of 4.1, $V_\mu$ is spanned by the vectors $u \cdot v^+$, where $u \in U_\mathbb{Z}^-$ is the product (in some fixed order) of the $Y_{\alpha_i}^{b_i}/b_i!$ and $\lambda - \mu = \sum b_i \alpha_i$. Write briefly $u = Y_B$ with $B = (b_1, \ldots, b_m)$. Similarly, we can form elements $X_A \in U_\mathbb{Z}^+$, where $A = (a_1, \ldots, a_m)$ and $\lambda - \mu = \sum a_i \alpha_i$.

With this notational setup it is clear that $X_A Y_B v^+$ is just an integral multiple $c_{AB} v^+$. Write $C := C^{(\lambda, \mu)}$ for the resulting $d \times d$ matrix, relative to some fixed ordering of the $d := \mathcal{P}(\lambda - \mu)$ permitted $m$-tuples of integers. For the Verma module $M = M(\lambda)_\mathbb{C}$, note that $C$ may be viewed as the matrix of a linear map $\varphi : M_\mu \to \mathbb{C}^d$, relative to the basis of $M_\mu$ consisting of all $Y_B v^+$. It is easy to see that $\text{Ker}\,\varphi$ is just the intersection of $M_\mu$ with the maximal submodule $M'$ of $M$. Thus the rank of $C$ is the desired weight multiplicity $\dim V_\mu$. It is computable over $\mathbb{Z}$ from a knowledge of the commutation relations in $U_\mathbb{Z}$. (First one has to fix the choice of the Chevalley basis by making compatible sign choices.)

In practice, one can limit the computation to the case where $\mu$ is dominant by taking advantage of the Weyl group invariance of weight multiplicities in $V(\lambda)_\mathbb{C}$. Note too that $C$ can be made *symmetric* by choosing the order of roots in the products $X_A$ to be the reverse of that chosen for $Y_B$.

Once everything is set up over $\mathbb{Z}$, one can in turn study the transition from a Weyl module $V(\lambda) = K \otimes V(\lambda)_\mathbb{Z}$ to its simple quotient $L(\lambda)$ in characteristic $p$.

Fixing $\mu$ as above, $\dim L(\lambda)_\mu$ may be computed as the rank of the matrix obtained from $C$ by reduction modulo $p$.

Looking back at the discussion of the contravariant form in 3.1, we can see another, more suggestive, way to think about the computational approach just described. The matrix $C$ attached to the pair of weights $(\lambda, \mu)$ may be interpreted as the matrix of a contravariant form on the $\mu$-weight space of the Verma module $M$. This form is typically degenerate, but induces the earlier nondegenerate contravariant form on the corresponding weight space of $V(\lambda)_{\mathbb{C}}$. The form takes integer values on $V(\lambda)_{\mathbb{Z}}$ and thus permits reduction modulo $p$. The $p$-rank of $C$ then equals $\dim L(\lambda)_\mu$.

## 4.3. An Easy Example: $SL(2, K)$

Implementation of the algorithm just outlined normally requires a computer, apart from the easy case of $\mathbf{G} = SL(2, K)$. Here the matrix $C$ is $1 \times 1$.

Denote the standard Chevalley basis of $\mathfrak{g}$ by $(X, H, Y)$. Dominant weights $\lambda$ may be identified with nonnegative integers. Over $\mathbb{C}$ or $\mathbb{Z}$ the simple module of highest weight $\lambda$ has a basis $(v_0, v_1, \ldots, v_\lambda)$ on which $\mathfrak{g}$ acts as in [**215**, §7]:

$$
\begin{aligned}
H \cdot v_i &= (\lambda - 2i)v_i \\
Y \cdot v_i &= (i + 1)v_{i+1} \\
X \cdot v_i &= (\lambda - i + 1)v_{i-1}
\end{aligned}
$$

with the convention $v_{-1} = 0 = v_{\lambda+1}$.

Now one has to use a standard identity [**215**, (26.2)] in $U_{\mathbb{Z}}$:

$$
\frac{X^c}{c!} \frac{Y^a}{a!} = \sum_{k=0}^{\min(a,c)} \frac{Y^{a-k}}{(a-k)!} \binom{H - a - c + 2k}{k} \frac{X^{c-k}}{(c-k)!}.
$$

In the special case $c = a$, this readily yields

$$
\frac{X^a}{a!} \frac{Y^a}{a!} \cdot v_0 = \binom{\lambda}{a} v_0.
$$

This becomes 0 in $K$ just when $p$ divides the binomial coefficient on the right. In particular, we recover the fact that $V(\lambda) = L(\lambda)$ for $0 \le \lambda < p$.

## 4.4. Computational Algorithms

Here is a brief survey of computational algorithms based on either the contravariant form theory outlined above or more direct characteristic $p$ methods:

- The original papers by Burgoyne and Williamson cited above contain a number of relatively small tables computed by their method for the root systems of rank $\le 4$. There is one table for each coset of the root lattice in the weight lattice. Table 1 below extracts a small sample to illustrate their format in the case of type $C_3$. This follows the numbering of simple roots in Bourbaki [**56**] in which the first two are short. The weight $a\varpi_1 + b\varpi_2 + c\varpi_3$ is denoted by $abc$. The last three columns of the table give weight multiplicities in $L(\lambda)$ for the weight $\lambda$ in the left column. These are valid for all $p$ except as indicated in the third column.

| weight | dim $L(\lambda)$ | primes | 100 | 001 | 110 |
|--------|------------------|--------|-----|-----|-----|
| 100 | 6 | | 1 | | |
| 001 | 14 | | 1 | 1 | |
| | 8 | $p = 2$ | 0 | 1 | |
| 110 | 64 | | 4 | 2 | 1 |
| | 44 | $p = 3$ | 3 | 1 | 1 |
| | 58 | $p = 7$ | 3 | 2 | 1 |

TABLE 1. Some weight multiplicities for $C_3$ [**67**]

- Gilkey and Seitz [**186**] modify the Burgoyne–Williamson algorithm to make it more efficient. For example, when the highest weight has a non-trivial stabilizer in the Weyl group, they can reduce to a substantially smaller matrix computation. To illustrate their method, they exhibit some weight multiplicities for each of the exceptional groups. In Table 2 below, adapted from their table, some results are summarized for the group **G** of type $F_4$. In particular, they confirm earlier partial results for fundamental weights by Veldkamp [**425**] when $p = 2$ and by Wong [**437**].

  The numbering of simple roots follows the convention in Bourbaki [**56**], with the first two being long (the reverse of Wong's convention). Coordinates of a dominant weight are then written briefly as a string like 0011. Here the ten weights $\lambda_i$ (abbreviated $i$) index rows and columns, with the $(i, j)$ entry giving the multiplicity of the weight $\lambda_j$ in $L(\lambda_i)$. Only those primes for which $\lambda_i$ is *restricted* are considered. This is indicated by notation such as $(p > 2)$, followed by the generic data for this range of primes and then by the finite number of exceptions for special values of $p$.

- Koppinen [**280**] develops a substantially different (and faster) computational approach, in which $p$ is fixed at the outset. His method is similar in spirit to that of Burgoyne, but he works directly in characteristic $p$ with the hyperalgebra $\mathrm{Dist}(\mathbf{G}_r)$ of the $r$th Frobenius kernel $\mathbf{G}_r$. Here $r$ is chosen so that a given highest weight $\lambda$ is $p^r$-restricted. Generalizing an idea of G.M. Nielsen in the case $r = 1$, he writes down an explicit (not very complicated) generator for the simple module $L(\lambda)$ inside $\mathbf{u}_r$. Using the commutator relations which come from the Kostant $\mathbb{Z}$-form $U_{\mathbb{Z}}$, he can then make explicit computations of weight space dimensions.

  Koppinen gives a careful summary of his algorithm. But the actual Pascal programs are written down in an unpublished document, where it is indicated that the algorithm was checked thoroughly in the complicated case of type $G_2$.

- Purely combinatorial algorithms are developed for $\mathrm{SL}(n, K)$ and $\mathrm{Sp}(2n, K)$ by Pittaluga and Strickland [**399, 339, 340**], who allow $K$ to be any infinite field. In our notation, they study $L(\lambda)$ as the image of the module homomorphism $V(\lambda) \to H^0(\lambda)$ (the latter called a "Schur module" in this context). This can all be written down explicitly over $\mathbb{Z}$. The weights are parametrized by certain Young diagrams, while bases over $\mathbb{Z}$ are given in terms of tableaux, as in the work of De Concini and others. This permits an algorithmic approach which ultimately reduces to rank computations

| weight | $\dim L(\lambda_i)$ | primes | 1 | 2 | 3 | 4 | 5 | 6 | 7 | 8 | 9 | 10 |
|---|---|---|---|---|---|---|---|---|---|---|---|---|
| $\lambda_1 = 0000$ | 1 | | 1 | | | | | | | | | |
| $\lambda_2 = 0001$ | 26 | | 2 | 1 | | | | | | | | |
| | 25 | $p = 3$ | 1 | 1 | | | | | | | | |
| $\lambda_3 = 1000$ | 52 | | 4 | 1 | 1 | | | | | | | |
| | 26 | $p = 2$ | 2 | 0 | 1 | | | | | | | |
| $\lambda_4 = 0010$ | 273 | | 9 | 5 | 2 | 1 | | | | | | |
| | 246 | $p = 2$ | 6 | 4 | 2 | 1 | | | | | | |
| | 196 | $p = 3$ | 4 | 3 | 1 | 1 | | | | | | |
| $\lambda_5 = 0002$ | 324 | $(p > 2)$ | 12 | 5 | 3 | 1 | 1 | | | | | |
| | 298 | $p = 7$ | 10 | 4 | 3 | 1 | 1 | | | | | |
| | 323 | $p = 13$ | 11 | 5 | 3 | 1 | 1 | | | | | |
| $\lambda_6 = 1001$ | 1053 | | 21 | 14 | 6 | 4 | 1 | 1 | | | | |
| | 676 | $p = 2$ | 4 | 8 | 2 | 3 | 0 | 1 | | | | |
| $\lambda_7 = 0100$ | 1274 | | 26 | 13 | 10 | 4 | 3 | 1 | 1 | | | |
| | 246 | $p = 2$ | 6 | 0 | 4 | 0 | 2 | 0 | 1 | | | |
| | 1222 | $p = 3$ | 22 | 12 | 9 | 4 | 3 | 1 | 1 | | | |
| $\lambda_8 = 2000$ | 1053 | $(p > 2)$ | 21 | 9 | 8 | 3 | 3 | 1 | 1 | 1 | | |
| | 755 | $p = 7$ | 11 | 5 | 5 | 2 | 2 | 1 | 1 | 1 | | |
| $\lambda_9 = 0011$ | 4096 | | 64 | 40 | 24 | 14 | 8 | 4 | 2 | 0 | 1 | |
| | 2404 | $p = 3$ | 28 | 20 | 12 | 8 | 5 | 3 | 1 | 0 | 1 | |
| | 2991 | $p = 7$ | 39 | 25 | 17 | 10 | 7 | 3 | 2 | 0 | 1 | |
| | 3773 | $p = 13$ | 53 | 35 | 21 | 13 | 7 | 4 | 2 | 0 | 1 | |
| $\lambda_{10} = 0003$ | 2652 | $(p > 3)$ | 36 | 24 | 13 | 9 | 5 | 3 | 1 | 0 | 1 | 1 |
| | 2651 | $p = 7$ | 35 | 24 | 13 | 9 | 5 | 3 | 1 | 0 | 1 | 1 |

TABLE 2. Some weight multiplicities for $F_4$ [186]

for matrices. The authors remark that they have implemented their algorithm on a computer. (For some related work on choosing spanning sets for modules in the case of $GL(n, K)$, see Cliff and Stokke [110].)

- Lübeck [306] refines the method of Gilkey–Seitz, in order to get more comprehensive results on degrees of "small" representations: see 4.7 below.

## 4.5. Fundamental Modules for Symplectic Groups

When **G** is of classical type, representations have often been studied by methods which stand apart from Lie theory. Instead of focusing on Weyl modules and their composition factors, one can start with the natural module for **G** and study the structure of the resulting exterior or symmetric powers. The initial constructions here are essentially characteristic-free but acquire richer internal structure in characteristic $p$. (This is closely related to questions about "tilting" modules and fusion rules: see [**RAGS**, II.E].) General and special linear groups are considered in more detail below in Chapter 19.

Simple modules with a fundamental highest weight tend to arise naturally from exterior powers. We call these **fundamental modules** for short. Here we look briefly at the symplectic group $Sp(2\ell, K)$.

Premet and Suprunenko [345] determine the composition factors of Weyl modules having fundamental highest weights, by reducing the problem to known properties of symmetric group representations involving 2-part partitions. Their multiplicity and dimension results, while explicit, are given in terms of combinatorial algorithms rather than closed formulas. Recently McNinch [318] has recovered these results in a more direct way, by exploiting tilting modules together with Howe dual pairs: known facts about groups of rank 1 are used to explore groups of higher rank.

Meanwhile Gow [191] shows how to use the classical contraction operators on exterior powers to pin down various kernels and images. When $p \leq \ell - 1$, he thereby constructs $p - 1$ fundamental modules inside the $k$th exterior powers (with $k = \ell - p + 2, \ldots, \ell$). Their dimensions can be worked out explicitly by a combinatorial algorithm. This provides another alternative to the Premet–Suprunenko computation for the corresponding fundamental weights $\varpi_{\ell-p+2}, \ldots \varpi_{\ell}$.

Baranov–Suprunenko [32] explore the structure of exterior powers further, working out branching rules relative to the subgroup $\mathrm{Sp}(2\ell-2, K)$ of $\mathrm{Sp}(2\ell, K)$.

Further results on the fundamental modules for small primes are sketched by Foulle [176], using ideas from the theory of tilting modules.

### 4.6. Small Weights and Small Characteristics

Outside the framework of Lusztig's Conjecture, a number of special cases have been worked out by various methods. Typically these involve particular Lie types and are limited to "small" dominant weights (such as fundamental weights) or "small" primes (often smaller than the Coxeter number). Here is a brief survey of calculations in the literature beyond those already discussed:

- Springer [382] makes the character data explicit for $G_2$ when $p = 3$, using the special isogeny as in Steinberg [393].
- Hagelskjaer [201] computes the character data for $G_2$ when $p = 5$, thereby showing consistency with Lusztig's approach even though $p$ is smaller than the Coxeter number 6.
- Mertens [319] works out many aspects of the representation theory of $G_2$, including some detailed computations for small $p$: see Chapter 18 below.
- Xu and Ye [438], Ye and Zhou [449, 450, 451, 452] compute the irreducible characters of low-rank algebraic groups: types $A_4, A_5, A_6, B_3, C_3$ when $p = 2$ and types $B_3, C_3, A_4, D_4$ when $p = 3$. They use results of N. Xi coming from a comparison with quantum groups, along with computer assistance.
- Seitz [364, §6] determines, for a simple algebraic group $\mathbf{G}$, those $p$ and restricted $\lambda$ for which all weight spaces of $L(\lambda)$ are 1-dimensional. (He does this under mild restrictions on $\lambda$ when there are two root lengths and the ratio of squared lengths is $p$.)
- Krcmar [288] determines the smallest dimension of a nontrivial simple module for each group of Lie type; these correspond to certain fundamental weights.

### 4.7. Small Representations

The computations in special cases just surveyed contribute to the search for "small" representations. This search has interested finite group theorists for some time in connection with the classification and internal structure of simple groups: see Kleidman–Liebeck [**274**, 5.4]. There are similar applications to the study of the subgroup structure of reductive algebraic groups: see for example Brundan [**63**, 1.13], whose Table 2 is based on Liebeck [**298**], and related work of Liebeck–Seitz on maximal subgroups.

Broadly speaking, finite group theorists look for all low-dimensional irreducible projective representations of arbitrary finite simple groups over fields of arbitrary characteristic. Since the universal groups of Lie type are (with a handful of exceptions in characteristics $2, 3$) universal central extensions of their simple quotients, attention shifts in this case to ordinary representations of the universal groups.

Several specific questions overlap here in the setting of a group $G = \mathbf{G}^F$ of Lie type in the defining characteristic:

(1) For a given $G$, determine the smallest nontrivial simple $KG$-module.

(2) For a given bound $M$, determine all simple $KG$-modules of dimension $\leq M$.

(3) If $\mathbf{G}$ has rank $\ell$, find all simple $KG$-modules of dimension bounded by some reasonable function of $\ell$.

These questions can be viewed as questions about $\mathbf{G}$-modules, thanks to the Restriction Theorem and the Tensor Product Theorem. Since duals and Frobenius twists of simple modules are again simple of the same dimension, these are usually not mentioned explicitly in tables of the sort given by Brundan. Similarly, it is enough to list tensor-indecomposable simple modules: those which are not tensor products of smaller nontrivial simple modules. So the case of $p$-restricted highest weights is the most crucial one.

As noted above, Question 1 is treated by Krcmar [**288**]; but it is also covered by Question 2. For Question 2, a popular bound (for the more general finite group problem above) has been $M = 250$, which for example takes in the adjoint module for $E_8$. But much larger bounds are often feasible for groups of Lie type, while for classical types Question 3 is a better formulation.

By refining the Liebeck–Seitz computational algorithm and combining it with a variety of other case-by-case arguments and known results (many of them due to Liebeck), Lübeck [**306**] has worked out satisfying answers to these questions for all Lie types including the groups of Suzuki and Ree. His computations are summarized in many tables. (Note that his conventions for numbering of vertices in the Dynkin diagram differ from those of Bourbaki.) For example, for the group of type $E_8$, the bound $M = 100000$ is used. There are at most five $p$-restricted weights yielding simple modules of dimension $< M$, typically of dimensions $1, 248, 3875, 27000, 30380$. (Some of these dimensions shrink for $p = 2, 3, 5, 7, 31$.)

Lübeck organizes his results in two ways. For each root system of rank $\leq 11$ he chooses a realistic degree bound (ranging from 300 to 100000) for his list of $p$-restricted weights. But for a classical type of rank $\ell > 11$ he can use $\ell^3$ as an upper bound (or $\ell^3/8$ in the case of type $A_\ell$). Ultimately his results determine all simple modules of dimension $\leq 250$ for groups of Lie type in the defining characteristic, thus completing earlier work of Hiss, Malle, and others on simple groups over arbitrary fields.

# CHAPTER 5

# Other Aspects of Simple Modules

So far we have focused attention mainly on the classification and description of simple modules for a group of Lie type. This chapter assembles a number of more-or-less independent topics which emerge naturally from this study:

- In 5.1 we look at how Frobenius maps permute simple **G**-modules of arbitrary highest weight. This contributes in turn to a determination of minimal splitting fields (5.2).
- When there are two root lengths, with $p$ equal to the squared ratio of long to short, special isogenies exist between algebraic groups with corresponding dual root systems. This leads to a refined tensor product factorization of their simple modules. For the nonisomorphic groups of types $B_\ell$ and $C_\ell$ with $\ell > 2$ in characteristic 2, we then find a dimension-preserving bijection between the two collections of simple modules (5.3–5.4).
- After recalling some standard facts about Grothendieck rings and Brauer characters (5.5), we compare formal characters for **G**-modules with Brauer characters for $\mathbf{G}^F$ (5.6).
- Several sections explore a recurring theme: the behavior of **G**-modules on restriction to a finite group of Lie type. Examples include simple modules $L(\lambda)$ for arbitrary $\lambda \in X^+$, tensor products of simple modules, Weyl modules. Further instances will arise later in the study of projective modules, Ext groups, etc.

When dealing with an arbitrary $KG$-module $M$ we use the notation $[M : L]_G$ for the number of composition factors of $M$ isomorphic to a given simple module $L$, and similarly with $G$ replaced by the algebraic group **G**.

## 5.1. Restriction of Frobenius Maps

Recall from 2.5 two types of Frobenius endomorphisms $F : \mathbf{G} \to \mathbf{G}$. In one case $F$ is the standard Frobenius map relative to $p^r$ (and may be denoted $F_r$ to emphasize $r$). Its group of fixed points is a Chevalley group over $\mathbb{F}_q$, where $q = p^r$. In the other case $F$ is the composite of $F_r$ and a (commuting) graph automorphism $\pi$ of order 2 or 3. Its fixed points comprise a twisted group, contained in a corresponding Chevalley group over $\mathbb{F}_{q^2}$ or $\mathbb{F}_{q^3}$.

It is natural to ask how an arbitrary simple **G**-module $L(\lambda)$ with $\lambda \in X^+$ behaves on restriction to one of these finite subgroups. In view of the Tensor Product Theorem and the fact that Frobenius maps commute with tensoring, the key question is how $L(\lambda)^{[r]}$ restricts to $G = \mathbf{G}^F$ when $F$ is defined relative to $q = p^r$ and $\lambda \in X_1$. Iteration then makes it easy to treat higher Frobenius twists. (See Kleidman–Liebeck [**274**, Prop. 5.4.2] and Liebeck [**298**].)

PROPOSITION. *Let $G = \mathbf{G}^F$ over a field of $p^r$ elements, and fix $\lambda \in X_1$.*

(a) If $F = F_r$ is the standard Frobenius map relative to $p^r$, then the **G**-module $L(\lambda)^{[r]}$ is isomorphic as $KG$-module to $L(\lambda)$.

(b) If $F = F_r \circ \pi$ for a nontrivial graph automorphism $\pi$, then $L(\lambda)^{[r]}$ is isomorphic as $KG$-module to the twist by $\pi$ of the $KG$-module $L(\lambda)$.

PROOF. In either case let $\varphi : \mathbf{G} \to \mathrm{GL}(L(\lambda))$ be the representation in question, whose $r$th Frobenius twist is $\varphi \circ F_r$. If $G$ is a Chevalley group, this map coincides on $G$ with $\varphi$. But if $G$ is twisted and $g \in G$, we have $\varphi(F_r(g)) = \varphi(\pi^{-1}\pi F_r(g)) = \varphi(\pi^{-1}(g))$. □

## 5.2. Splitting Fields

The strategy in Chapter 2 was to look first at the simple **G**-modules over the algebraically closed field $K$, then to show that those having $p^r$-restricted highest weights remain simple and nonisomorphic for $G = \mathbf{G}^F$ when $F$ is a Frobenius map relative to a field of order $p^r$. From the algebraic group side comes the further refinement in the description of $L(\lambda)$ as a twisted tensor product. This development bypasses the question of splitting fields, which we now address.

Recall that for an arbitrary finite group $G$, a field $k$ is called a **splitting field** for a simple $kG$-module if the module remains simple under any extension of scalars. In turn, $k$ is a splitting field for $G$ if it is a splitting field for every simple $kG$-module. Equivalently, if $k$ is a subfield of the algebraically closed field $K$, all simple $KG$-modules can be realized by matrix representations over $k$. When $k$ has characteristic $p$, it is sufficient (but often not necessary) for $k$ to contain all $e$th roots of 1, where $e$ is the $p'$-part of the exponent of $G$.

A simplification in prime characteristic is that all Schur indices are 1, due to Wedderburn's theorem on finite division algebras. This insures for a simple $KG$-module $M$ that the subfield of $K$ generated by all traces of the representing matrices is already a splitting field for $M$, indeed a minimal splitting field. A practical test for a finite extension $k$ of $\mathbb{F}_p$ to be a splitting field for $M$ is that the Galois group of $k$ should fix all traces of elements of $G$ acting on $M$.

For this and other background information see for example Curtis–Reiner [**125**, §70] and Gorenstein–Lyons–Solomon [**190**, 2.8].

The following basic result is implicit in Steinberg [**393**] and spelled out more fully in his lectures [**394**, §13].

THEOREM. Let $G = \mathbf{G}^F$ over a field of $q = p^r$ elements.

(a) If $G$ is a Chevalley group, then $\mathbb{F}_q$ is a splitting field for $G$.

(b) If $G$ is of type $^2\mathrm{A}_\ell, ^2\mathrm{D}_\ell$, or $^2\mathrm{E}_6$, then $\mathbb{F}_{q^2}$ is a splitting field for $G$.

(c) If $G$ is of type $^3\mathrm{D}_4$, then $\mathbb{F}_{q^3}$ is a splitting field for $G$.

How one proves this theorem depends on the approach taken to the classification and construction of simple modules. For example, Steinberg [**394**] uses the $BN$-pair setting of Curtis–Richen to construct simple modules for $G$, where it becomes clear that all representations can be realized within the group algebra over the field specified in the theorem. In effect, a splitting field for the finite torus $\mathbf{T}^F$ is also a splitting field for $G$.

On the other hand, the approach via algebraic groups yields a somewhat different argument: When $\lambda \in X_r$, the Tensor Product Theorem coupled with Proposition 5.1 show for a Chevalley group $G$ that the $KG$-module $L(\lambda)$ is isomorphic to the Frobenius twist $L(\lambda)^{[r]}$. Thus the associated matrix representations lead to the same traces, i.e., the traces are invariant under raising to the $q$th power. This just says that the traces lie in the field $\mathbb{F}_q$. For twisted groups the idea is similar, but here one has to substitute $\mathbb{F}_{q^2}$ or $\mathbb{F}_{q^3}$ for $\mathbb{F}_q$ when $\pi$ has order 2 or 3.

This type of reasoning also pins down the minimal splitting field for each individual simple module $L(\lambda)$. For example, when $G$ is a Chevalley group over $\mathbb{F}_{p^2}$ and $\lambda \in X_1$, we see that $L(\lambda) \otimes L(\lambda)^{[1]}$ is isomorphic as a $KG$-module to its Frobenius twist and thus has splitting field $\mathbb{F}_p$. See the discussion in Kleidman–Liebeck [274, §5.4], where minimal splitting fields of individual modules are determined.

Similarly, one shows that a Suzuki or Ree group (of type $^2B_2, {}^2G_2, {}^2F_4$) over a field whose order is an odd power of 2 or 3 has that field as a splitting field.

## 5.3. Special Isogenies

In his 1956–58 Paris seminar, Chevalley discovered (in the course of classifying semisimple algebraic groups up to isomorphism) some special isogenies between groups having dual root systems $\Phi, \Phi^\vee$. These maps exist only when $p = 2$ or 3 and there are squared root lengths having the corresponding ratio 2 or 3. In the cases $B_2 = C_2, F_4$, and $G_2$, where $\Phi$ and $\Phi^\vee$ are isomorphic, the isogenies are closely related to the existence of Suzuki and Ree groups: see 1.3 and Chapter 20.

After reviewing Chevalley's construction we shall explore, following Steinberg [393, §10–11], how special isogenies lead to a further refinement of the Tensor Product Theorem 2.7. This is especially interesting when $p = 2$ and the isogeny involves the simply connected algebraic groups of types $B_\ell$ and $C_\ell$. This isogeny is actually an isomorphism of abstract groups (though not of algebraic groups), inducing an isomorphism between the corresponding finite groups over $\mathbb{F}_q$. In particular, both finite groups have the same collection of simple modules coming by restriction from the two collections of simple modules for the algebraic groups having $q$-restricted highest weights. So the dimensions of the latter must match up. In 5.4 we make this comparison precise, as a byproduct of Steinberg's refined factorization.

To fix the ideas, consider $\mathbf{G} = \mathbf{G}'$ of type $G_2$ when $p = 3$. Taking $\alpha$ and $\beta$ to be simple roots (with $\alpha$ short), the corresponding fundamental weights are $\varpi_\alpha = 2\alpha + \beta$ (the highest short root) and $\varpi_\beta = 3\alpha + 2\beta$ (the highest long root). In the dual root system $\Phi^\vee$, long and short just get reversed in these descriptions. There is a special isogeny $\varphi : \mathbf{G} \to \mathbf{G}$, whose comorphism $\varphi^* : X(\mathbf{T}) \to X(\mathbf{T})$ maps the long $\alpha^\vee$ to $3\alpha$ and the short $\beta^\vee$ to $\beta$. As a result, the dual fundamental weight $\varpi_{\alpha^\vee} = 2\alpha^\vee + 3\beta^\vee$ maps to $3\varpi_\alpha$ while $\varpi_{\beta^\vee} = \alpha^\vee + 2\beta^\vee$ maps to $\varpi_\beta$.

In general we have a pair of simple, simply connected algebraic groups $\mathbf{G}$ and $\mathbf{G}'$ with dual root systems $\Phi$ and $\Phi^\vee$. Assume that $p = 2$ or 3, while $\Phi$ has roots for which the ratio of squared lengths is $p$. Denote by $\varphi : \mathbf{G} \to \mathbf{G}'$ the associated special isogeny: it maps $x_\alpha(c) \mapsto x_{\alpha^\vee}(c)$ when $\alpha$ is long and $x_\alpha(c) \mapsto x_{\alpha^\vee}(c^p)$ when $\alpha$ is short. Note that if $\varphi' : \mathbf{G}' \to \mathbf{G}$ is defined similarly, the fact that $\mathbf{G}$ is generated by its root subgroups implies that the composite $\varphi' \circ \varphi : \mathbf{G} \to \mathbf{G}$ is just the standard Frobenius map relative to $p$.

When $\varphi$ is restricted to a map of maximal tori $\mathbf{T} \to \mathbf{T}'$, the comorphism $\varphi^* : X(\mathbf{T}') \to X(\mathbf{T})$ sends $\alpha^\vee \mapsto \alpha$ when $\alpha$ is long and sends $\alpha^\vee \mapsto p\alpha$ when $\alpha$ is short. Corresponding fundamental weights are mapped similarly, as illustrated above in the case when both $\mathbf{G}$ and $\mathbf{G}'$ are of type $G_2$.

Denote by $X(\mathbf{T})_L$ and $X(\mathbf{T})_S$ respectively the set of $\lambda$ for which $\langle \lambda, \alpha^\vee \rangle = 0$ whenever $\alpha$ is short (resp. long). (Similar notation is used for $\mathbf{T}'$.) We say that

$\lambda \in X(\mathbf{T})_L$ has *long support*, since $\lambda$ is a $\mathbb{Z}$-linear combination of fundamental weights corresponding to long simple roots. Similarly, $\lambda \in X(\mathbf{T})_S$ has *short support*. In this language, $\varphi^*$ sends weights with short support to corresponding weights with long support but sends those with long support to $p$ times the corresponding weights with short support.

REMARK. When $\varphi : \mathbf{G} \to \mathbf{G}'$ is a special isogeny, the ideal $\mathfrak{g}_S := \operatorname{Ker} d\varphi$ in $\mathfrak{g}$ is generated by the root vectors corresponding to short roots and is stable under the adjoint representation. For example, when $\mathbf{G} = \mathbf{G}'$ has type $G_2$ and $p = 3$, the $\mathbf{G}$-module $\mathfrak{g}/\mathfrak{g}_S$ is simple of dimension 7 and has as highest weight the highest root $3\alpha + 2\beta$ (which for other primes is the highest weight of the 14-dimensional adjoint representation). As a restricted Lie algebra, $\mathfrak{g}_S$ corresponds to a group scheme which plays a role here analogous to that of the first Frobenius kernel $\mathbf{G}_1$.

## 5.4. Steinberg's Refined Factorization

Now we can explain the refined version of the Tensor Product Theorem found by Steinberg [393]:

THEOREM. *Let $\varphi : \mathbf{G} \to \mathbf{G}'$ be a special isogeny. With notation as in 5.3, we have:*

(a) *Composition of $\varphi$ with an irreducible representation of $\mathbf{G}'$ yields an irreducible representation of $\mathbf{G}$. If the highest weight $\lambda'$ lies in $X(\mathbf{T}')_S$, the resulting representation of $\mathbf{G}$ has the same dimension and has highest weight $\varphi^*(\lambda')$ lying in $X(\mathbf{T})_L$. If $\lambda' \in X(\mathbf{T}')_L$, then $\varphi^*(\lambda')$ lies in $X(\mathbf{T})_S$ and is divisible by $p$. Thus $\varphi$ induces a natural dimension-preserving bijection between simple modules for $\mathbf{G}$ with highest weight having long support and simple modules for $\mathbf{G}'$ with highest weight having short support.*

(b) *Write $\lambda \in X(\mathbf{T})^+$ uniquely as a sum of dominant weights $\mu \in X(\mathbf{T})_L$ and $\nu \in X(\mathbf{T})_S$. Then $L(\lambda) \cong L(\mu) \otimes L(\nu)$.*

PROOF. Part (a) follows almost immediately from the observations made about $\varphi$ in the preceding section. Here the surjectivity of $\varphi$ insures that simple modules for $\mathbf{G}'$ pull back to simple modules for $\mathbf{G}$.

Thanks to the Tensor Product Theorem it is enough in (b) to deal with the case when $\lambda, \mu, \nu$ are $p$-restricted weights.

Steinberg's proof works directly with the corresponding Lie algebra representations and the derived map $d\varphi$ of Lie algebras. Instead we focus on the parallel situation with finite groups over $\mathbb{F}_p$, exploiting the Restriction Theorem along with the Tensor Product Theorem. Note that since the $p$th power map fixes elements of $\mathbb{F}_p$, the isogeny $\varphi$ induces an isomorphism from $G = \mathbf{G}(\mathbb{F}_p)$ onto $G' = \mathbf{G}'(\mathbb{F}_p)$. So pullbacks under $\varphi$ of simple $KG'$-modules are simple $KG$-modules.

Using (a), write $\mu = \varphi^*\mu'$ and $p\nu = \varphi^*\nu'$ for some $p$-restricted $\mu' \in X(\mathbf{T}')_S$ and $\nu' \in X(\mathbf{T}')_L$. Then the pullback to $\mathbf{G}$ of the representation of $\mathbf{G}'$ with highest weight $\lambda' := \mu' + \nu'$ has as highest weight $\lambda_\circ := \mu + p\nu$. Moreover, this $\mathbf{G}$-module $L(\lambda_\circ)$ remains simple on restriction to $KG$.

On the other hand, the Tensor Product Theorem says that the $\mathbf{G}$-module $L(\lambda_\circ)$ is isomorphic to $L(\mu) \otimes L(\nu)^{[1]}$. On restriction to $G$ this becomes the ordinary tensor product $L(\mu) \otimes L(\nu)$ (while being simple for $KG$). But as a $\mathbf{G}$-module, $L(\mu) \otimes L(\nu)$ involves a composition factor $L(\mu + \nu)$. Because the tensor product is simple as a $KG$-module, we must have $L(\mu) \otimes L(\nu) \cong L(\mu + \nu) = L(\lambda)$ as $\mathbf{G}$-modules. $\square$

It is especially interesting to see how the theorem applies to the non-isomorphic algebraic groups of types $B_\ell$ and $C_\ell$ when $\ell > 2$. Here there are special isogenies in both directions. By applying part (a) of the theorem to each isogeny, we get natural bijections between the collections of simple **G**-modules with highest weight having long (resp. short) support and simple **G′**-modules with highest weight having short (resp. long) support. In turn part (b) yields a natural bijection (preserving dimension) between arbitrary simple **G**-modules and simple **G′**-modules having corresponding highest weights.

This is of course compatible with the fact that the restriction of either special isogeny to the group of rational points over $\mathbb{F}_p$ is an isomorphism of finite groups.

Direct computations of the type described in Chapter 4 may give some insight into how these simple modules emerge from the corresponding Weyl modules, whose dimensions are usually quite different for types $B_\ell$ and $C_\ell$. Consider for example the rank 3 groups. Recall that the first two simple roots are long for $B_3$ but short for $C_3$; here the Bourbaki numbering respects the duality of the root systems.

For $B_3$ we have $\dim V(\varpi_i) = 7, 21, 8$ for $i = 1, 2, 3$, whereas for $C_3$ these dimensions are $6, 14, 14$. In each case $\varpi_1$ is the highest weight of the standard representation. When $p = 2$ the 0 weight (linked to $\varpi_1$ in type $B_3$) is lost, leading to $\dim L(\varpi_1) = 6$ as in type $C_3$. At the other extreme, $V(\varpi_3)$ for $C_3$ has a composition factor $L(\varpi_1)$, leading to $\dim L(\varpi_3) = 8$ as in type $B_3$.

Finally, for $C_3$ we get $V(\varpi_2) = L(\varpi_2)$, whereas for $B_3$ the corresponding Weyl module has extra composition factors $L(\varpi_1)$ and $L(0)$. So $\dim L(\varpi_2) = 14$ as in type $C_3$.

For a related discussion involving the spin representations in characteristic 2, see Smith [**378**, p. 398].

## 5.5. Brauer Characters and Grothendieck Rings

Although we rely mainly on the language of module theory, it is sometimes useful to invoke the formalism of **Brauer characters**. This especially comes into play when comparing representations in characteristic 0 and in characteristic $p$, which we take up only later in Chapter 16. The underlying idea is straightforward, but needs to be adapted carefully when reduction modulo $p$ via a suitably chosen ring is involved. For details see Curtis–Reiner [**125**, §82] or [**126**, §17B], noting that their notational conventions differ somewhat in the two treatments and differ even more from our notation.

Let $G$ be an arbitrary finite group. The subset $G_{\mathrm{reg}}$ consisting of $p$-regular elements (elements of order not divisible by $p$) is a union of conjugacy classes. In turn the set of $\mathbb{C}$-valued class functions on $G_{\mathrm{reg}}$ has a natural structure of algebra over $\mathbb{C}$. The characteristic functions on classes can be taken as a basis.

Denote by $e$ the $p'$-exponent of $G$: the least common multiple of orders of $p$-regular elements of $G$. Given a representation $\varphi : G \to \mathrm{GL}(V)$ over $K$ and a $p$-regular element $g \in G$, the eigenvalues of $\varphi(g)$ are $e$th roots of unity. The usual character of $\varphi$ at $g$ is obtained by summing these eigenvalues. To get the Brauer character we simply replace these roots of unity by corresponding ones in $\mathbb{C}$. This requires, however, a fixed (but arbitrary) isomorphism between the cyclic group of $e$th roots of 1 in $K^\times$ and the corresponding subgroup of $\mathbb{C}^\times$. Say $c$ generates the first group and $\gamma$ the second. Writing the trace of $\varphi(g)$ as $c^{a_1} + \cdots + c^{a_n}$, we then define the Brauer character value by

$$(\operatorname{br}\varphi)(g) := \gamma^{a_1} + \cdots + \gamma^{a_n}.$$

As a matter of notation, we may write $\operatorname{br} V$ in place of $\operatorname{br}\varphi$ here. The standard properties of the Brauer character are these:

PROPOSITION. *Let $G$ be any finite group, and fix an isomorphism as above between the eth roots of $1$ in $K^\times$ and the eth roots of $1$ in $\mathbb{C}^\times$.*

(a) *The resulting Brauer characters are class functions on $G_{reg}$.*

(b) *The Brauer character of a tensor product of two $KG$-modules is the product of the corresponding Brauer characters.*

(c) *Two representations of $G$ over $K$ have the same Brauer character if and only if they have the same composition factors, counting multiplicities. In particular, an irreducible representation is determined up to equivalence by its Brauer character.*

(d) *The Brauer characters afforded by a full set of nonisomorphic irreducible representations of $G$ over $K$ form a basis of the $\mathbb{C}$-valued class functions on $G_{reg}$.*

Note that part (d) requires Theorem 2.9.

The proposition implies that the ring of class functions on $G_{reg}$ is identifiable with the complexification of the **Grothendieck ring** of $G$ over $K$. We call the latter ring $R_K(G)$, following Serre [**366**, §14] as opposed to the Curtis–Reiner notation $G_0(KG)$ [**126**, §16].

REMARK. If $K$ is taken to be an algebraic closure of $\mathbb{F}_p$, one can make an all-purpose identification of $K^\times$ with the group of roots of unity in $\mathbb{C}^\times$ having order relatively prime to $p$. This is worked out in Carter [**89**, 3.1.3].

## 5.6. Formal Characters and Brauer Characters

Now return to the case of a finite group $G$ of Lie type over a field of $p^r$ elements. Using the language of Brauer characters, we can recast some features of simple $KG$-modules and related modules. This requires a transition from formal characters, expressed as $\mathbb{Z}$-linear combinations of the basis elements $e(\lambda)$ of $\mathbb{Z}[X]$ (see 2.1). Here we review some of the development in Wong [**436**, §3] and Ballard [**29**, §2], both of whom deal just with Chevalley groups. The set-up here will be used to study projective modules in Chapter 9.

As in the previous section we let $e$ be the $p'$-exponent of $G$ and fix an isomorphism between the group of $e$th roots of unity in $K^\times$ and the corresponding group in $\mathbb{C}^\times$. Each $KG$-module $V$ then has a Brauer character $\operatorname{br} V$ defined on the set $G_{reg}$ of $p$-regular elements. If $V$ is actually a **G**-module with formal character $\chi \in \mathcal{X} = \mathbb{Z}[X]^W$, we can also write $\operatorname{br}\chi$. More generally, we can define a "formal Brauer character" $\operatorname{br}\chi$ for any $\chi \in \mathcal{X}$ by writing $\chi$ uniquely as a $\mathbb{Z}$-linear combination of the $\chi_p(\lambda)$ (with $\lambda \in X^+$).

Another $\mathbb{Z}$-basis of $\mathcal{X}$ consists of the orbit sums $s(\lambda) := \sum e(w\lambda)$ for $\lambda \in X^+$, the sum taken over coset representatives of $W/W_\lambda$ (where $W_\lambda$ is the stabilizer of $\lambda$). The corresponding Brauer characters are easy to describe. It is well known that each $x \in G_{reg}$ is conjugate to an element $t \in \mathbf{T}$, with all possible choices for $t$ being $W$-conjugate. So it makes sense to define $s_\lambda : G_{reg} \to \mathbb{C}$ by assigning to $x$

the sum of complex roots of unity corresponding to the eigenvalues of all $(w\lambda)(t)$ for a fixed $t$ conjugate to $x$. Clearly $s_\lambda = \operatorname{br} s(\lambda)$.

Here are some useful properties of the $s_\lambda$:

PROPOSITION. *Let* $G = \mathbf{G}^F$, *and let* $\lambda, \mu \in X^+$.

(a) *If* $G$ *is a Chevalley group, then* $s_{q\lambda} = s_\lambda$.

(b) *If* $G$ *is a twisted group, then* $s_{q\lambda} = s_{\widetilde{\lambda}}$, *where the graph automorphism* $\pi$ *sends* $\lambda \mapsto \widetilde{\lambda}$.

(c) $s_\lambda s_\mu = s_{\lambda+\mu} + \mathbb{Z}^+$-*linear combination of various* $s_\nu$ *with* $\nu < \lambda + \mu$.

PROOF. Parts (a) and (b) follow the reasoning of Proposition 5.1 above, while (c) is a consequence of the corresponding formula for the product $s(\lambda)s(\mu)$. $\quad\square$

We have to relate this formalism to the characters of $\mathbf{G}$-modules and $KG$-modules. Since the formal characters $\chi_p(\lambda)$ for $\lambda \in X^+$ form a $\mathbb{Z}$-basis of $\mathcal{X}$, every Brauer character $\operatorname{br}\chi$ of a $\mathbf{G}$-module (restricted to $G$) is a $\mathbb{Z}$-linear combination of the various $\operatorname{br}\chi_p(\lambda)$. The weights (with multiplicity) of any $\mathbf{G}$-module are permuted by the Weyl group $W$, so in turn the formal character can be written as a $\mathbb{Z}^+$-linear combination of the various $s(\mu)$. In particular, since the highest weight of the simple $\mathbf{G}$-module $L(\lambda)$ has multiplicity 1, its formal character can be written as:

$$(1) \qquad \chi_p(\lambda) = s(\lambda) + \sum_{\mu < \lambda,\ \mu \in X^+} a_\mu\, s(\mu),\ \text{with } a_\mu \in \mathbb{Z}^+.$$

Write $\varphi_\lambda = \operatorname{br} L(\lambda)$ when $\lambda \in X_r$. Combining the above observations, we get

$$(2) \qquad \varphi_\lambda = s_\lambda + \mathbb{Z}^+\text{-linear combination of various } s_\mu,\ \text{with } \mu \in X^+,\ \mu < \lambda.$$

This involves however an unavoidable complication: Some of the dominant $\mu < \lambda$ in the usual partial ordering of weights may fail to lie in $X_r$. To rewrite the sums inductively in terms of restricted weights (in the following section) we need a refinement of this ordering, which essentially takes into account the way in which twisted tensor products of $\mathbf{G}$-modules restrict to $G$ as ordinary tensor products. Define $\mu \leq_{\mathbb{Q}} \lambda$ to mean that $\lambda - \mu$ is a $\mathbb{Q}^+$-linear combination of positive roots. For example, it is well-known that each fundamental weight $\varpi$ is a positive $\mathbb{Q}$-linear combination of simple roots ([**215**, 13.2]), so $0 <_{\mathbb{Q}} \varpi$.

For twisted groups we have to settle for a less precise comparison of the relative size of weights. Define for any $\lambda \in X^+$

$$|\lambda| := \langle \lambda, \alpha_0^\vee \rangle.$$

Since we are in type $A_\ell$, $D_\ell$, or $E_6$, the highest short root $\alpha_0$ is actually the highest root. In particular, $\langle \alpha, \alpha_0^\vee \rangle \geq 0$ for all $\alpha \in \Delta$, because $\alpha_0 + \alpha$ cannot be a root. This implies

$$\mu < \lambda \Longrightarrow |\mu| \leq |\lambda|.$$

But easy examples show that the converse can fail.

We also need the fact that for a *fundamental* weight $\varpi$, the weight $\widetilde{\varpi}$ is again fundamental and satisfies $|\varpi| = |\widetilde{\varpi}|$. This equality reflects the fact that when $\alpha_0^\vee$ is expressed as a $\mathbb{Z}^+$-linear combination of simple coroots, the coefficients for a pair of coroots related by $\pi$ are equal. (See [**215**, 12.2, Table 2].) As a result, $|\lambda| = |\widetilde{\lambda}|$ for all $\lambda \in X^+$.

## 5.7. Rewriting Formal Sums

Now we can adapt some of Wong's arguments in [**436, 437**]. (Note however that he considers only Chevalley groups and uses a total lexicographic ordering rather than the partial ordering $<_\mathbb{Q}$.)

PROPOSITION. *Let $G = \mathbf{G}^F$ over a field of $p^r$ elements.*

(a) *Let $G$ be a Chevalley group. If $\lambda \in X^+$, then $s_\lambda$ is a $\mathbb{Z}$-linear combination of various $s_\mu$ with $\mu \in X_r$ and $\mu \leq_\mathbb{Q} \lambda$.*

(b) *Let $G$ be a twisted group. If $\lambda \in X^+$, then $s_\lambda$ is a $\mathbb{Z}$-linear combination of various $s_\mu$ with $\mu \in X_r$ and $|\mu| \leq |\lambda|$.*

PROOF. (a) Use induction on the partial ordering $\leq_\mathbb{Q}$. If $\lambda$ fails to lie in $X_r$, we can find a fundamental weight $\mu$ for which $\lambda - q\mu$ is also dominant. Since $G$ is a Chevalley group, $s_{q\mu} = s_\mu$. Proposition 5.6(c) allows us to write

$$s_\lambda = s_{\lambda-q\mu+q\mu} = s_{\lambda-q\mu}\, s_\mu + \mathbb{Z}\text{-linear combination of } s_\nu,$$

with $\nu < (\lambda - q\mu + \mu) <_\mathbb{Q} \lambda$. But similarly we have

$$s_{\lambda-q\mu}\, s_\mu = s_{\lambda-(q-1)\mu} + \mathbb{Z}\text{-linear combination of } s_\nu,$$

again with $\nu <_\mathbb{Q} \lambda$. Combining steps, $s_\lambda$ can be replaced by a $\mathbb{Z}$-linear combination of various $s_\nu$ with $\nu <_\mathbb{Q} \lambda$. By induction, all $\nu$ can be assumed to lie in $X_r$.

(b) When $G$ is twisted, we have to replace induction on the ordering by induction on the "size" of weights. Otherwise the steps in the argument are similar. Using $s_{q\mu} = s_{\widetilde{\mu}}$ in place of $s_{q\mu} = s_\mu$ above, we can write

$$s_\lambda = s_{\lambda-q\mu+q\mu} = s_{\lambda-q\mu}\, s_{\widetilde{\mu}} + \mathbb{Z}\text{-linear combination of } s_\nu,$$

with $\nu < (\lambda - q\mu + \widetilde{\mu})$. In turn,

$$s_{\lambda-q\mu}\, s_{\widetilde{\mu}} = s_{\lambda-q\mu+\widetilde{\mu}} + \mathbb{Z}\text{-linear combination of } s_\nu.$$

Here we recall that $|\mu| = |\widetilde{\mu}|$. Thus we have $|\nu| < |\lambda - q\mu + \widetilde{\mu}| < |\lambda|$, allowing the induction to proceed. □

COROLLARY. *The collection of formal Brauer characters $\{s_\lambda \mid \lambda \in X_r\}$ is a basis of the space of $\mathbb{C}$-valued class functions on $G_{reg}$.*

PROOF. The $q^\ell$ Brauer characters $\varphi_\lambda$ ($\lambda \in X_r$) form a basis of the $\mathbb{C}$-valued class functions on $G_{reg}$ (Proposition 5.5(d)). Each $\varphi_\lambda$ is in turn the sum of $s_\lambda$ and various $s_\mu$ with $\mu < \lambda$. Thanks to the proposition, $\varphi_\lambda$ can be expressed as a $\mathbb{Z}$-linear combination of the $q^\ell$ formal Brauer characters $s_\nu$ ($\nu \in X_r$). So these $s_\nu$ must also form a basis. □

## 5.8. Restricting Highest Weight Modules to Finite Subgroups

In Chapter 2 we saw that the restriction of a simple $\mathbf{G}$-module $L(\lambda)$ to $G$ remains simple when $\lambda$ is $p^r$-restricted. Otherwise $L(\lambda)$ becomes a tensor product for $G$ of two or more $L(\mu)$ with $\mu \in X_r$.

We can ask more generally how an arbitrary $\mathbf{G}$-module $M$ restricts to $G$ if it has a unique highest weight $\lambda$ (in the usual partial ordering): all weights $\mu$ of $M$ must satisfy $\mu \leq \lambda$. One example is a highest weight module, generated by a weight vector which is fixed by $\mathbf{U}$; such a module is always a quotient of some Weyl module (3.3). Another important example is a tensor product of the form $L(\lambda) \otimes L(\mu)$, with $\lambda, \mu \in X^+$.

In order to do the necessary bookkeeping with weights on restriction to $G$, we again rely on the partial ordering $\leq_Q$ (for Chevalley groups) or the weaker comparison of size $|\mu| \leq |\lambda|$ (for twisted groups).

PROPOSITION. *Let $M$ be a $\mathbf{G}$-module with a unique highest weight $\lambda \in X^+$, and let $\nu \in X_r$ with $[M : L(\nu)]_G > 0$.*
   (a) *If $G$ is a Chevalley group, then $\nu \leq_Q \lambda$.*
   (b) *If $G$ is a twisted group, then $|\nu| \leq |\lambda|$.*
*In either case, if $\lambda \in X_r$, then $[M : L(\lambda)]_G = \dim M_\lambda$.*

PROOF. This follows readily from Proposition 5.7, coupled with the obvious analogue for $M$ of 5.6(1). ☐

## 5.9. Restriction of Weyl Modules

Here we consider what can be said about the restriction of Weyl modules $V(\lambda)$ to $G$. When $\lambda \in X_r$, we can show that $V(\lambda)$ remains *indecomposable* on restriction to $G$, even though there will usually be some non-restricted weights below $\lambda$. For the proof we need a few preliminaries.

In order to manipulate some tensor products, we invoke the standard vector space isomorphism $\mathrm{Hom}(X \otimes Y^*, Z) \cong \mathrm{Hom}(X, Y \otimes Z)$. If $X, Y, Z$ are $KH$-modules for an arbitrary group $H$, this implies:

(*) Assuming $X$ is a simple module and $Z$ is a quotient of $X \otimes Y^*$, then $X$ embeds in $Y \otimes Z$.

From the theory of $\mathbf{G}$-modules, we need a consequence of the universal property of Weyl modules (Theorem 3.3). Note that Wong [**437**, (1C)] argues instead by reduction modulo $p$, using admissible lattices to define Weyl modules.

PROPOSITION. *If $\lambda, \mu \in X^+$, then $V(\lambda + \mu) \otimes V(\mu)^*$ has a quotient isomorphic to $V(\lambda)$.*

PROOF. From the vector space isomorphism above we get an isomorphism between $\mathrm{Hom}_{\mathbf{G}}(V(\lambda + \mu) \otimes V(\mu)^*, V(\lambda))$ and $\mathrm{Hom}_{\mathbf{G}}(V(\lambda + \mu), V(\lambda) \otimes V(\mu))$. This last tensor product has a maximal vector of weight $\lambda + \mu$. So the universal property of Weyl modules gives a $\mathbf{G}$-homomorphism from $V(\lambda + \mu)$ into the tensor product. This translates into a $\mathbf{G}$-homomorphism $V(\lambda + \mu) \otimes V(\mu)^* \to V(\lambda)$ whose image contains a vector of weight $(\lambda + \mu) - \mu$ (where $-\mu$ is the lowest weight of $V(\mu)^*$), hence is all of $V(\lambda)$. ☐

We also have to use the Steinberg module. Letting $\sigma = (p^r - 1)\rho$, recall from 3.7 that the $r$-th Steinberg module $\mathrm{St}_r = V(\sigma) = L(\sigma)$ is simple for $G$ as well as $\mathbf{G}$.

Following Wong [**437**, (2D)], we can now prove:

THEOREM. *Let $G = \mathbf{G}^F$ over a field of $p^r$ elements. If $\lambda \in X_r$, then $L(\lambda)$ is the unique simple $G$-quotient of the Weyl module $V(\lambda)$. In particular, $V(\lambda)$ is an indecomposable $KG$-module.*

PROOF. We assume that $G$ is a Chevalley group, leaving to the reader the adaptation to twisted groups. Suppose $L(\nu)$ occurs as a quotient of the $G$-module $V(\lambda)$. Proposition 5.8 says that $\nu \leq_Q \lambda$.

To get $\lambda \leq_Q \nu$ we apply the above proposition with $\mu = \sigma - \lambda$, which is in $X^+$ since $\lambda \in X_r$. This yields an epimorphism of $\mathbf{G}$-modules (restricted to $G$):

$\mathrm{St}_r \otimes V(\sigma - \lambda)^* \to V(\lambda)$. Compose this with the assumed $G$-module epimorphism $V(\lambda) \to L(\nu)$. Now we can invoke $(*)$ above, with $\mathrm{St}_r$ in the role of $X$, to get a $G$-module embedding of $\mathrm{St}_r$ into $V(\sigma - \lambda) \otimes L(\nu)$. But this tensor product module is just the restriction to $G$ of the corresponding one for $\mathbf{G}$, whose highest weight is $\sigma - \lambda + \nu$. Using Proposition 5.8 again, we conclude $\sigma \leq_{\mathbb{Q}} \sigma - \lambda + \nu$, or $\lambda \leq_{\mathbb{Q}} \nu$.  $\square$

### 5.10. Restriction of Simple Modules to Levi Subgroups

When studying simple modules for any group, it is natural to ask how these modules behave on restriction to various subgroups. Typically such questions are hard to answer in the modular setting. But in our situation, one type of restriction is easy to analyze.

Start with a simple module $L(\lambda)$ for a group $G$ of Lie type and consider a parabolic subgroup $P$ of $G$. If $P$ has Levi factor $L$ and unipotent radical $V$, the subspace $L(\lambda)^V$ of $V$-invariants is automatically stable under $L$ (since $V$ is normal in $P$). Smith [**377**] observed that $L(\lambda)^V$ is then a simple $KL$-module. Notice that the two extreme cases are obvious: If $P = G$, then $L = G$ and $L(\lambda)^V = L(\lambda)$. If $P$ is a Borel subgroup $B = TU$, the space of $U$-invariants is one-dimensional (multiples of a maximal vector).

Smith actually deduces the result for $G$ from the corresponding result for $\mathbf{G}$, as we do here. His approach is concrete, though it is not especially efficient and requires a separate treatment of twisted groups. A more unified argument is given in the context of split $BN$-pairs, applying simultaneously to $\mathbf{G}$ and $G$, by Cabanes [**73**]; this avoids weight arguments at the cost of some transparency. His method also yields the more precise statement that the restriction of $L(\lambda)$ to $L$ has the simple module $L(\lambda)^V$ as a *direct summand*, which we incorporate below. Independently but somewhat later, Timmesfeld [**415**] gives a more concise formulation of Smith's argument which covers twisted and untwisted groups simultaneously (see Gorenstein–Lyons–Solomon [**190**, 2.8.11]).

THEOREM. *Let* $\mathbf{P} = \mathbf{LV}$ *be a parabolic subgroup of* $\mathbf{G}$, *with Levi subgroup* $\mathbf{L}$ *and unipotent radical* $\mathbf{V}$. *For any* $\lambda \in X^+$, *the fixed point subspace of* $\mathbf{V}$ *in* $L(\lambda)$, *denoted* $L(\lambda)^{\mathbf{V}}$, *is a simple* $\mathbf{L}$-*module.*

PROOF. Given a simple system $\Delta$ for the root system $\Phi$ of $\mathbf{G}$, the parabolic subgroup $\mathbf{P}$ is determined by a subset $\theta$ of $\Delta$, which is a simple system for the root system $\Theta$ of $\mathbf{L}$ (or of its semisimple derived group). Then $\mathbf{V}$ is the product of root groups $\mathbf{U}_\alpha$ for $\alpha \in \Phi^+ - \Theta$.

We need some related subgroups. Let $\mathbf{U}$ and $\mathbf{U}^-$ be the unipotent radicals of the opposite Borel subgroups $\mathbf{B}$ and $\mathbf{B}^-$. Let $\mathbf{V}^-$ be the unipotent radical of the parabolic subgroup opposite to $\mathbf{P}$, and denote by $\mathbf{U}_{\Theta^+}$ and $\mathbf{U}_{\Theta^-}$ the products of root groups $\mathbf{U}_\alpha$ for $\alpha \in \Theta^+$ or $\Theta^-$. Thus $\mathbf{U} = \mathbf{U}_{\Theta^+}\mathbf{V}$ and similarly for $\mathbf{U}^-$.

Let $v^+$ be a maximal vector in $L(\lambda)$. Now observe that $L(\lambda) = M \oplus M'$, where $M$ is the sum of all weight spaces $L(\lambda)_\mu$ for which $\lambda - \mu$ lies in $\mathbb{Z}\Theta^+$ and $M'$ is the sum of all other weight spaces. In particular, $v^+ \in M$. It is easy to check that $\mathbf{L}$ stabilizes both $M$ and $M'$, while $\mathbf{V}$ fixes $M$ pointwise. If the space of $\mathbf{V}$-invariants in $M'$ was nonzero, it would clearly be stable under the action of $L$ and hence contain a nonzero vector invariant under $\mathbf{U} = \mathbf{U}_{\Theta^+}\mathbf{V}$. This contradicts the fact that all maximal vectors in $L(\lambda)$ are multiples of $v^+ \in M$. Therefore $M = L(\lambda)^{\mathbf{V}}$.

Since $L(\lambda)$ is the span of $\mathbf{U}^- v^+ = \mathbf{V}^- \mathbf{U}_{\Theta^-} v^+$, it is clear that $M$ is the span of $\mathbf{U}_{\Theta^-} v^+$. Now it follows that $M$ is a simple $\mathbf{L}$-module: A proper submodule would contain a vector fixed by $\mathbf{U}_{\Theta^+}$ but also fixed by $\mathbf{V}$, hence by $\mathbf{U}$. But all such vectors are multiples of $v^+$ and hence generate $M$. $\qquad\square$

COROLLARY. *Let $G = \mathbf{G}^F$ over a field of $p^r$ elements. Let $P = LV$ be a parabolic subgroup of $G$. If $\lambda \in X_r$, the fixed point space $L(\lambda)^V$ is a simple $KL$-module, whose highest weight is the restriction of $\lambda$.*

PROOF. Assume that $G$ is a Chevalley group; the argument in the twisted case is quite similar.

Since $\lambda \in X_r$, it is clearly enough to show that $\mathbf{V}$ and $V$ have the same fixed point space $M$ in $L(\lambda)$. Then the simple $\mathbf{L}$-module $M$ restricts to a simple $L$-module. But $V$ fixes $M$ pointwise and by an argument like that above cannot fix any nonzero vector in $M'$. $\qquad\square$

## 5.11. Restriction to Elementary Abelian $p$-Subgroups

Another question about restriction of simple modules to subgroups comes up naturally in connection with recent work on cohomology and support varieties (see 15.2 below): How does such a module restrict to an elementary abelian $p$-subgroup $H$ of $G$? In particular, when is the module $KH$-free?

Alperin and Mason [10, 11] first look carefully at $G = \mathrm{SL}(2, q)$ over the field $k = \mathbb{F}_q$. Here there is a precise statement: A simple $kG$-module $M$ has dimension a power of $p$ if and only if the restriction of $M$ to some elementary abelian $p$-subgroup affords the regular representation. Thanks to the Tensor Product Theorem, the only such modules are tensor products of distinct Frobenius twists of the basic $p$-dimensional Steinberg module $L((p-1)\rho)$.

When $G$ is a universal Chevalley group of type A, D or E, the rank 1 case is used to obtain a general criterion. Starting with an arbitrary simple $kG$-module $M$, use the Tensor Product Theorem to write $M = R \otimes S$, where $S$ involves all the twists of the basic Steinberg module occurring in $M$. (They call $S$ a "partial Steinberg module".)

THEOREM. *Let $G$ be a (universal) Chevalley group over $k = \mathbb{F}_p$. Assume all roots in $\Phi$ are of equal length and that $M$ is a simple $kG$-module. Let $H$ be any elementary abelian $p$-subgroup lying in some root subgroup. Then the restriction of $M$ to $kH$ is free if and only if the restriction of the partial Steinberg module $S$ to $kH$ is free.*

CHAPTER 6

# Tensor Products

Apart from their intrinsic interest, tensor products of simple modules—or related modules such as Weyl modules—play a significant role in the study of projective modules and extensions: see Chapters 9, 10, and 12 below.

We begin by outlining some natural problems in 6.1. To deal with these, one typically has to study tensor products first for $\mathbf{G}$ and then adapt the results to $G = \mathbf{G}^F$ using Proposition 5.1. In practice only some special highest weights lead to reasonable computations. In 6.2 we consider briefly the algorithmic computation of composition factor multiplicities, with reference to some special cases worked out in the literature. We also look at several more narrowly focused questions on tensor products, e.g., when is a tensor product of two (nontrivial) simple modules also simple (6.3) or semisimple (6.4)?

In the remainder of the chapter, we derive a precise formula (Theorem 6.9) for the multiplicity of the Steinberg module in a tensor product of simple $KG$-modules. This depends only on standard character data for $\mathbf{G}$ and will be an essential ingredient in the later study of projective modules in Chapters 9–10.

### 6.1. Tensor Products of Simple Modules

Given two simple $KG$-modules $L(\lambda)$ and $L(\mu)$, what can be said about the structure of $L(\lambda) \otimes L(\mu)$? Here are a few natural questions, for a pair of weights $\lambda, \mu \in X_r$.

- What are the multiplicities of the composition factors of the $KG$-module $L(\lambda) \otimes L(\mu)$?
- For a fixed $\nu$, what is the multiplicity of $L(\nu)$ in $L(\lambda) \otimes L(\mu)$?
- When is the $KG$-module $L(\lambda) \otimes L(\mu)$ simple? semisimple?
- What is the **socle** (the sum of all simple submodules) of $L(\lambda) \otimes L(\mu)$ as a $\mathbf{G}$-module or $KG$-module?
- How does $L(\lambda) \otimes L(\mu)$ decompose as a direct sum of indecomposable $KG$-modules?
- Which indecomposable $KG$-modules occur as summands of tensor products of simple modules?
- What is the submodule structure of $L(\lambda) \otimes L(\mu)$?

Some of these matters have been explored reasonably well for algebraic groups, along with the related tensor products of Weyl modules. Here there is a deep result: any $\mathbf{G}$-module $V(\lambda) \otimes V(\mu)$ has a filtration with quotients isomorphic to Weyl modules having highest weights dictated by the character formula. Dually, the tensor product of induced modules has a filtration with quotients of the form $H^0(\nu)$. Such filtrations are called respectively **Weyl filtrations** and **good filtrations**: see [**RAGS**, II.4.16–4.21].

In general the full **G**-module or $KG$-module structure of $L(\lambda) \otimes L(\mu)$ may be expected to look arbitrarily complicated. This can be seen already in the relatively simple case of $\mathbf{G} = \mathrm{SL}(2, K)$, where weights of simple modules all have multiplicity 1. Here Doty and Henke [**150**] work out explicitly the structure of indecomposable **G**-summands in $L(\lambda) \otimes L(\mu)$. Adapting their results to $\mathrm{SL}(2, q)$ would involve many further steps. In higher rank such explicit computations are usually out of reach.

Even with standard character data for **G** in hand, the determination of composition factor multiplicities $[L(\lambda) \otimes L(\mu) : L(\nu)]_G$ (where $\lambda, \mu, \nu \in X_r$) is already highly nontrivial. Typically one has to go back and forth between the finite group and the algebraic group in studying such questions, which already pose serious challenges in the setting of **G**-modules. Some of the problems just outlined probably have no reasonable solution, but others can be dealt with in interesting special cases.

The Tensor Product Theorem (2.7) illustrates well the conceptual distance between the classical case and characteristic $p$: in characteristic 0, a tensor product of two simple modules can never be simple unless one of them is the trivial module. Steinberg's theorem also helps to motivate the decomposition problem. Starting with an arbitrary simple module $L(\lambda)$ for **G**, written as a twisted tensor product, the restriction of this module to the finite group $G$ becomes an ordinary tensor product.

## 6.2. Multiplicities

In principle the knowledge of all formal characters of simple **G**-modules (or equivalently, knowledge of composition factor multiplicities for all Weyl modules) would permit one to write the formal character of a tensor product $L(\lambda) \otimes L(\mu)$ as a sum of various ch $L(\nu)$. One method is to start with the computable tensor product multiplicities in characteristic 0, which determine the multiplicities of composition factors in $V(\lambda) \otimes V(\mu)$. Then plug in the data on multiplicities of composition factors of Weyl modules.

Even more than in characteristic 0, this naive approach leads to a formidable amount of computation when the dimensions of the modules are large. While there do exist some reasonable formulas in characteristic 0 for tensor product multiplicities, one cannot expect a simple closed formula (like Weyl's) in characteristic $p$ for the characters of simple modules. So it is unrealistic to expect closed formulas (or easy algorithms) for tensor product multiplicities.

For later reference we recall briefly some of the standard results in characteristic 0 (over $\mathbb{C}$ or other algebraically closed field). Probably the nicest closed formula for the multiplicities in $V(\lambda)_{\mathbb{C}} \otimes V(\mu)_{\mathbb{C}}$ is the one found by Steinberg (see [**215**, 24.4]). But this involves a complicated double sum over the Weyl group, which is impractical for computational purposes. Other methods, generalizing for example the Littlewood–Richardson Rule, continue to be explored: crystal graphs, Littelmann's path model, ...

An old formula of Brauer, developed also by Klimyk, shows conceptually where the difficulty lies (see [**215**, §24], [**RAGS**, II.5.8]). The proof by Kumar [**293**] of the "PRV Conjecture" brings out more subtle features. These results are summarized in the two parts of the following theorem. The essential problem for arbitrary

$\lambda, \mu \in X^+$ is to find the coefficients $a^\nu_{\lambda\mu} \in \mathbb{Z}^+$ in

(1) $$\chi(\lambda)\chi(\mu) = \sum_{\nu \in X^+} a^\nu_{\lambda\mu}\chi(\nu).$$

THEOREM. *Let* $\lambda, \mu \in X^+$, *and define for any* $\nu \in X$ *the formal character* $\chi(\nu)$ *as in* 3.9.

(a) *Write* $\chi(\mu) = \sum_{\nu \in X} m^\nu_\mu\, e(\nu)$ *with* $m^\nu_\mu \in \mathbb{Z}$. *Then*

$$\chi(\lambda)\chi(\mu) = \sum_{\nu \in X} m^\nu_\mu\, \chi(\lambda + \nu).$$

*Moreover,* $a^{\lambda+\nu}_{\lambda\mu} \le m^\nu_\mu$ *for all* $\nu$.

(b) *For each* $w \in W$, *let* $\nu$ *be the dominant weight* $W$-*conjugate to* $\lambda + w\mu$. *Then* $\chi(\nu)$ *occurs at least once in the decomposition* (1).

For semisimple groups in characteristic $p$, only a small amount of explicit multiplicity data for tensor products is found in the literature. Typically this is worked out in low ranks when $p$ is small, in connection with projective modules and extensions.

In the case of finite Chevalley groups, Holmes [208] introduces a directed graph (following earlier work of Chastkofsky and Feit) to organize the calculation of multiplicities: these correspond to lengths of paths emanating from a vertex associated with a specific tensor product of simple $KG$-modules. (He applies this approach to projectives for $\mathrm{SL}(4, 2^r)$ for small $r$.)

A considerable amount of special tensor product data has been compiled as part of the computation of extensions between simple modules: see the survey in 12.11 below. (There is also some parallel work by Pillen [333] in the setting of Lie algebras.)

## 6.3. Simple Tensor Products

As remarked earlier, the Tensor Product Theorem shows that a tensor product of two nontrivial simple **G**-modules may again be simple. Moreover, Steinberg's study of special isogenies (see 5.3–5.4) reveals some additional ways in which to express certain simple **G**-modules with a restricted highest weight as tensor products of smaller simple modules. Seitz [364, (1.6)] has shown that only in this way can a tensor product of simple modules with (nonzero) $p$-restricted highest weights actually be simple:

PROPOSITION. *Let* $\lambda \in X^+$. *The* **G**-*module* $L(\lambda)$ *is isomorphic to a tensor product* $L(\mu) \otimes L(\nu)$ *for nonzero* $\mu, \nu \in X_1$ *if and only if* $\lambda \in X_1$ *and* **G** *is of type* $B_\ell, C_\ell, F_4$ *with* $p = 2$ *or type* $G_2$ *with* $p = 3$, *while* $\langle \mu, \alpha^\vee \rangle = 0$ *for all long simple roots* $\alpha$ *and* $\langle \nu, \alpha^\vee \rangle = 0$ *for all short simple roots* $\alpha$ *(or vice versa).*

Note that it is not hard to see why $\lambda = \mu + \nu$ must be restricted here: otherwise there must exist a *simple* root $\alpha$ such that both $\dim L(\mu)_{\mu-\alpha} = 1$ and $\dim L(\nu)_{\nu-\alpha} = 1$, leading quickly to the contradiction $\dim L(\lambda)_{\lambda-\alpha} \ge 2$.

This result together with the Tensor Product Theorem is then applied systematically to more general weights for finite groups of Lie type by Magaard–Tiep [313, p. 240].

## 6.4. Semisimple Tensor Products

Though it is comparatively rare for $L(\lambda) \otimes L(\mu)$ to be simple as a **G**-module, there are many situations in which this module is semisimple provided $p$ is large enough. In order to prove that "generic" Cartan invariants exist for finite groups of Lie type (see 11.9 below), Humphreys [**232**, Lemma] develops the following criterion.

The essential point is that $\lambda$ should lie far enough inside a $p$-alcove to insure that adding the various weights of $L(\mu)$ to $\lambda$ produces weights inside the same alcove. (This can be expressed in terms of distance to walls in the euclidean space $\mathbb{R} \otimes_{\mathbb{Z}} X$ as the requirement that the distance from $\lambda$ to any wall is at least half the diameter of the convex hull of the weight diagram of $\mu$.) Then Theorem 6.2, together with linkage and translation functors, yields the desired character formula.

PROPOSITION. *Let $\lambda, \mu \in X^+$, and suppose that $\lambda$ lies inside a p-alcove far enough from all walls so that $\lambda + \nu$ belongs to the same alcove for all weights $\nu$ of $L(\mu)$. Then $L(\lambda) \otimes L(\mu)$ splits into the direct sum of simple **G**-modules $L(\lambda + \nu)$, where $\nu$ runs over all weights of $L(\mu)$ counted with multiplicity.*

This adapts immediately to the case of a finite group $G$ of Lie type over a field of order $p^r$: whenever $\lambda, \mu \in X_r$ satisfy the hypothesis of the proposition, $L(\lambda) \otimes L(\mu)$ is semisimple as a $KG$-module.

## 6.5. Dimensions Divisible by $p$

One recurring question in modular representation theory is the extent to which powers of $p$ divide the dimensions of simple (or, more generally, indecomposable) modules. See for example the discussion of blocks and defect groups in Chapter 8. For a group of Lie type we have already noted the special role of the Steinberg module, whose dimension is the full $p$-power dividing $|G|$.

For an arbitrary finite group there are some interesting connections between tensor products and divisibility by $p$, summarized in the following theorem of Benson and Carlson [**50**, 2.1–2.2]. This generalizes earlier work of Landrock.

THEOREM. *Let $G$ be any finite group.*

(a) *Suppose $M$ and $N$ are indecomposable $KG$-modules. Then the trivial module $K$ occurs as a direct summand of $M \otimes N$ if and only if $M \cong N^*$ and $p$ does not divide $\dim M$, in which case $K$ occurs with multiplicity one.*

(b) *Suppose $M$ is an indecomposable $KG$-module, with $\dim M$ divisible by $p$. Then for any $KG$-module $N$, $p$ divides the dimension of every indecomposable summand of $M \otimes N$.*

Benson and Carlson work more generally over an arbitrary field $k$ of characteristic $p$, with "indecomposable" replaced by "absolutely indecomposable" in the sense that $\mathrm{End}_{kG}(M)/\mathrm{Rad}\,\mathrm{End}_{kG}(M) \cong k$.

Part (b) of the theorem is a simple corollary of part (a), as follows. Take $U$ to be an indecomposable summand of $M \otimes N$. If $p$ does not divide $\dim U$, part (a) implies that $K$ is a direct summand of $U \otimes U^*$, hence also of $(M \otimes N) \otimes U^* \cong M \otimes (N \otimes U^*)$. Applying (a) again, we find that $M^*$ must be isomorphic to a summand of $N \otimes U^*$ and in turn that $p$ cannot divide $\dim M$, contrary to hypothesis.

REMARK. For Benson and Carlson, the theorem is a tool in the study of the **Green ring** $A(G)$ of a finite group $G$. The idea is to take the isomorphism classes of indecomposable $KG$-modules as a basis of a free $\mathbb{Z}$-module $a(G)$, with addition given by direct sum. Then use the tensor product to define a commutative ring structure. This **representation ring** $a(G)$ generalizes in a natural way the Grothendieck ring of $\mathbb{C}G$ in characteristic 0. The Green ring is defined by $A(G) := \mathbb{C} \otimes a(G)$. (See Benson [**45**, 2.2], [**47**, 5.1].)

While the study of Green rings has become important for arbitrary finite groups, it is probably unrealistic to expect a unified theory for groups of Lie type. For example, SL$(2, p)$ has only finitely many isomorphism classes of indecomposable modules, whereas this fails for SL$(2, p^r)$ when $r > 1$. But Lie theory should cast light on those indecomposable modules which arise directly from the group algebra, including projectives. Green rings have been studied only in special situations for groups of Lie type: see Alperin [**5**] on SL$(2, 2^r)$; Kouwenhoeven [**284**] on the Green ring of GL$(2, p)$; Sin [**368**] on the use of the Green ring to study indecomposable summands of the module induced from a Sylow $p$-subgroup.

## 6.6. Formal Characters and Multiplicities

In the remainder of this chapter we investigate systematically (following Jantzen [**251**]) the multiplicity of the Steinberg module St$_r$ for $G$ as a $KG$-composition factor in an arbitrary **G**-module such as a tensor product of simple modules. For this we work with formal characters, as in 5.6.

In this chapter and later, we need to work with several types of multiplicity. We have already denoted the composition factor multiplicity of $L(\lambda)$ in a **G**-module $V$ by the symbol $[V : L(\lambda)]_\mathbf{G}$. This is the same as the coefficient of $\chi_p(\lambda)$ when ch $V$ is written as a $\mathbb{Z}^+$-linear combination of the basis elements $\chi_p(\lambda)$ of $\mathcal{X}$. More generally, whenever $\chi \in \mathcal{X}$, we denote the coefficient of $\chi_p(\lambda)$ in $\chi$ by $[\chi : \chi_p(\lambda)]_\mathcal{X}$ or by $[\chi : L(\lambda)]_\mathcal{X}$.

It is also necessary to keep track of multiplicities for the finite group $G$. If $V$ is a $KG$-module (for example, the restriction to $G$ of a **G**-module) and $\lambda \in X_r$, denote by $[V : L(\lambda)]_G$ the composition factor multiplicity of $L(\lambda)$ in $V$. In particular, if $\mu \in X^+$ and $\lambda \in X_r$, then $[L(\mu) : L(\lambda)]_G$ makes sense. To generalize a bit further, define for any $\chi \in \mathcal{X}$ and any $\lambda \in X_r$ an integer

$$(1) \qquad [\chi : L(\lambda)]_G := \sum_{\mu \in X^+} [\chi : L(\mu)]_\mathcal{X} \, [L(\mu) : L(\lambda)]_G.$$

## 6.7. Twisting by the Frobenius

The $p^r$th power map acts in a natural way on $\mathbb{Z}[X]$, by sending a basis element $e(\lambda)$ to $e(p^r\lambda)$. We write $\chi \mapsto \chi^{[r]}$ for the restriction of this action to $\mathcal{X}$. For example, when $\chi = \mathrm{ch}\, L(\lambda)$, we have $\chi^{[r]} = \mathrm{ch}\, L(\lambda)^{[r]}$.

We are working with a Frobenius map $F$ which is the composite of the $p^r$th power map and a possibly nontrivial graph automorphism $\pi$ (viewed as an automorphism of **G** which leaves **T** invariant). Suppose $\pi \neq 1$. If we follow $\pi^{-1}$ with a rational representation $\mathbf{G} \to \mathrm{GL}(V)$, we get a twisted representation $\mathbf{G} \to \mathrm{GL}(V^\pi)$. Clearly ch $V^\pi = (\mathrm{ch}\, V)^\pi$, where the induced action of $\pi$ on $\mathbb{Z}[X]$ is denoted by $e(\lambda) \mapsto e(\widetilde{\lambda})$. Similarly the action on $\mathcal{X}$ is denoted by $\chi \mapsto \widetilde{\chi}$. As in Proposition 5.1, the restriction of $\pi^{-1}$ to $G$ agrees with the restriction of the standard Frobenius map relative to $p^r$, so $V^\pi$ is isomorphic to $V^{[r]}$ as a $KG$-module. In particular, we get:

LEMMA. *Let $G = \mathbf{G}^F$ over a field of $p^r$ elements. If $\lambda \in X_r$ and $\mu \in X^+$, then for all $\nu \in X_r$,*

$$[\chi_p(\lambda)\chi_p(\mu)^{[r]} : L(\nu)]_G = [\chi_p(\lambda)\chi_p(\widetilde{\mu}) : L(\nu)]_G,$$

*with $\widetilde{\mu}$ replaced by $\mu$ if $\pi = 1$. In general, if $\chi \in \mathcal{X}$ and $\nu \in X_r$, then*

$$[\chi^{[r]} : L(\nu)]_G = [\widetilde{\chi} : L(\nu)]_G,$$

*with $\widetilde{\chi}$ replaced by $\chi$ if $\pi = 1$.*                                              $\square$

## 6.8. Rewriting Multiplicities

For the main theorem we will need a more devious rewriting of multiplicities, still in the spirit of Lemma 6.7 but not referring explicitly to $G$. Note that in the following lemma, the factor $\psi$ is just carried along through the steps. This factor is needed to give some flexibility in the way the lemma is applied. Here and later, the tilde decorating some weights should be ignored when $F$ involves no graph automorphism.

LEMMA. *For all $\chi, \psi \in \mathcal{X}$ and $\lambda \in X_r$,*

$$\sum_{\nu \in X^+} [\chi^{[r]}\psi\chi_p(\widetilde{\nu}) : L(\lambda + p^r\nu)]_{\mathcal{X}}$$

*is unchanged if $\chi^{[r]}$ is replaced by $\widetilde{\chi}$.*

PROOF. First write the given sum in a more complicated way:

$$\sum_{\mu,\nu \in X^+} \sum_{\lambda' \in X_r} [\psi\chi_p(\widetilde{\nu}) : L(\lambda' + p^r\mu)]_{\mathcal{X}} [\chi^{[r]}\chi_p(\lambda')\chi_p(\mu)^{[r]} : L(\lambda + p^r\nu)]_{\mathcal{X}}.$$

Next substitute

$$\chi_p(\mu)^{[r]}\chi^{[r]}\chi_p(\lambda') = \sum_{\mu' \in X^+} [\chi_p(\mu)\chi : L(\mu')]_{\mathcal{X}} \chi_p(\lambda' + p^r\mu')$$

to obtain successively:

$$\sum_{\mu,\nu \in X^+} [\psi\chi_p(\widetilde{\nu}) : L(\lambda + p^r\mu)]_{\mathcal{X}} [\chi_p(\mu)\chi : L(\nu)]_{\mathcal{X}}$$

$$= \sum_{\mu,\nu \in X^+} [\psi\chi_p(\widetilde{\nu}) : L(\lambda + p^r\mu)]_{\mathcal{X}} [\chi_p(\widetilde{\mu})\widetilde{\chi} : L(\widetilde{\nu})]_{\mathcal{X}}$$

$$= \sum_{\mu \in X^+} [\widetilde{\chi}\psi\chi_p(\widetilde{\mu}) : L(\lambda + p^r\mu)]_{\mathcal{X}}.$$

The last step uses the substitution:

$$\widetilde{\chi}\chi_p(\widetilde{\mu}) = \sum_{\nu \in X^+} [\widetilde{\chi}\chi_p(\widetilde{\mu}) : L(\widetilde{\nu})]_{\mathcal{X}} \, \chi_p(\widetilde{\nu}).$$

In the next-to-last step we use the fact that applying $\pi$ leaves multiplicities unchanged.                                              $\square$

## 6.9. Multiplicity of the Steinberg Module

Now we are in a position to obtain a key formula (see Jantzen [**251**, Satz 1.5]) expressing the $G$-multiplicity of $\mathrm{St}_r = L(\sigma_r)$ in any rational **G**-module (for example, a tensor product). The salient feature of this formula is that its right side involves only standard character data for **G**, without reference to $G$.

THEOREM. *Let* $G = \mathbf{G}^F$ *over a field of* $p^r$ *elements. For any* $\chi \in \mathcal{X}$, *we have*

(1) $$[\chi : \mathrm{St}_r]_G = \sum_{\nu \in X^+} [\chi \chi_p(\tilde{\nu}) : \chi_p(\sigma_r + p^r \nu)]_{\mathcal{X}}.$$

PROOF. Note that the sum here is actually finite: to get a nonzero multiplicity on the right side, some maximal weight $\theta$ occurring in $\chi$ must satisfy

$$\sigma_r + p^r \nu \le \theta + \tilde{\nu}, \text{ or } p^r \nu - \tilde{\nu} \le \theta - \sigma_r,$$

possible for only finitely many dominant $\nu$.

For the proof, we use induction on the maximum value of $|\tau| = \langle \tau, \alpha_0^\vee \rangle$ as $\tau$ runs over the weights of $\chi$. Both sides of (1) being linear with respect to $\chi$, it is clearly enough to consider $\chi = \chi_p(\lambda + p^r \mu)$ with $\mu \in X^+$ and $\lambda \in X_r$. Here the maximum value involved is $\langle \lambda + p^r \mu, \alpha_0^\vee \rangle$. This is 0 just when $\lambda = \mu = 0$, in which case both sides are clearly 0.

If $\lambda$ is arbitrary, the case $\mu = 0$ is easy to analyze directly: in the sum over $\nu$, any nonzero term yields

$$\sigma_r + p^r \nu \le \lambda + \tilde{\nu} \le \sigma_r + \tilde{\nu},$$

forcing $p^r \nu \le \tilde{\nu}$ and then $p^r |\nu| \le |\tilde{\nu}| = |\nu|$ (see 5.6). Since $\nu \in X^+$, we must have $\nu = 0$. Now (1) reduces to $[\chi_p(\lambda) : \mathrm{St}_r]_G = [\chi_p(\lambda) : \mathrm{St}_r]_G$ for $\lambda \in X_r$, an immediate consequence of the Restriction Theorem 2.11.

For the induction step we may suppose $\mu \ne 0$. In order to appeal to the induction hypothesis, we want to replace $\lambda + p^r \mu$ by a smaller weight. For this we apply Lemma 6.7 to $\chi = \chi_p(\lambda) \chi_p(\mu)^{[r]}$, getting $[\chi : \mathrm{St}_r]_G = [\chi' : \mathrm{St}_r]_G$, where $\chi' := \chi_p(\lambda) \chi_p(\tilde{\mu})$. We can use Lemma 6.8 to make a similar transformation of the sum on the right side of (1):

$$\sum_{\nu \in X^+} [\chi \chi_p(\tilde{\nu}) : \chi_p(\sigma_r + p^r \nu)]_{\mathcal{X}} = \sum_{\nu \in X^+} [\chi' \chi_p(\tilde{\nu}) : \chi_p(\sigma_r + p^r \nu)]_{\mathcal{X}}.$$

Now it is enough to see that the induction hypothesis already takes care of $\chi'$. The maximum value in question for $\chi$ is $|\lambda| + p^r |\mu|$, while for $\chi'$ it is $|\lambda| + |\tilde{\mu}| = |\lambda| + |\mu|$. Since $|\mu| > 0$ ($\mu$ being dominant but not 0), we see that induction does apply. $\square$

# CHAPTER 7

# $BN$-Pairs and Induced Modules

Thus far we have looked at simple modules from the algebraic group viewpoint. In this chapter we digress briefly to survey the more intrinsic approach involving finite groups with a split $BN$-pair (1.7). Here there are several natural ways to realize the simple modules as quotients or submodules of explicitly constructed $KG$-modules.

- Curtis and Richen work inside $KG$ itself, finding a reasonable substitute for highest weights (7.1).
- Induction of 1-dimensional $B$-modules to $G$ provides an alternative "principal series" framework for the classification of simple modules, which can be characterized in terms of images of intertwining maps (7.2). The approach by Carter and Lusztig parallels in many respects the study of maps between cohomology groups of line bundles on the flag variety of $\mathbf{G}$.
- Further work by Sawada and others has clarified the nature of the indecomposable summands in the related module induced from the trivial character of the Sylow $p$-subgroup $U$ (7.4).
- From a more geometric viewpoint, Ronan and Smith have studied the 0th homology of the Tits complex (based on parabolic subgroups of $G$) relative to coefficient systems coming from the module theory of $KG$ (7.5). Schmidt makes a further comparison between the top homology modules and the principal series modules (7.6).

While the techniques involved in these various constructions are concrete and self-contained, they do not as yet yield more detailed information or conjectures about the Brauer characters of simple modules. For this reason we do not provide too many details. To facilitate reference to the original papers, we follow in most cases the special notation (such as $\Delta$ for the Tits complex and $n$ rather than $\ell$ for the rank of a $BN$-pair) used in those papers. Note that groups of Lie type are treated in varying generality: sometimes $G$ is assumed to be a Chevalley group, while some papers exclude the groups of Suzuki and Ree.

## 7.1. Weights for Groups with a Split $BN$-Pair

We have seen in Chapter 2 how to approach the study of simple $KG$-modules systematically via the study of highest weight modules for the ambient algebraic group $\mathbf{G}$. But a finite group theorist might reasonably ask for a more intrinsic development based just on properties of the group $G$—preferably those properties encoded in the axioms for a split $BN$-pair (1.7). Some experimentation by Curtis [**121**] led him and his student Richen to a self-contained construction and classification of the modules, which we outline here following [**122**] (see also Richen [**349, 350**]). To avoid trivialities one can assume that the rank $n$ is $\geq 2$, in which case the classification of simple groups with a split $BN$-pair assures a good fit with the finite groups of Lie type.

It is tempting to imitate the "highest weight" theory, but the example $G = \mathrm{SL}(2,p)$ already shows that a naive approach cannot work. Here the finite torus has only $p-1$ characters $\chi : T \to K^{\times}$, not enough to distinguish the $p$ simple modules: the trivial module and the Steinberg module have the same "highest weight". Even so, if we start with a nonzero $U$-fixed vector $v$ affording a character $\chi$ of $T$, we can ask a more refined question: What is the largest parabolic subgroup of $G$ stabilizing the line $Kv$? This leads to a precise parametrization of the simple modules, though it takes the full force of the axioms for a split $BN$-pair to carry out the program.

The key observation of Curtis and Richen is that certain elements of $KG$ will act by distinct scalars on $U$-fixed vectors in two distinct simple modules having the same $T$-character $\chi$. The scalar is 0 in the case of the trivial module and $-1$ in the case of the Steinberg module, for example. (The reader can check the recipe below directly for $\mathrm{SL}(2,p)$.)

Formally, one proceeds as follows. The split $BN$-pair set-up produces root subgroups $U_{\alpha}$ for the various simple roots $\alpha$. Choose a representative $n_{\alpha} \in N$ for the corresponding simple reflection $s_{\alpha}$ in $W$. (How the choice is made eventually has to be refined.) Conjugating $U_{\alpha}$ by $n_{\alpha}$ then produces a group $U_{-\alpha}$. Together $U_{\alpha}$ and $U_{-\alpha}$ generate a group with $BN$-pair of rank 1, having a "torus" $T_{\alpha} \subseteq T$.

Using a bar to denote the sum in $KG$ over a subgroup, one finds that $\overline{U_{\alpha}}n_{\alpha}$ just acts by a scalar $c_{\alpha}$ on a "highest weight vector". Define a **weight** to be an $(n+1)$-tuple $(\chi, c_1, \ldots, c_n)$, where $\chi : T \to K^{\times}$ is a character and where the $c_i \in K$ correspond to a fixed ordering of simple roots $\alpha_1, \ldots, \alpha_n$. Define a **highest weight vector** of this weight in a $KG$-module to be a (nonzero) vector $v$ on which $U$ acts trivially and $T$ acts by the character $\chi$, while for each simple root $\alpha$, we have $\overline{U_{\alpha}}n_{\alpha} \cdot v = c_{\alpha}v$.

THEOREM. *Let $G$ be a finite group with a split BN-pair of characteristic $p$. Fix as above an ordering of the simple roots $\alpha_1, \ldots, \alpha_n$, and abbreviate $T_{\alpha_i}$ by $T_i$.*

(a) *Each nonzero $KG$-module $M$ has a highest weight vector $v$ of some weight $(\chi, c_1, \ldots, c_n)$, and $KGv = KU^{-}v$.*

(b) *Every simple $KG$-module has a unique line fixed pointwise by $U$, which therefore contains all highest weight vectors.*

(c) *Two simple $KG$-modules are isomorphic if and only if they have the same weight.*

(d) *For a simple $KG$-module, all $c_i \in \{0, -1\}$.*

(e) *For a weight $(\chi, c_1, \ldots, c_n)$ with all $c_i \in \{0, -1\}$, there exists a simple $KG$-module having this weight if and only if $\chi|_{T_i} = 1$ whenever $c_i = -1$.*

(f) *In case $T$ is the direct product of its subgroups $T_i$, the number of nonisomorphic simple $KG$-modules is $\prod(|T_i| + 1)$.*

By bringing in the parabolic subgroup in the $BN$-pair which stabilizes the line through a highest weight vector, the classification in the theorem can be recast in the spirit of Corollary 2.14 above.

REMARKS. (1) The Curtis–Richen approach described in [**122**] requires a precise way of choosing representatives in $N$ for simple reflections. For this they need to assume that the split $BN$-pair is "restricted", a technical condition later shown by Richen [**350**] to be implied just by the original axioms.

(2) In his Yale lectures Steinberg [**394**, §13] gives a self-contained treatment of the classification of simple modules for the finite groups, along lines suggested by Curtis (but not using the $BN$-pair axiomatics explicitly). This concrete approach in $KG$ also exhibits the splitting field in a direct way: one only has to work over a splitting field for the abelian $p'$-group $T$.

## 7.2. Principal Series and Intertwining Operators

While it is natural to look inside the group algebra of a finite group for simple modules, another equally natural approach is to analyze the structure of various modules induced from subgroups. Carter and Lusztig [**93**] provide a new perspective on the Curtis–Richen theory, emphasizing the role of parabolic subgroups, induced modules, and intertwining operators.

As before, the group $G$ has a split $BN$-pair of rank $n$ and characteristic $p$. (But we use $T$ in place of $H$ to denote $B \cap N$.) The induced module constructions in [**93**] make sense over any field, in particular a field of characteristic 0 (see Chapter 16 below); but we look only at $KG$-modules. The field here need only be a splitting field for $T$; it then turns out to be a splitting field for $G$.

The trivial character of $U$ induces a $KG$-module of dimension $[G : U] = |G|_{p'}$. Rather than describe it as $KG \otimes_{KU} K$, Carter and Lusztig work with an equivalent function space description like that used in the theory of algebraic groups. Their space $\mathcal{F}$ is defined to be the space of $K$-valued functions on $U \backslash G$, or

$$\mathcal{F} := \{f : G \to K \mid f(ug) = f(g) \text{ for all } g \in G, u \in U\}.$$

Here $x \in G$ acts by right translation: $(x \cdot f)(g) = f(gx)$.

In turn, it is easy to see that $\mathcal{F} = \bigoplus_\chi \mathcal{F}_\chi$, where

$$\mathcal{F}_\chi := \{f \in \mathcal{F} \mid f(tg) = \chi(t)f(g) \text{ for all } g \in G, t \in T\}$$

and $\chi$ runs over all linear characters $T \to K^\times$. The subspace $\mathcal{F}_\chi$ is a $KG$-submodule, naturally isomorphic to the induced module $KG \otimes_{KB} K_\chi$, where $\chi$ defines in the usual way a 1-dimensional $KB$-module $K_\chi$. Its dimension is $[G : B]$. The modules $\mathcal{F}_\chi$ are often referred to as **principal series** modules.

Denote by $I$ the set of $n$ simple reflections generating $W$. To each $i \in I$ corresponds a subgroup $T_i \subset T$ coming from a rank 1 subgroup. Define for each $\chi$ a subset $J_0(\chi) := \{i \in I \mid \chi(T_i) = 1\}$. This set is often empty and is equal to $I$ just when $\chi$ is the trivial character.

In this setting Carter and Lusztig define for each $i \in I$ an explicit $KG$-endomorphism $T_i$ of $\mathcal{F}$. On the submodule $\mathcal{F}_\chi$, one gets quadratic relations with a Hecke algebra flavor: $T_i^2 = -T_i$ when $i \in J_0(\chi)$ but is 0 otherwise. Using a reduced expression for $w \in W$ and a fixed $J \subset I$, they next define an endomorphism $\Theta_w^J$ as a composite of the $T_i$ (or sometimes $I + T_i$) corresponding to the simple reflections involved. (It takes considerable work to show that the result is independent of the reduced expression.)

In their main theorem Carter and Lusztig recover the Curtis–Richen classification:

THEOREM. *Let $G$ be a finite group with a split $BN$-pair of characteristic $p$. Denote by $w_\circ$ the longest element in its Weyl group $W$.*

(a) *For each character $\chi : T \to K^\times$ and each subset $J \subset J_0(\chi)$, the image of the map $\Theta^J_{w_0} : \mathcal{F}_\chi \to \mathcal{F}_{w_0(\chi)}$ is a simple $KG$-module. It has a unique $U$-invariant line on which $T$ acts by $\chi$ and whose stabilizer in $G$ is the parabolic subgroup $P_J$.*

(b) *The simple modules constructed in this way are pairwise nonisomorphic and exhaust the isomorphism classes of simple $KG$-modules.*

(c) *The isomorphism classes of simple $KG$-modules are in natural bijection with the pairs $(J, \chi)$ such that $J \subset J_0(\chi)$.*

(d) *The submodules of $\mathcal{F}_{w_0(\chi)}$ obtained in this way for subsets $J \subset J_0(\chi)$ are the only simple submodules.*

REMARKS. (1) A character computation in 9.7 below reveals indirectly that the $KG$-modules $\mathcal{F}_\chi$ and $\mathcal{F}_{w\chi}$ for any $w \in W$ have the same composition factors, which is implicit in the work of Carter and Lusztig.

(2) These modules will reappear (in different notation) in the setting of finite groups of Lie type. Principal series modules in characteristic $p$ are studied by Pillen [336] in conjunction with his work on Loewy series of projective modules for Chevalley groups: see 13.10 below. Jantzen [253] explores further the submodule structure of the principal series modules, in relation to the Deligne–Lusztig character theory over $\mathbb{C}$. When $G$ is a Chevalley group, he constructs good analogues of his Weyl module filtrations, with a computable "sum formula": see 17.13 below.

## 7.3. Examples

When $G = \mathrm{SL}(2, p)$, it is not too difficult to work out the full structure of the principal series modules, all of which have dimension $p + 1$. Characters $\chi$ of $T$ (restrictions of weights of $\mathbf{T}$) can be identified naturally with integers between 0 and $p - 2$. When $\chi = 0$, the module $\mathcal{F}_0$ is just the direct sum of the trivial module and the Steinberg module, corresponding to the respective sets $J = I$ and $J = \emptyset$. When $\chi > 0$, the module $\mathcal{F}_\chi$ has a unique simple submodule, of dimension $p - \chi$, and a simple quotient of dimension $\chi + 1$.

The study of principal series modules starts to get more complicated when $G = \mathrm{SL}(3, p)$. Carter and Lusztig illustrate the methods of their paper concretely in this case. Images and kernels of the various endomorphisms reveal a considerable amount (though not all) of the submodule lattice in $\mathcal{F}_\chi$. (Note that Figure 7 in §8 is not entirely accurate, as observed in Jantzen [253, p. 473] and Schmidt [359, 3.3].)

Here the analogy with the structure of Weyl modules $V(\lambda)$ for $\mathrm{SL}(3, K)$ with highest weights weights lying in the lowest $p^2$-alcove for $W_p$ is striking. There are two typical patterns of composition factors, depending on which type of alcove $\lambda$ lies in. In "generic" cases $V(\lambda)$ has 9 composition factors, with multiplicity 1, lying in a certain pattern of alcoves. (This is pictured in 3.10.) The same kind of generic behavior occurs for principal series modules.

## 7.4. Summands of Principal Series Modules

The construction of induced modules outlined above leaves quite a few unanswered questions. For example, it is reasonable to ask how each principal series module decomposes into a direct sum of indecomposable modules. The prototype of this investigation is the study in characteristic 0 of the character induced from the trivial character of a Borel subgroup. There the Hecke algebra formalism coming from the study of intertwining operators plays a key role: see Curtis–Reiner [**126**, §11D].

In characteristic $p$ there is a substantial literature on such questions, usually related explicitly to the Curtis–Richen framework (but independent of the work of Carter–Lusztig). See in particular the papers by Cabanes [**74, 75, 76**], Green [**196**], Khammash [**271, 272, 273**], Richen [**349, 350**], Sawada [**354, 355, 356**], Tinberg [**416, 417, 419**].

One important step is taken by Sawada [**354**]. He denotes by $Y$ the $KG$-module induced from the trivial module of $U$. This can be constructed concretely inside the group algebra as the left ideal generated by the sum over the elements of $U$. Write $Y = Y_1 \oplus \cdots \oplus Y_s$ as a sum of indecomposable submodules, and set $E := \operatorname{End}_{KG}(Y)$.

THEOREM. *Let $G$ be a finite group with split $BN$-pair of rank $n$ and characteristic $p$. Then there are natural bijections between the sets*

(a) *isomorphism classes of simple $KG$-modules,*
(b) *indecomposable modules $Y_1, \ldots, Y_s$,*
(c) *equivalence classes of irreducible representations of the algebra $E$.*

*Moreover, the representations of $E$ are all 1-dimensional, each $Y_i$ has a unique simple submodule, and the socle of $Y$ is multiplicity-free.*

In more detail, he labels each $Y_i$ by an admissible pair $(J, \chi)$ as in Curtis–Richen, so that the socle of $Y_i$ is labelled by the "opposite" pair $({}^{w_\circ}J, {}^{w_\circ}\chi)$. From the viewpoint of Carter–Lusztig [**93**], the sum of all $Y_i$ sharing a common $\chi$ will be the principal series module denoted there by $\mathcal{F}_\chi$.

Other authors cited above go on to rework and expand this framework. For example, Tinberg [**416**] works out an explicit combinatorial formula for the dimension of each indecomposable summand of $Y$, as a sum of various computable powers of $p$, and describes the vertex of each summand as a Sylow intersection.

## 7.5. Homology Representations

If a finite group $G$ acts on a simplicial complex (or algebraic variety), associated homology or cohomology groups may afford interesting representations of $G$. When $G$ has a $BN$-pair, it is natural to study its action on the associated **Tits complex**, here denoted $\Delta$. This is a finite simplicial complex of dimension 1 less than the rank of $G$, based on the collection of parabolic subgroups: vertices correspond to maximal parabolics, etc. In characteristic 0 the top homology group of $\Delta$ yields an elegant construction of the Steinberg representation, as explored by Solomon and Curtis–Lehrer–Tits (see Curtis–Reiner [**127**, §66]).

In characteristic $p$ Ronan–Smith [**351**] study homology representations based on $\Delta$ as part of their more general investigations into geometries and group actions. They ask in particular to what extent a $KG$-module can be reconstructed from homology, which depends on "local" data coming from parabolic subgroups. In good cases, an algorithmic procedure leads back to the module.

Here we sketch a few ideas from [**351**], referring also to related papers by Smith [**379, 380, 381**] and Fisher [**170, 171**]. Results in [**351**] are usually formulated for universal Chevalley groups, assumed to have rank $n \geq 2$ in order to avoid trivialities; but the techniques involved are fairly general.

Associated with any $KG$-module $M$ is a coefficient system (presheaf) $\mathcal{F}_M$, computed in terms of fixed point spaces in $M$ of unipotent radicals of parabolic subgroups of $G$ (cf. Smith [**377**] and 5.10 above). The focus is on the $KG$-module $H_0(\mathcal{F}_M)$ rather than on the top homology; intermediate homology groups may be nonzero in this setting, but their nature is less clear. As an example, one sees (because St is a projective module) that $H_0(\mathcal{F}_{\mathrm{St}}) \cong \mathrm{St}$. In general:

THEOREM. *Let $G$ be a universal Chevalley group of rank $n \geq 2$ over $\mathbb{F}_q$. If $M$ is a simple $KG$-module, then $H_0(\mathcal{F}_M)$ has a unique simple quotient, isomorphic to $M$, and is therefore indecomposable.*

In some very special cases, especially when $p = 2$, it is possible to be more precise. For example, in [**351**, (4.1)] it is shown that for a simple module $M$ whose highest weight is *minimal* in the partial ordering of dominant weights, $H_0(\mathcal{F}_M) \cong M$ except in type $\mathrm{C}_n$. Here the symplectic module $M$ has dimension $2n$, whereas $H_0(\mathcal{F}_M)$ is isomorphic to the natural module for type $\mathrm{B}_n$ of dimension $2n + 1$. (See also the discussion of the spin modules in characteristic 2 by Smith [**378**, 2.3].)

Another interesting example in [**351**, p. 329] exhibits the various dimensions for $\mathrm{SL}(4, 2)$.

## 7.6. Comparison with Principal Series

One might ask how homology representations of the type discussed above compare with the principal series modules discussed earlier. In his thesis, Schmidt [**359**] explores this question from the dual viewpoint of the top homology of the Tits complex. His main results are formulated in the setting of split $BN$-pairs used by Carter and Lusztig.

Schmidt's notation is a hybrid of that found in previous papers, with some variations. The Tits complex is again called $\Delta$, while a $KG$-module $M$ determines a coefficient system called $S_M$ on $\Delta$. Simple modules are denoted $L(\lambda, J)$ for characters $\lambda : T \to K^\times$ and subsets $J \subset J_0(\lambda)$. If $n$ denotes the rank of $G$, the top homology group $H_{n-1}(\Delta, S_M)$ is the main object of study. For example, as in characteristic 0 the trivial module $M$ yields the Steinberg module. When $M$ is simple, the main theorem of Kapitel 1 compares the associated homology module with the principal series module determined by its "weight".

THEOREM. *Let $G$ be a finite group with split $BN$-pair of rank $n$ and characteristic $p$. If $L = L(\lambda, J)$ is a simple $KG$-module, then $H_{n-1}(\Delta, S_L)$ has a simple socle and can be embedded in the principal series module corresponding to $L$. The socle has the form $L(w_\circ\lambda, J')$ for a specified set $J'$.*

The embedding can in fact be described fairly explicitly in terms of kernels and images of the Carter–Lusztig intertwining maps.

In the remainder of his thesis, Schmidt studies the groups of type A in more depth, looking for example at the case when $L$ is the "natural" module. By introducing additional intertwining maps in type A, he is able to give a complete description of the composition series of principal series for $\mathrm{SL}(3, p)$ (while correcting a few errors in [**93**]).

# CHAPTER 8

# Blocks

Block theory plays a fundamental role in modular representation theory of finite groups (and has been imitated profitably in other non-semisimple situations). For a finite group $G$, the blocks of $KG$ can be defined straightforwardly to be the indecomposable two-sided ideals; but their structure may be quite complicated when $p$ divides $|G|$. Each simple (or, more generally, indecomposable) $KG$-module "belongs to" a unique block. There are also important connections with the representations of $G$ in characteristic 0. Brauer theory shows that any reduction modulo $p$ of a simple $\mathbb{C}G$-module yields well-defined composition factors over $K$, all of which belong to the same block. (This theme is pursued in Chapter 16 below.)

There are many ways to approach the determination of blocks of $KG$ for a particular group $G$, depending on what is already known about the ordinary and modular representations. Here we review some basic theory involving the defect of a block (8.2) and the associated defect groups (8.4), then apply it to a group of Lie type in the defining characteristic. It turns out in this case that $KG$ has blocks of only two extreme types. One type involves just the Steinberg module, while the other type has "defect group" equal to a Sylow $p$-subgroup (8.5). In particular, when $G$ is *simple* it has precisely two $p$-blocks. This result can also be placed in the more general setting of "vertices" of simple modules (8.8).

To round out the chapter we look at the related issue of representation type of a block (8.9). Here the analysis for groups of Lie type is fairly straightforward, but not very enlightening.

## 8.1. Blocks of a Group Algebra

When $G$ is an arbitrary finite group, the indecomposable two-sided ideals of $KG$ are called **blocks**, or $p$-**blocks**. They are algebras in their own right, called **block algebras**. The identity element of $KG$ can be written uniquely as a sum of commuting primitive idempotents, each of which is the identity element of a block algebra. A standard reference for the basic theory is Curtis–Reiner [**125**, §85–87].

It is common in the literature to denote a block by $B$ (or sometimes $b$), which creates a notational conflict for groups of Lie type. We use instead the notation $\mathcal{B}$.

PROPOSITION. *Let $G$ be any finite group. If $M$ is a nonzero indecomposable $KG$-module, exactly one block algebra $\mathcal{B}$ acts nontrivially on $M$, and then each composition factor of $M$ belongs to $\mathcal{B}$.*

We say that $\mathcal{B}$ "contains" $M$ or that $M$ "belongs to" $\mathcal{B}$. The block of $KG$ containing the trivial one-dimensional module is called the **principal block** or just the 1-**block**.

At first sight it seems an impossible task to say anything definite about the total number of $p$-blocks of a given group $G$ of Lie type in the defining characteristic. As a first step in this direction we introduce a numerical invariant of a block, called the "defect", and try to count the blocks having each possible defect.

## 8.2. The Defect of a Block

With $G$ still an arbitrary finite group, let $p^e$ be the full power of $p$ dividing $|G|$.

First we define the **defect** of a conjugacy class $C$ of $G$ to be $d$ if $p^d$ is the full power of $p$ dividing $|C_G(x)|$ for one (hence any) $x \in C$. In other words, $p^d$ is the order of a Sylow $p$-subgroup of $C_G(x)$. For example, the defect of a one-element class in $Z(G)$ is $e$.

Next we define the **defect** of a block $\mathcal{B}$ of $KG$ to be $d$ if $p^{e-d}$ divides the dimension of every indecomposable $KG$-module in $\mathcal{B}$ while no higher power of $p$ does so. (To compute $d$ it is in fact enough to consider just the simple modules in the block.) For example, the 1-block has highest possible defect $e$. The defect of $\mathcal{B}$ measures in some sense how far $\mathcal{B}$ is from being a semisimple algebra: indeed, it turns out that $\mathcal{B}$ is a semisimple algebra precisely when its defect is 0.

REMARK. When we consider the interaction between ordinary and modular representations in Chapter 16, we will observe that the ordinary characters of $G$ may also be assigned to blocks relative to any prime $p$ dividing $|G|$. The defect of a block can then be characterized equally well in terms of powers of $p$ dividing the degrees of the characters in a block.

The relationship between the two notions of "defect" just defined is not at all obvious, but by studying closely the way the center of $KG$ acts on $KG$-modules one can prove:

THEOREM. *Let $G$ be any finite group, with $p^e$ being the full power of $p$ dividing $|G|$.*

(a) *The number of blocks of $KG$ of highest defect $e$ is equal to the number of $p$-regular conjugacy classes of $G$ of defect $e$.*

(b) *The number of blocks of defect $\geq d$ is at most the number of $p$-regular conjugacy classes of defect $\geq d$.*

(c) *A block has defect $0$ if and only if it contains a simple module $M$ which is also projective. Then $M$ is unique and $p^e | \dim M$.*

Counting blocks of lower defect than $e$ is a subtle matter. There is, for example, no easy way to decide whether a finite group $G$ has such a block (in particular, whether it has a block of defect 0) relative to a given prime dividing $|G|$. For *simple* groups the study of groups of Lie type in all characteristics is central to the study of such questions: see Brockhaus–Michler [**62**], Michler [**323**], Willems [**433**].

## 8.3. Groups of Lie Type

The general theory just summarized leads quickly to a determination of the number of blocks of highest and lowest defect for a finite group of Lie type. (But more work is required to complete the picture.) Here we have to anticipate one feature of the Steinberg module to be discussed more fully in Chapter 9.

Let $G = \mathbf{G}^F$ over a field of $p^r$ elements. From the formulas of Chevalley and Steinberg for the group order, one knows that the highest power of $p$ dividing $|G|$ is $p^{rm}$, where $m = |\Phi^+|$: see 1.5. So $rm$ is the highest possible defect of a block of

$KG$. The principal block, which contains the trivial module $L(0)$, always has this defect.

THEOREM. *Let* $G = \mathbf{G}^F$ *over a field of* $p^r$ *elements. Then:*

(a) *The number of blocks of highest defect in* $KG$ *is the order of* $Z(G)$.

(b) *There is a unique block of defect* 0 *in* $KG$, *containing just the Steinberg module* $\mathrm{St}_r = L((p^r - 1)\rho)$ *of dimension* $p^{rm}$.

PROOF. (a) Theorem 8.2(a) says that the number of blocks of highest defect is the number of $p$-regular (i.e., semisimple) classes in $G$ having this defect. Clearly each $s \in Z(G)$ determines such a class. Conversely, let $s \in G$ be a semisimple element for which $C_G(s)$ contains a Sylow $p$-subgroup of $G$. Then $C_{\mathbf{G}}(s)$ is $F$-stable and $C_G(s) = C_{\mathbf{G}}(s)^F$. It is well known that the centralizer in $\mathbf{G}$ of any semisimple element is a reductive subgroup of $\mathbf{G}$ having maximal rank (and is connected since $\mathbf{G}$ is simply connected): see for example Carter [**89**, 3.5]. Therefore $C_{\mathbf{G}}(s)$ is generated by an $F$-stable maximal torus together with root groups corresponding to a closed symmetric $F$-stable subsystem $\Psi \subset \Phi$. A positive system $\Psi^+$ determines an $F$-stable maximal unipotent subgroup $\mathbf{V}$ of $C_{\mathbf{G}}(s)$. Then the finite group $\mathbf{V}^F$ is a Sylow $p$-subgroup of $C_G(s)$ and has order $q^{|\Psi^+|}$. By assumption this equals $q^m$, forcing $\Psi = \Phi$ and $C_{\mathbf{G}}(s) = \mathbf{G}$. Thus $s \in Z(\mathbf{G})^F \subset Z(G)$.

(b) We saw in 3.7 that the Steinberg module $\mathrm{St}_r$ has dimension $p^{rm}$, and that no other simple module has as large a dimension. This already shows that there is at most one candidate for a block of defect 0, thanks to Theorem 8.2(c). As we will explain in 9.3 below, $\mathrm{St}_r$ is also a *projective* module and therefore lies by itself in a block of defect 0. □

REMARK. The assumption that $\mathbf{G}$ is simply-connected insures that (with just a few exceptions listed in 1.5) $G$ is equal to its derived group. It is for this reason that the Steinberg block is the unique one of defect 0. For related groups such as $\mathrm{PGL}(n, q)$, the situation gets a little more complicated: there are additional modules of dimension $p^{rm}$ analogous to the Steinberg module and hence additional blocks of defect 0. The number of such blocks is in fact precisely the index of the derived group in $G$. This follows readily from Clifford's Theorem and the fact that this index is relatively prime to $p$. In the case of $\mathrm{PGL}(n, q)$, for example, the number is the greatest common divisor of $n$ and $q - 1$.

## 8.4. Defect Groups

Letting $G$ again be an arbitrary finite group, Brauer theory assigns a conjugacy class of $p$-groups having order $p^d$ to a block $\mathcal{B}$ of $KG$ having defect $d$. Each such group is called a **defect group** of $\mathcal{B}$. Thus the trivial group is the defect group of a block of defect 0, while any Sylow $p$-subgroup is a defect group of a block of highest defect. (See Curtis–Reiner [**125**, §87].)

The original definition of defect groups is somewhat involved. Roughly speaking, one associates to a block $\mathcal{B}$ of defect $d$ a Sylow $p$-subgroup $D$ of order $p^d$ in the centralizer of a suitably chosen $p$-regular element (whose class has defect $d$). As an alternative, the notion of defect group can be subsumed into the more easily defined notion of "vertex" discussed in 8.8 below.

The requirement that $D$ be a Sylow group in a centralizer already restricts somewhat the possible $p$-subgroups of $G$ which can occur as a defect group of a block. But there are other strong constraints on $D$:

THEOREM. *Let $G$ be any finite group. If $\mathcal{B}$ is a block of $KG$ and $D$ a defect group of $\mathcal{B}$, then*

(a) *$D$ is the largest normal $p$-subgroup of $N_G(D)$.*
(b) *For any Sylow $p$-subgroup $S$ of $G$ containing $D$, there exists $c \in C_G(D)$ for which $D = S \cap S^c$. Here $S^c := c^{-1}Sc$.*

Note that (b) actually implies (a): For this only the weaker fact that $c \in N = N_G(D)$ is needed. By choosing $S$ to contain a Sylow $p$-subgroup $R$ of $N$ (which automatically includes $D$), one gets $D \subset R \cap R^c \subset S \cap S^c = D$. Thus $D$ is a Sylow intersection in $N$ as well as in $G$. Since any normal $p$-subgroup of $N$ lies in every Sylow subgroup of $N$, (a) follows.

Part (a) originates with Brauer [**60**, (10C)] and part (b) with Green [**194**, Thm. 3]. See also Curtis–Reiner [**127**, 57.31], Alperin [**6**, §13], Benson [**47**, 6.1.1].

REMARK. It has been a guiding philosophy since Brauer's early work that the study of ordinary and modular representations for an arbitrary finite group should be organized around the possible types of blocks and defect groups. After early success in the case of cyclic defect groups, recent work has focused on further conjectures about blocks with abelian defect groups. Groups of Lie type over fields of arbitrary characteristic have been a fruitful laboratory for study of such conjectures, but the defining characteristic case plays a relatively minor role here in view of Theorem 8.5 below and the fact that these groups rarely have abelian Sylow $p$-subgroups.

## 8.5. Defect Groups for Groups of Lie Type

For groups of Lie type, including the groups of Suzuki and Ree, the block structure of $KG$ turns out to be uncomplicated (in the defining characteristic):

THEOREM. *Let $G = \mathbf{G}^F$. Then all blocks of positive defect in $KG$ have a Sylow $p$-subgroup of $G$ as defect group. In particular, when $G$ is simple there are just two $p$-blocks: one of defect 0 and one of highest defect.*

This was first worked out for some classical groups by Dagger [**128**], after which a uniform proof was given by Humphreys [**214**]. (Study of the vertices of simple modules by Dipper then recovered the theorem indirectly, but at the cost of much more case-by-case work: see 8.8 below.)

The theorem implies that block theory has little to say directly about the behavior of $KG$-modules. By way of contrast, there is a rich block theory for primes other than $p$ dividing $|G|$: see Cabanes–Enguehard [**78**] and their references to the literature. In that situation the blocks interact in subtle ways with the Deligne–Lusztig generalized characters of $G$.

## 8.6. Steps in the Proof

We consider first how to prove Theorem 8.5 in the setting of finite subgroups $G = \mathbf{G}^F$ of algebraic groups. The proof can also be carried out in a suitably enriched $BN$-pair setting: see 8.7 below. For finite group theorists who prefer to avoid some of the algebraic group background, that is undoubtedly the most efficient proof.

Our starting point is Dagger's analysis [**128**] of Green's Theorem 8.4(b).

LEMMA. *Let $G = \mathbf{G}^F$ over a field of $p^r$ elements. Set $D := U \cap U^x$, and write $x = unu'$ in the Bruhat decomposition, with $u, u' \in U$ and $n \in N$ representing $w \in W$. If $x \in C_G(D)$, then $w\alpha > 0$ implies $w\alpha = \alpha$ for all $\alpha > 0$.*

Dagger assumed (as we do initially) that $G$ is a Chevalley group; the ideas carry over to twisted groups by substituting their "relative" roots and Weyl groups. To prove the lemma, one computes straightforwardly from the hypothesis that $nu'u$ centralizes $U \cap U^n$, which is conjugate to $D$ and can therefore replace $D$ as a defect group. In general $n^{-1}\mathbf{U}_\alpha n = \mathbf{U}_{w^{-1}\alpha}$. Thus $U \cap U^n$ is just the product of root subgroups $U_\alpha$ for those $\alpha > 0$ satisfying $w^{-1}\alpha > 0$. Using the commutation rules in $U$ and a standard ordering of positive roots, one sees that the hypothesis implies $w^{-1}\alpha = \alpha$ (and thus $w\alpha = \alpha$) for all roots involved in $U \cap U^n$.

With this lemma in hand, the proof of the theorem can be organized as follows. We streamline the arguments in [128] and [214] by making better use of a theorem of Borel and Tits.

(1) The problem is to pin down the possible defect groups of blocks of $KG$ having defect $> 0$. Start with such a defect group $D \neq 1$, which can be replaced by a conjugate when convenient. Since the Sylow $p$-subgroups of $G$ are the conjugates of $U = \mathbf{U}^F$ for a fixed choice of maximal unipotent subgroup in $\mathbf{G}$, Theorem 8.4(b) allows us to assume that $D = U \cap U^x$ for some $x \in C_G(D)$.

(2) Theorem 8.4(a) also says that $D$ is the largest normal $p$-subgroup of its normalizer $N_G(D)$. Since $D \neq 1$ this property implies, by a theorem of Borel–Tits (simplified by Burgoyne and Williamson [68] for algebraic groups over finite fields), that $N_G(D)$ is a parabolic subgroup $P$ of $G$.

(3) Replacing $D$ by a conjugate and using the assumption $D \neq 1$, we may assume that for some standard parabolic subgroup $\mathbf{P} = \mathbf{LV}$ of $\mathbf{G}$ defined by a proper subset $\Delta' \subset \Delta$, we have $N_G(D) = \mathbf{P}^F$ while $D = \mathbf{V}^F$. In particular, $D$ is just a product of certain root groups $U = \mathbf{U}_\alpha$ corresponding to those positive roots not in the $\mathbb{Z}$-span of $\Delta'$.

(4) Since $D = U \cap U^x$, the Bruhat decomposition $x = unu'$ shows immediately that we can take $x$ to be a representative $n$ of some Weyl group element $w$. So we may write $D = U \cap U^n$.

(5) By Dagger's lemma, the fact that $x = unu'$ centralizes $D$ translates into the statement that $w$ fixes all roots occurring. Moreover, $D = U \cap U^n$ implies that $w$ sends all roots occurring in $L = \mathbf{L}^F$ to negative roots (indeed, $w$ must be the longest element in the Weyl group of $\mathbf{L}$).

(6) Now we invoke the assumption that the root system of $\mathbf{G}$ is *irreducible*. If $D$ is a proper subgroup of $U$, there must be a root $\beta > 0$ of $L$ not orthogonal to some simple root $\alpha$ involved in $D$ and for which $w\beta = -\gamma \in -\Delta'$ is *simple*. This implies that $\alpha + \beta$ is a root. But $w(\alpha + \beta) = \alpha - \gamma$ cannot be a root, since both $\gamma$ and $\alpha$ are simple. This contradiction shows that $D = U$. $\square$

REMARK. This proof adapts to the Suzuki and Ree groups (Chapter 20), which Steinberg constructs as fixed point subgroups of certain groups $\mathbf{G}$ under a surjective endomorphism $\sigma$. As in the adaptation to twisted groups, the $BN$-pair structure of $G$ comes into play. For step (2), note that Burgoyne–Williamson [68] use Steinberg's $\sigma$-setup.

## 8.7. The $BN$-Pair Setting

It is natural to place Theorem 8.5 in the axiomatic setting of $BN$-pairs, as Dagger originally attempted to do. But this seems to require some enrichment of the axioms in order to imitate the above use of the Borel–Tits result for algebraic groups over finite fields. Here we outline briefly the approach presented by Cabanes–Enguehard [**78**, Part I]. (As in Digne–Michel [**134**], the groups of Suzuki and Ree are not explicitly treated in their book; but the arguments should extend to this case.)

They start with a finite group $G$ having a split $BN$-pair of characteristic $p$ (as described in 1.7 above). In particular $U$ is a Sylow $p$-subgroup of $G$. Each subset $I$ of the set $S$ of simple reflections in $W$ defines a subgroup $W_I$ with longest element $w_I$. Then the $BN$-pair is called **strongly split** if the subgroup $U_I := U \cap U^{w_I}$ is normal in $U$. This leads in particular to a good version of the commutation relations in $U$ familiar in groups of Lie type: see Genet [**185**]. (On the other hand, earlier work by Tinberg [**418**] showed how the normality statement can be obtained from the $BN$-pair axioms in conjunction with some of the arguments of Kantor–Seitz, without appeal to the commutation relations, yielding in turn the Levi decomposition for parabolic subgroups of $G$.)

Even more is needed, however, for the theorem on defect groups. Hypothesis 6.14 in [**78**] states:

> (∗) If a subgroup $V$ of $G$ is the largest normal $p$-subgroup of its normalizer, then there exists $g \in G$ and a set $I \subset S$ such that $V = gU_Ig^{-1}$.

This is checked for Chevalley groups and twisted groups (but not explicitly for Suzuki and Ree groups, where similar arguments apply).

In this framework, the following version of Theorem 8.5 is then proved:

THEOREM. *Let $G$ be a finite group with a strongly split $BN$-pair, satisfying* (∗) *above. Assume further that $W$ is irreducible and that $C_G(U) = Z(U)$, so in particular $Z(G) = 1$. Then the only non-principal $p$-blocks of $G$ are those of defect* 0.

The proof is similar to the one in 8.5, but requires a couple of technical arguments based on the axioms as a substitute for quoting Burgoyne–Williamson [**68**].

## 8.8. Vertices of Simple Modules

Following the early work of Brauer on defect groups, the idea of associating $p$-subgroups of a finite group $G$ with blocks of $KG$ was broadened considerably. Here we outline briefly this further development of the theory, which is presented in a number of texts, for example Alperin [**6**, §9–13], Benson [**47**, 6.1], Curtis–Reiner [**126**, §19] and [**127**, §57].

Given a $KG$-module $M$, one may ask for which subgroups $H \leq G$ it is possible to obtain $M$ as a direct summand of a module induced from $H$ to $G$, written briefly as $M \,|\, \mathrm{Ind}_H^G N$. If $H$ is *minimal* with this property, it is called a **vertex** of $M$, while $N$ is called a **source** of $M$. When $M$ is indecomposable, $H$ is a $p$-group, uniquely determined up to conjugacy in $G$, and $N$ can be taken to be indecomposable (with a suitable uniqueness property). Moreover, $M$ is a projective module iff its vertex is trivial. In general the vertex measures in some sense how far $M$ is from being projective.

The theory of vertices generalizes in an elegant way the earlier theory of defect groups. Let $G$ be any finite group. Regard $KG$ as a module for $G \times G$ via the action $(a, b) \cdot x := axb^{-1}$. Then the submodules are just the two-sided ideals, the indecomposable summands being the blocks. Let $\delta : G \to G \times G$ be the diagonal map. One shows that the block $\mathcal{B}$ has a vertex of the form $\delta(D)$ for some $p$-subgroup $D$ of $G$, and that moreover all choices for $D$ are conjugate in $G$. This turns out to agree with Brauer's earlier definition of defect group.

In this context it can further be shown that for a block $\mathcal{B}$ of defect $D$, every indecomposable module belonging to $\mathcal{B}$ has a vertex contained in $D$. Moreover, at least one such module has $D$ as a vertex. Thus $\mathcal{B}$ has a Sylow $p$-subgroup as its defect group iff it contains a module having a Sylow group as its vertex.

Now let $G$ be a finite group of Lie type. While there is no unified Lie-theoretic approach to the study of all indecomposable $KG$-modules, it is still reasonable to ask: What are the vertices of the simple modules $L(\lambda)$? Dipper [135, 137] answers this question in an elegant way:

THEOREM. Let $G = \mathbf{G}^F$ over a field of $p^r$ elements. Then the vertex of every non-projective simple $KG$-module is a Sylow $p$-subgroup of $G$.

The actual statement of Dipper's theorem involves the assumption that $G$ modulo its center is *simple*, which excludes just a few groups $\mathbf{G}^F$ with $\mathbf{G}$ simply connected. But he can check the non-simple groups case-by-case as well.

In view of the preceding remarks, the theorem immediately implies Theorem 8.5 on defect groups. However, Dipper's proof is developed separately for Chevalley groups and for twisted groups, while involving a considerable amount of case-by-case checking. A more unified proof would be desirable (encompassing also the groups of Suzuki and Ree). One obstacle to this is the lack of criteria like those given in Theorem 8.4 to limit the possible vertices.

We remark that in a related paper, Dipper [136] goes on to determine block structure and vertices of simple modules for the parabolic subgroups of Chevalley groups.

## 8.9. Representation Type of a Block

A standard theme in the study of representations of finite dimensional algebras (such as group algebras or block algebras) is the determination of **representation type**. The issue here is whether indecomposable modules can be classified in some meaningful sense. There turns out to be a useful trichotomy, as formulated for example in Benson [47, 4.4]. But note that definitions of "tame" vary somewhat in the literature.

An algebra $\mathcal{A}$ has **finite representation type** if it possesses only finitely many non-isomorphic indecomposable modules. (This is known to be equivalent to the apparently weaker requirement that there are only finitely many indecomposables of each dimension.) When this fails, $\mathcal{A}$ is said to have **tame representation type** if in a suitable sense all its indecomposable representations can be parametrized. The remaining possibility is that $\mathcal{A}$ has **wild representation type**.

For blocks in group algebras there is a clearcut way to sort things out in terms of defect groups:

THEOREM. *For an arbitrary finite group $G$, let $B$ be a $p$-block of $KG$ having defect group $D$. Then $B$ has finite representation type iff $D$ is cyclic, while $B$ has tame representation type iff $p = 2$ and $D$ is dihedral, semidihedral, or generalized quaternion. In all other cases $B$ has wild representation type.*

The proof evolved in a number of steps, starting with work of D.G. Higman (1954), who showed for an arbitrary finite group $G$ that $KG$ has finite representation type if and only if $G$ has a cyclic Sylow $p$-subgroup. The refinement involving blocks came later; see for example Benson [**47**, 6.3.5] for a proof. V.M. Bondarenko and Yu. A. Drozd (1977) found the criterion for $KG$ to have tame representation type, which was then adapted to blocks. In the tame case, the requirement that the defect group $D$ be dihedral, semidihedral, or generalized quaternion implies that any noncyclic abelian 2-subgroup of $D$ has order at most 4.

How does the theorem apply to a finite group $G$ of Lie type in the defining characteristic $p$? In view of Theorem 8.5 we need only consider the structure of the Sylow $p$-subgroup of $G$, for which $G$ itself may be assumed to be simple. This structure has been worked out in considerable detail for each type of group: see Gorenstein–Lyons–Solomon [**190**, 3.3].

The only simple groups of Lie type having a cyclic Sylow $p$-subgroup are the groups $PSL(2, p)$ for $p > 3$, along with the group ${}^2G_2(3)'$ for $p = 3$ (though the latter group is more commonly viewed as $SL(2, 8)$).

On the other hand, when the defining characteristic is $p = 2$, there are no groups of Lie type having semidihedral or generalized quaternion 2-Sylow subgroups. But there are a few small cases in which the Sylow 2-subgroups are dihedral: $SL(2, 4)$, $SL(3, 2)$, $B_2(2)' = C_2(2)'$ (isomorphic to the alternating group $A_6$).

REMARK. To $B$ is associated its stable *Auslander–Reiten quiver*, a directed graph which typically has many connected components. Each simple module in $B$ corresponds to a vertex of just one of these components. Kawata–Michler–Uno [**268**] study the case when a block in a finite group $G$ of Lie type has wild representation type. They prove the following result: *Suppose the $p$-block $B$ of $G$ has a Sylow $p$-subgroup as defect group. If $B$ is of wild representation type, then any simple $KG$-module in $B$ lies at the end of its Auslander–Reiten component.*

CHAPTER 9

# Projective Modules

In the study of a non-semisimple module category, projective modules are an essential ingredient. The general theory of these modules is fairly well-organized for any finite dimensional algebra, with additional features in the case of a group algebra $KG$. We begin by recalling some standard facts for an arbitrary $G$ in 9.1, then raise a number of questions in 9.2 concerning families of finite groups of Lie type.

A pivotal role is played by Steinberg modules (9.3): simple **G**-modules having highest weights of the form $(p^r - 1)\rho$, where $\rho$ is the sum of fundamental weights. Unlike other simple modules for the finite group $G$ over a field of $p^r$ elements, $L((p^r - 1)\rho)$ is its own projective cover. Tensoring arbitrary $KG$-modules with this one produces new projective modules, whose indecomposable summands turn out to exhaust the projective covers of all simple modules (9.4).

In the framework of Brauer characters (9.5–9.6), we see that the Steinberg character "divides" all characters of projective modules. Moreover, there is at least a rough lower bound (9.7) for the dimensions of indecomposable projectives, though this is usually far too small in practice. A more thorough study of projectives is deferred to the following chapter, where the parallel theory for Frobenius kernels comes into play.

Here we just take a detailed look at projective modules for $SL(2, p)$ (9.8). The data can be efficiently encoded in a "Brauer tree" (9.9). In this rather isolated case there are only finitely many indecomposable $KG$-modules (up to isomorphism), which can be described explicitly in terms of the projectives (9.10).

## 9.1. Projective Modules for Finite Groups

Let $G$ be any finite group. Since $KG$ is a Frobenius algebra, the notions of projective module and injective module coincide. Usually we opt for the language of projectives. If $p$ fails to divide $|G|$, all $KG$-modules are projective as well as semisimple; we are interested here only in the contrary case. For the following standard facts, see for example Curtis–Reiner [**125**, VIII, XII].

THEOREM. *Let $G$ be an arbitrary finite group. If $\{L_\lambda \mid \lambda \in \Lambda\}$ is a complete set of non-isomorphic simple $KG$-modules, denote the corresponding projective covers by $U_\lambda$. Then:*

(a) *The module $U_\lambda$ is indecomposable.*

(b) *$U_\lambda$ has a unique simple quotient and a unique simple submodule, each isomorphic to $L_\lambda$. Thus $U_\lambda$ is also the injective hull of $L_\lambda$.*

(c) *Every projective $KG$-module $Q$ is isomorphic to a direct sum of various $U_\lambda$. In any such decomposition, the number of summands isomorphic to $U_\lambda$ is equal to $\dim \operatorname{Hom}_{KG}(Q, L_\lambda)$.*

(d) *In any (left module) decomposition of $KG$ as a direct sum of copies of various $U_\lambda$, each $U_\lambda$ occurs exactly $\dim L_\lambda$ times. Thus*

(1)
$$|G| = \sum_{\lambda \in \Lambda} \dim L_\lambda \ \dim U_\lambda.$$

(e) *If $c_{\lambda\mu} = [U_\lambda : L_\mu]$ denotes the multiplicity of $L_\mu$ as a composition factor of $U_\lambda$, then $c_{\lambda\mu} = c_{\mu\lambda}$.*

(f) *If $P$ is a Sylow $p$-subgroup of $G$, then all projective $KP$-modules are free. Moreover, all $KG$-modules $M$ are $(G, P)$-projective: if $M$ is $KP$-projective, then it is also $KG$-projective. As a result, if $p^e$ is the full power of $p$ dividing $|G|$, then $p^e$ divides $\dim U_\lambda$ for all $\lambda \in \Lambda$.*

(g) *The tensor product of a projective $KG$-module and an arbitrary $KG$-module is again projective.*

Traditionally the projective cover of a simple module is called a **principal indecomposable module**, abbreviated PIM. The multiplicities $c_{\lambda\mu}$ are called the **Cartan $p$-invariants** of $KG$. They form a **Cartan matrix** indexed by $\Lambda \times \Lambda$, which is *symmetric* thanks to (e). This reflects the fact that $KG$ is a "symmetric" algebra in the sense of Nesbitt. (As a finite dimensional Hopf algebra, $KG$ is automatically a Frobenius algebra, but the symmetry property is stronger.)

Since each indecomposable $KG$-module belongs to a single block (8.1), the PIMs as well as simple modules distribute naturally into blocks. In particular, for an ordering of $\Lambda$ compatible with the block decomposition of $KG$, the Cartan matrix has the corresponding block form.

## 9.2. Groups of Lie Type

Return now to the setting of finite groups of Lie type, with notation as in 1.8. We have seen in Chapter 2 that the simple $KG$-modules $L(\lambda)$ are the restrictions of those simple **G**-modules having highest weights $\lambda \in X_r$. Denote by $U_r(\lambda)$ the projective cover (= injective hull) of $L(\lambda)$. Here the subscript is needed, because for a fixed $\lambda \in X_r$ the module $L(\lambda)$ remains simple for a group over a larger finite field but then has a larger projective cover. When $r = 1$ we may write just $U(\lambda)$ rather than $U_1(\lambda)$.

In types $A_\ell, D_\ell, E_6$, the simple modules for a Chevalley group and its twisted analogue are actually the same. But the projective covers may differ, in which case there is a further notational ambiguity.

Many questions arise at this point about the nature of projective modules for groups of Lie type:

- What is the dimension of $U_r(\lambda)$?
- What are the Cartan invariants of $G$, and how do they depend on $p$?
- If $L(\lambda)$ with $\lambda \in X_r$ is written as a twisted tensor product involving various $L(\mu)$ with $\mu \in X_1$, how is $U_r(\lambda)$ related to the various $U_1(\mu)$?
- In types $A, D, E_6$, how are the PIMs for a Chevalley group related to those for its twisted analogue?
- What are the possible Loewy structures, such as radical series and socle series?
- Which features of the PIMs are independent of $p$?

This and the following chapters will describe how much is currently known about these questions, most of which are still only partially answered.

REMARK. Though the internal structure of a PIM may be quite complicated, the projective modules tend to play the role of trivial objects in the overall module category for $KG$. (See for example 19.8 below.) Recent study of the "stable module category" raises a number of questions about what happens when projective summands are discarded. If $M$ is any $KG$-module, consider the related $KG$-module $\mathrm{End}(M) \cong M^* \otimes M$. In case $\mathrm{End}(M)$ decomposes as the direct sum of the trivial module and a projective module, $M$ is called **endotrivial**. When $G$ is a finite group of Lie type, such modules are studied by Carlson–Mazza–Nakano [**85**].

## 9.3. The Steinberg Module

How does one get started with the study of projective modules when $G$ is of Lie type over $\mathbb{F}_q$ (with $q = p^r$)? We know from 3.7 that $\mathrm{St}_r = L((q-1)\rho)$ has dimension $q^m$ ($m = \dim \mathbf{U} = |\Phi^+|$), whereas all other simple $KG$-modules have smaller dimensions. Thanks to part (f) of Theorem 9.1, $\mathrm{St}_r$ is therefore the only simple $KG$-module which could also be projective.

THEOREM. *Let $G = \mathbf{G}^F$ over a field of $q = p^r$ elements. The Steinberg module $\mathrm{St}_r = L((q-1)\rho)$ is a projective and injective $KG$-module, of dimension $q^m$ (the order of a Sylow $p$-subgroup of $G$).*

PROOF. This can be approached in more than one way, depending on what one already knows.

The most standard line of reasoning is this: Starting with the existence of the **Steinberg character** of $G$ over $\mathbb{C}$ (see 16.6), whose degree is $q^m$, one can appeal to Brauer's theory of blocks of defect 0 (as defined in 8.2). This theory shows that the reduction modulo $p$ of a representation affording this character is simultaneously irreducible and projective. This module must in turn be isomorphic to $\mathrm{St}_r$, the only simple $KG$-module of dimension $q^m$. (This is the proof given by Steinberg [**393**].)

Since we are deferring consideration of ordinary characters, we instead look directly at how the Sylow $p$-subgroup $V = U^-$ of $G$ acts on $\mathrm{St}_r$. For any simple $KG$-module $L(\lambda)$, with highest weight vector $v^+$, Proposition 2.15 shows that

$$KV \cdot v^+ = KG \cdot v^+ = L(\lambda).$$

Thus the $KV$-linear map $KV \to KV \cdot v^+ \hookrightarrow L(\lambda)$ is *surjective*. Since both $\mathrm{St}_r$ and $KV$ have dimension $q^m$, the associated map $KV \to \mathrm{St}_r$ is actually an isomorphism of $KV$-modules. In other words, $\mathrm{St}_r$ is projective (hence free) as a $KV$-module. Now we can invoke part (f) of Theorem 9.1 to conclude that $\mathrm{St}_r$ is also projective as a $KG$-module. $\qquad\qquad\square$

REMARKS. (1) The latter argument is similar in spirit to the original construction given by Steinberg [**392**], which can be carried out over fields of arbitrary characteristic such as $K$. Working inside the group algebra $KG$, he defines an element $e$ which generates a right ideal affording the desired irreducible representation. (We work instead with left modules.) It then has to be shown that this $KG$-module is also projective (or injective). The construction shows that it affords the regular representation of the Sylow $p$-subgroup $V$, so it is already $KV$-projective, as above.

(2) For a detailed survey of the Steinberg representation in its various guises, see Humphreys [**231**].

## 9.4. Tensoring with the Steinberg Module

Having shown that the Steinberg module is $KG$-projective, our basic strategy is to tensor it with arbitrary simple modules (or other highest weight modules) and then try to sort out which PIMs occur as direct summands. First we have to assemble some notation.

As before we abbreviate $\sigma_r := (p^r - 1)\rho$. To compare weights when $G$ is a Chevalley group, recall from 5.6 the refined partial ordering $\mu \leq_{\mathbb{Q}} \lambda$. For twisted groups we use instead the length function $|\lambda| = \langle \lambda, \alpha_0^{\vee} \rangle$.

As usual, $w_o$ denotes the longest element of $W$, interchanging positive and negative roots. For $\lambda \in X^+$, write $\lambda^* := -w_o\lambda$ for the highest weight of the dual module $L(\lambda)^*$ (see 2.2). Define an involution $\lambda \mapsto \lambda^{\circ}$ of $X_r$ by the recipe

$$\lambda^{\circ} := \sigma_r + w_o\lambda.$$

Note that $(\lambda^{\circ})^* = (\lambda^*)^{\circ} = \sigma_r - \lambda$.

THEOREM. *Let* $G = \mathbf{G}^F$ *over a field of* $p^r$ *elements, and let* $\lambda, \mu \in X_r$.

(a) *In any decomposition of* $L(\lambda^{\circ}) \otimes \mathrm{St}_r$ *as a direct sum of PIMs, the number* $m_{\mu}$ *of summands isomorphic to* $U_r(\mu)$ *is the composition factor multiplicity* $[L(\mu) \otimes L(\lambda^{\circ}) : \mathrm{St}_r]_G$. *In particular, we have* $m_{\lambda} = 1$.

(b) *If* $G$ *is a Chevalley group and* $m_{\mu} > 0$, *then* $\lambda \leq_{\mathbb{Q}} \mu$.

(c) *If* $G$ *is a twisted group and* $m_{\mu} > 0$, *then* $|\lambda| \leq |\mu|$.

PROOF. On the level of $\mathbf{G}$-modules we have

$$\mathrm{Hom}_{\mathbf{G}}(L(\lambda^{\circ}) \otimes \mathrm{St}_r, L(\mu)) \cong \mathrm{Hom}_{\mathbf{G}}(\mathrm{St}_r, L(\mu) \otimes L(\lambda^{\circ})^*)$$
$$= \mathrm{Hom}_{\mathbf{G}}(\mathrm{St}_r, L(\mu) \otimes L(\sigma_r - \lambda)).$$

When $\mu = \lambda$, this last tensor product has highest weight $\sigma_r$, with multiplicity 1, and therefore has a (unique) $\mathbf{G}$-composition factor isomorphic to $\mathrm{St}_r$.

Now replace $\mathbf{G}$ by $G$ in the calculation. By Theorem 9.1(c)) the dimension of the Hom space on the left side is $m_{\mu}$. Since $\mathrm{St}_r$ is both simple and projective, the dimension of the Hom space on the right side is equal to the number of times $\mathrm{St}_r$ occurs as a direct summand (or equivalently, composition factor) of the $G$-module $L(\mu) \otimes L(\lambda^{\circ})^*$.

The highest weight of any $\mathbf{G}$-module $L(\nu)$ occurring as a composition factor of $L(\mu) \otimes L(\sigma_r - \lambda)$ must satisfy $\nu \leq \mu + \sigma_r - \lambda$. In turn, $\mathrm{St}_r$ can only occur as a $G$-composition factor in the tensor product if it occurs as a composition factor of such an $L(\nu)$. If $G$ is a Chevalley group, Proposition 5.8 forces $\sigma_r \leq_{\mathbb{Q}} \nu \leq \mu + \sigma_r - \lambda$. Thus $\lambda \leq_{\mathbb{Q}} \mu$. Moreover, equality can hold here only if $\mathrm{St}_r$ occurs as the unique $\mathbf{G}$-composition factor having the highest possible weight in the tensor product.

If $G$ is a twisted group, the argument is similar.                    $\square$

This line of argument suggests an algorithmic approach to the decomposition of $L(\lambda^{\circ}) \otimes \mathrm{St}_r$ into PIMs, based on Theorem 6.9 (which depends only on standard character data for $\mathbf{G}$). However, knowing just the multiplicities of indecomposable summands in $L(\lambda^{\circ}) \otimes \mathrm{St}_r$ would not immediately reveal much about the PIMs, e.g., their dimensions. For this a more careful recursion (based on a partial ordering of weights) is required, starting with the obvious case $\lambda^{\circ} = 0$. We turn next to some computations with Brauer characters which facilitate such bookkeeping.

In the following chapter we shall investigate less directly the way $U_r(\lambda)$ and other PIMs occur in tensoring with $\mathrm{St}_r$, by comparing $KG$-modules with projective modules for Frobenius kernels. This indirect approach turns out to be more efficient for low rank computations. It also provides more insight into the "generic" behavior of PIMs.

## 9.5. Brauer Characters and Orthogonality Relations

A more classical viewpoint on the process of tensoring with the Steinberg module is expressed in terms of Brauer characters (see 5.5).

First let $G$ be any finite group. As is common in the literature, we list the distinct irreducible Brauer characters as $\varphi_1, \ldots, \varphi_n$. To each of these corresponds the Brauer character of the corresponding PIM, denoted $\eta_1, \ldots, \eta_n$. Generalizing the usual orthogonality relations for ordinary characters, one has a modular version: see Curtis–Reiner [125, §84] or [126, §18C]. Here the inner product of characters is replaced by

$$(\chi, \theta)_G := \frac{1}{|G|} \sum \chi(g)\theta(g^{-1}),$$

with the sum taken only over the set $G_{\text{reg}}$ of $p$-regular elements of $G$ and $\chi, \theta$ assumed to be $\mathbb{C}$-valued class functions on $G_{\text{reg}}$. (Alternatively, one can extend Brauer characters to class functions on $G$ by defining them to be 0 outside $G_{\text{reg}}$ and then using the usual inner product formula.)

THEOREM. *Let $G$ be any finite group. With the above notation, we have* $(\varphi_i, \eta_j)_G = \delta_{ij}$.

## 9.6. Brauer Characters of PIMs

Return now to the case where $G$ is of Lie type over a field of $q = p^r$ elements. In the setting of 9.5 it is not difficult to translate the proof of Theorem 9.4 above. Here we recover some of the results of Ballard [29]. But note that he (like Wong [436, 437]) treats only Chevalley groups and works with a lexicographic ordering of weights.

Ballard shows in effect that the Steinberg module "divides" all PIMs for $G$. Independently, Lusztig [308] (also in the context of Chevalley groups) approaches the divisibility from a different direction, having first extended the definition of the Steinberg representation to a more general class of algebraic groups. Yet another approach will be taken in the following chapter, by comparing $G$-modules with modules for Frobenius kernels of $\mathbf{G}$ in the spirit of 9.4.

Following 5.6, denote by $\varphi_\lambda = \operatorname{br} L(\lambda)$ the Brauer character of $L(\lambda)$, abbreviating by $\varphi$ the Brauer character of the Steinberg module $\operatorname{St}_r = L(\sigma_r)$. Write $\eta_\lambda = \operatorname{br} U_r(\lambda)$, the Brauer character of the PIM corresponding to $\lambda \in X_r$. Recall the formal Brauer characters $s_\mu$ coming from $W$-orbits $s(\mu)$, where $\mu \in X^+$. These are building blocks for the $\varphi_\lambda$, as shown in Proposition 5.7:

$$(1) \qquad \varphi_\lambda = s_\lambda + \sum a_\mu s_\mu \ (a_\mu \in \mathbb{Z}, \mu \in X_r).$$

Here the sum is taken over $\mu <_{\mathbb{Q}} \lambda$ if $G$ is a Chevalley group, or $|\mu| < |\lambda|$ if $G$ is a twisted group. As in 5.7 and 5.8, we proceed in slightly different ways for $G$ of these two types. We shall write the details here just for Chevalley groups, leaving the modifications in the twisted case to the reader.

Recall from 9.4 the involution $\lambda \mapsto \lambda^\circ$ on $X_r$ defined by

$$\lambda^\circ := \sigma_r + w_\circ \lambda.$$

Note that since $w_\circ$ interchanges positive and negative roots, we have $\lambda \leq_{\mathbb{Q}} \mu$ if and only if $\mu^\circ \leq_{\mathbb{Q}} \lambda^\circ$. (For the twisted case, note instead that $|\lambda| \leq |\mu|$ if and only if $|\mu^\circ| \leq |\lambda^\circ|$.)

THEOREM. *Let $G = \mathbf{G}^F$ over a field of $p^r$ elements. Let $\lambda \in X_r$, with $\varphi$ as above the Brauer character of $\mathrm{St}_r$. Then $\eta_\lambda = s_{\lambda^\circ} \varphi + \mathbb{Z}$-linear combination of various products $s_{\mu^\circ} \varphi$, with $\mu \in X_r$ and $\lambda <_{\mathbb{Q}} \mu$ (in the twisted case, $|\lambda| < |\mu|$). Thus the Steinberg character divides all Brauer characters of PIMs for $KG$.*

PROOF. Assume that $G$ is a Chevalley group. We use downward induction on $\lambda$, starting with $\varphi = s_0\varphi$.

Start by translating Theorem 9.4 into the language of Brauer characters:

$$\varphi_{\lambda^\circ}\varphi = \eta_\lambda + \mathbb{Z}\text{-linear combination of } \eta_\mu \text{ for } \lambda <_{\mathbb{Q}} \mu.$$

Then use (1) above to substitute on the left:

$$\varphi_{\lambda^\circ} = s_{\lambda^\circ} + \sum_{\nu^\circ <_{\mathbb{Q}} \lambda^\circ} a_{\nu^\circ} s_{\nu^\circ}$$

with coefficients in $\mathbb{Z}$ and with $\nu^\circ \in X_r$. Here the condition $\nu^\circ <_{\mathbb{Q}} \lambda^\circ$ is equivalent to $\lambda <_{\mathbb{Q}} \nu$.

Apply the induction hypothesis to replace the terms $\eta_\mu$ with $\mathbb{Z}$-linear combinations of various $s_{\nu^\circ}\varphi$ with $\mu <_{\mathbb{Q}} \nu$. Since $\lambda <_{\mathbb{Q}} \mu$ in each case, this yields the desired expression for $\eta_\lambda$.      □

REMARK. In his development, Ballard instead uses the Orthogonality Relations to derive various preliminary results which feed into his proof of the theorem. He shows for example that if $\lambda \in X_r$, then $(s_\lambda, \varphi)_G = \delta_{\lambda, \sigma_r}$.

## 9.7. A Lower Bound for Dimensions of PIMs

A rough lower bound for dimensions of PIMs is implicit in Ballard's calculations with Brauer characters, reflecting a type of Weyl group symmetry. We already know from Theorem 9.1(f) that $\dim U_r(\lambda)$ is a multiple of $q^m = |U|$, so that $\mathrm{St}_r$ achieves the smallest possible dimension. Following [**29**, §5], we can go somewhat further, at least in the case of Chevalley groups. (We leave it to the reader to adapt the argument to twisted groups.)

The idea is to restrict a PIM to the Borel subgroup $B$ of $G$, where it decomposes into a direct sum of PIMs for $KB$. These turn out to be very easy to describe, since $B = TU$ with $U$ a normal $p$-subgroup and $T$ commutative of order prime to $p$. Each simple $T$-module (coming from a weight $\lambda$) is also projective, so the resulting induced module is $KB$-projective and of dimension $q^m = |U|$. By Theorem 9.1 again, the induced module is itself a PIM for $KB$. Call it $I_r(\lambda)$. Since the simple $KB$-modules are all of dimension 1, the dimension formula 9.1(1) applied to $B$ shows at once that we obtain all PIMs in this way.

In the language of Brauer characters, we are starting with a weight $\lambda \in X$ and associating with it a Brauer character $\zeta_\lambda$ of $T$. The resulting induced character $\widehat{\zeta_\lambda} := \mathrm{Ind}\,\zeta_\lambda$ is then the Brauer character of the PIM $I_r(\lambda)$ for $KB$. Note that $\zeta_\lambda = \zeta_\mu$ if and only if $\lambda \equiv \mu \pmod{q-1}$. Moreover, the Weyl group acts naturally on the characters of $T$, with $(w\zeta_\lambda)(t) = \zeta_\lambda(t^w)$, i.e., $w\zeta_\lambda = \zeta_{w\lambda}$. (Here the normalizer action of $W$ on $T$ is written for convenience on the right.)

Now consider the restriction of a PIM $U_r(\lambda)$ to $KB$. As the injective hull of $L(\lambda)$, $U_r(\lambda)$ contains a one-dimensional $KB$-submodule on which $T$ acts via $\lambda$. The corresponding PIM $I_r(\lambda)$ therefore occurs at least once as a $KB$-summand of $U_r(\lambda)$.

THEOREM. *Let $G$ be a Chevalley group. If $\lambda \in X_r$, each $I_r(w\lambda)$ for $w \in W$ is a direct summand of the restriction of $U_r(\lambda)$ to $KB$. Therefore*

$$\dim U_r(\lambda) \geq |W\zeta_\lambda| q^m.$$

PROOF. In the simply connected group $\mathbf{G}$ the centralizer of a semisimple element $t \in \mathbf{T}$ can easily be described: it is a connected reductive group, generated by $\mathbf{T}$ together with various pairs of root groups $\mathbf{U}_\alpha$ and $\mathbf{U}_{-\alpha}$. As a result, $C_{\mathbf{G}}(t) \cap \mathbf{B}$ is a Borel subgroup of $C_{\mathbf{G}}(t)$ (see for example Carter [**89**, 3.5.5]).

We apply this to elements $t$ and $t^w$ of $T$ in $G$ (with $w \in W$). Their centralizers in $\mathbf{G}$ are conjugate, so the respective Borel subgroups $C_{\mathbf{G}}(t) \cap \mathbf{B}$ and $C_{\mathbf{G}}(t^w) \cap \mathbf{B}$ are also conjugate. Taking rational points, we see that $t$ and $t^w$ have centralizers (hence conjugacy classes) in $B$ of the same size. Write $|\operatorname{cl}(t)|$ for the size of the class of $t$ in $B$. Moreover, each semisimple element of $B$ is conjugate (in $B$) to a unique element of $T$. These facts enable us to rewrite the Brauer character pairing for $B$ as follows when a character $\eta_\lambda$ is restricted to $KB$ and $\mu \in X$. Here the sums can be taken over $t \in T$, since the Brauer characters are defined just on $p$-regular classes:

$$
\begin{aligned}
(\operatorname{Res}\eta_\lambda, \widehat{\zeta_\mu})_B &= |B|^{-1} \sum |\operatorname{cl}(t)|\, \eta_\lambda(t^{-1})\, \zeta_\mu(t) \\
&= |B|^{-1} \sum |\operatorname{cl}(t^w)|\, \eta_\lambda(t^{-w})\, \zeta_\mu(t^w) \\
&= |B|^{-1} \sum |\operatorname{cl}(t)|\, \eta_\lambda(t^{-1})\, \zeta_{w\mu}(t) \\
&= (\operatorname{Res}\eta_\lambda, \widehat{\zeta_{w\mu}})_B
\end{aligned}
$$

The proposition follows from the case $\mu = \lambda$.  □

It is always true that $|W\zeta_\lambda| \leq |W\lambda|$, the size of the orbit of $\lambda$ in $X$. The inequality is strict just when $\lambda$ has one or more coordinates equal to $q - 1$ (as in the case of $\operatorname{St}_r$).

When $r = 1$, an interesting class of weights for which $\dim U(\lambda) = |W|p^m$ will be discussed in 10.7 below. The next section offers a brief preview in rank 1.

## 9.8. Projective Modules for SL(2, $p$)

When $G = \mathrm{SL}(2, p)$ it is easy to fill in the following outline: (1) Working first in the framework of $\mathbf{G}$-modules, tensor each simple module with St and extract the "block" summand involving the unique highest weight. This is limited by the Linkage Principle. (2) Observe that this summand is actually indecomposable, a special feature of rank 1. (3) Next restrict to $G$. This yields a projective module, usually of dimension $2p$. (4) Finally, compare dimensons to see that these modules are almost all indecomposable and to determine the exceptions.

Here one can avoid the recursive steps needed in higher ranks. To work out the details, we assume $p$ is odd, leaving the easy case $p = 2$ to the reader. Weights may be identified here with integers.

Fix $\lambda \in X_1$ and recall from 9.4 that $\lambda^\circ = (p-1)\rho + w_\circ\lambda = (p-1) - \lambda$. Here we just write $\mu = \lambda^\circ$. The $\mathbf{G}$-module $L(\mu)\otimes\mathrm{St}$ has $\mu+(p-1) = 2(p-1)-\lambda$ as its unique highest maximal weight, occurring with multiplicity 1. So there is a unique indecomposable summand involving that weight: call it $Q(\lambda)$. Thanks to the Linkage Principle, this summand lies in the "block" summand involving composition factors $L(\nu)$ with $\nu$ linked to $\lambda$ by the affine Weyl group $W_p$ relative to $p$.

In our situation, $L(\mu) = V(\mu)$, a Weyl module. To work out the **G**-composition factors in the tensor product, start with the composition factors of Weyl modules given by the classical Clebsch–Gordan formula. When $\mu = 0$, the block summand is just $\mathrm{St} = Q(p-1)$. When $0 < \mu \leq p-1$, the block summand involves two copies of $L(\lambda)$ along with one other composition factor coming from the Weyl module $V(\mu + p - 1)$, namely $L(\mu - 1) \otimes L(1)^{[1]}$. Here the block summand has dimension $2p$, which puts an upper bound on $\dim Q(\lambda)$.

The restriction of any summand of $L(\mu) \otimes \mathrm{St}$ to $KG$ is projective. Moreover, the multiplicities of composition factors can be worked out by passing from twisted tensor products for **G** to ordinary tensor products for $G$.

From Theorem 9.1(f) we know that $p$ divides each $\dim U(\lambda)$. Moreover,

$$(1) \qquad |G| = p^3 - p = \sum_{\lambda} \dim L(\lambda) \dim U(\lambda) = \sum_{\lambda=0}^{p-1} (\lambda + 1) \dim U(\lambda).$$

Two extreme cases are easy to dispose of: When $\lambda = p - 1$, we have $U(\lambda) = Q(\lambda) = \mathrm{St}$. When $\lambda = 0$, the **G**-module $Q(\lambda)$ has composition factors $L(0)$ (twice) and $L(p-2) \otimes L(1)^{[1]}$. On restriction to $G$ the tensor product yields composition factors $L(p-1) = \mathrm{St}$ and $L(p-3)$, with the PIM St splitting off as a direct summand. Thus $\dim U(0) = p$. Now it is an easy exercise to see that (1) forces $\dim U(\lambda) = 2p = \dim Q(\lambda)$ whenever $\lambda \neq 0, p-1$. This in turn allows one to work out the composition factors explicitly: Typically there are 4 of them, two copies of $L(\lambda)$ together with $L(p - \lambda - 1)$ and $L(p - \lambda - 3)$. (The latter module disappears when $\lambda = p - 2$.)

A similar argument applies to $G = \mathrm{SL}(2, q)$ when $q = p^r$, using twisted tensor product expressions for $L(\mu)$ and $\mathrm{St}_r$. (See Jeyakumar [261] and Humphreys [216].) Write $L(\mu) \cong L(\mu_0) \otimes L(\mu_1)^{[1]} \otimes \cdots \otimes L(\mu_{r-1})^{[r-1]}$ and $\mathrm{St}_r \cong \mathrm{St} \otimes \mathrm{St}^{[1]} \otimes \cdots \otimes \mathrm{St}^{[r-1]}$. Then we can reorganize

$$L(\mu) \otimes \mathrm{St}_r \cong (L(\mu_0) \otimes \mathrm{St}) \otimes (L(\mu_1) \otimes \mathrm{St})^{[1]} \otimes \cdots \otimes (L(\mu_{r-1}) \otimes \mathrm{St})^{[r-1]}.$$

There is an obvious summand involving the highest weight:

$$Q(\lambda) := Q(\lambda_0) \otimes Q(\lambda_1)^{[1]} \otimes \cdots \otimes Q(\lambda_{r-1})^{[r-1]}.$$

On restriction to $G$ this must in turn have $U(\lambda)$ as a summand. Now a simple dimension comparison like the one above for $\mathrm{SL}(2, p)$ shows that $Q(\lambda) = U(\lambda)$ for $\lambda \neq 0$, whereas $Q(0) = U(0) \oplus \mathrm{St}_r$.

We leave it as an exercise for the reader to compute the dimension of each $U(\lambda)$ based on the $p$-adic expansion of $\lambda$. (On the other hand, it is a much harder combinatorial problem to make explicit the composition factor multiplicities; this is discussed in 11.5 below.)

The line of argument just used reflects indirectly the role of the higher Frobenius kernels and suggests a partial analogue of the Tensor Product Theorem for PIMs (to be developed in 10.14). But for groups of higher rank, the precise dimensions of PIMs are not so readily extracted from the tensor product construction.

## 9.9. Brauer Trees for $SL(2,p)$

In the special case $G = SL(2,p)$ (or $PSL(2,p)$) the composition factors of PIMs can be determined without invoking Lie theory, as in the early work of Brauer–Nesbitt. For example, since the Sylow $p$-subgroups here are cyclic, the well-developed theory of blocks with cyclic defect groups may be applied. (See Alperin [6, V], Curtis–Reiner [127, §62].) For the moment we are ignoring the role of ordinary representations in this theory (see 16.7 below). So what we see at first is just a graphical encoding in a **Brauer tree** of the incidence of simple modules as composition factors of PIMs.

Leaving aside the trivial case $p = 2$, there are two similarly-behaved blocks in $KG$ of highest defect containing the simple modules with even and odd highest weights between 0 and $p - 2$. (See Figures 1 and 2.) In each case the Brauer tree

$$\underset{}{\circ}\!\overset{0}{\rule{1cm}{0.4pt}}\!\underset{}{\circ}\overset{p-3}{\rule{1cm}{0.4pt}}\underset{}{\circ}\overset{2}{\rule{1cm}{0.4pt}}\underset{}{\circ}\overset{p-5}{\rule{1cm}{0.4pt}}\underset{}{\circ}\overset{4}{\rule{1cm}{0.4pt}}\underset{}{\circ}\overset{p-7}{\rule{1cm}{0.4pt}}\underset{}{\circ}\cdots\underset{}{\circ}\overset{(p-1)/2}{\rule{1cm}{0.4pt}}\bullet$$

$$\underset{}{\circ}\overset{p-2}{\rule{1cm}{0.4pt}}\underset{}{\circ}\overset{1}{\rule{1cm}{0.4pt}}\underset{}{\circ}\overset{p-4}{\rule{1cm}{0.4pt}}\underset{}{\circ}\overset{3}{\rule{1cm}{0.4pt}}\underset{}{\circ}\overset{p-6}{\rule{1cm}{0.4pt}}\underset{}{\circ}\overset{5}{\rule{1cm}{0.4pt}}\underset{}{\circ}\cdots\underset{}{\circ}\overset{(p-3)/2}{\rule{1cm}{0.4pt}}\bullet$$

FIGURE 1. Brauer trees for $SL(2,p)$, $p \equiv 1 \pmod 4$

$$\underset{}{\circ}\overset{0}{\rule{1cm}{0.4pt}}\underset{}{\circ}\overset{p-3}{\rule{1cm}{0.4pt}}\underset{}{\circ}\overset{2}{\rule{1cm}{0.4pt}}\underset{}{\circ}\overset{p-5}{\rule{1cm}{0.4pt}}\underset{}{\circ}\overset{4}{\rule{1cm}{0.4pt}}\underset{}{\circ}\overset{p-7}{\rule{1cm}{0.4pt}}\underset{}{\circ}\cdots\underset{}{\circ}\overset{(p-3)/2}{\rule{1cm}{0.4pt}}\bullet$$

$$\underset{}{\circ}\overset{p-2}{\rule{1cm}{0.4pt}}\underset{}{\circ}\overset{1}{\rule{1cm}{0.4pt}}\underset{}{\circ}\overset{p-4}{\rule{1cm}{0.4pt}}\underset{}{\circ}\overset{3}{\rule{1cm}{0.4pt}}\underset{}{\circ}\overset{p-6}{\rule{1cm}{0.4pt}}\underset{}{\circ}\overset{5}{\rule{1cm}{0.4pt}}\underset{}{\circ}\cdots\underset{}{\circ}\overset{(p-1)/2}{\rule{1cm}{0.4pt}}\bullet$$

FIGURE 2. Brauer trees for $SL(2,p)$, $p \equiv 3 \pmod 4$

turns out to be an undirected graph without loops or branch points. Its vertices will later be labelled with ordinary characters (16.2). Each of the $(p - 1)/2$ edges is labelled with a weight $\lambda$ in such a way that $U(\lambda)$ has as composition factors two copies of $L(\lambda)$ together with simple modules labelled by the (one or two) weights on neighboring edges. The edge at the end with a dark vertex is exceptional: the simple module here occurs *three* times in its PIM.

Over fields of order $p^r$ everything becomes more complicated. But over the prime field we will see in 17.16 higher-dimensional analogues of the Brauer tree for higher-rank groups of Lie type.

## 9.10. Indecomposable Modules for $SL(2,p)$

The groups $SL(2,p)$ are unique among Chevalley groups in having cyclic Sylow $p$-subgroups. By the theorem of Higman quoted in 8.9, this is equivalent to $KG$ being of finite representation type: up to isomorphism, there are only finitely many indecomposable $KG$-modules. (For this $K$ just needs to be a splitting field of characteristic $p$.)

Janusz [259, 260] explores in detail how these finitely many indecomposables can be constructed, starting with PIMs. In the first paper, he takes $G$ to be a finite group with cyclic Sylow $p$-subgroup $P$. Using Green's theory of vertices and sources for indecomposable modules, he derives an exact formula for the number of (non-isomorphic) indecomposables:

THEOREM. *Let $G$ be a finite group with a cyclic Sylow $p$-subgroup. Then the number of nonisomorphic indecomposable $KG$-modules is*

$$n_p(G) + \sum_Q |Q| \left(1 - \frac{1}{p}\right) n_p(N_G(Q)),$$

*where $Q$ runs over all nontrivial subgroups $Q$ of $P$ and $n_p(H)$ denotes the number of $p$-regular conjugacy classes in a group $H$.*

*In particular, when $G = \mathrm{PSL}(2,p)$, the number of indecomposable modules is $1 + p(p-1)/2$, all but the Steinberg module St occurring in the principal block.*

Janusz goes on to construct the number of indecomposables for $\mathrm{PSL}(2,p)$ specified by his formula. These are recognized to be non-isomorphic by looking at their socles and numbers of composition factors. The idea is to start by ordering the maximal submodules of PIMs in a special way, so that two consecutive modules share a unique simple quotient. Then take the direct sum of any number of consecutive modules in the ordering. Such a module has two obvious maps onto one of the simple quotients; the submodule where these maps agree is indecomposable, with a multiplicity-free socle.

In his second paper (which doesn't refer explicitly to the first), Janusz studies the structure of an individual block having a cyclic defect group. Here he uses the Brauer tree as a tool in the intricate bookkeeping required. It turns out that only certain features of the block algebra $A$ are needed: one can replace it by any split symmetric algebra $A$ of finite representation type whose Cartan matrix comes from a "Brauer graph". In this generality, he obtains a concise formula for the number of indecomposable $A$-modules: $e(em + 1)$, where $e$ is the number of edges in the Brauer graph ($e + 1$ being the number of vertices) and $m$ is the multiplicity of the exceptional vertex.

When $G = \mathrm{PSL}(2,p)$ and we consider the Brauer tree of the 1-block as above, we have $e = (p-1)/2$ and $m = 2$, thus recovering the number $p(p-1)/2$.

REMARK. An application of the classification of indecomposable modules for $\mathrm{PSL}(2,p)$ occurs in work of Schoen [361]. To paraphrase his result: If $G = \mathrm{PSL}(2,p)$ acts faithfully on a smooth, irreducible curve $C$ defined over an algebraically closed field of characteristic different from $p$, then $G$ acts naturally on the étale cohomology $H^1(C, \mathbb{F}_p)$. When $p$ is odd, it is shown how to decompose this module into indecomposable summands, using the explicit description above of all indecomposables. For example, let $C = X(p)$ be the complete, complex algebraic curve which compactifies the parameter space for isomorphism classes of elliptic curves with a symplectic level $p$-structure. Here $G$ acts on $C$ via its natural action on the moduli problem. The cohomology module then splits into a direct sum of certain projectives along with certain submodules of PIMs having dimension $p - 1$, which can be specified explicitly.

## 9.11. Dimensions of PIMs in Low Ranks

To round out this chapter we look at some cases in which explicit formulas have been worked out for dimensions of PIMs using the direct methods outlined earlier.

Without the machinery to be developed in the following chapter, one has to analyze the unique $KG$-summand in a tensor product $L(\mu) \otimes \mathrm{St}_r$ (with $\mu \in X_r$) involving the highest weight $\mu + (p^r - 1)\rho$. This presupposes some knowledge of the formal or Brauer character of $L(\mu)$. The algebraic group action on such a tensor product leads mainly to a determination of the summand (usually much too large to be a PIM for $KG$) belonging to the linkage class of the highest weight. The program is extremely difficult to carry out in practice except when the rank is small, and even then mainly in cases like $p = 2, 3$.

On the other hand, in such small cases the examination of tensor products may also reveal a considerable amount of structural information about PIMs, along lines explored in Chapters 11, 12, and 13.

- For $G = \mathrm{SL}(2, p^r)$ the methods of Jeyakumar and Humphreys described above in 9.8 make it easy to write down explicit dimension formulas. Identifying $X^+$ with $\mathbb{Z}^+$, each $\lambda \in X^+$ has a $p$-adic expansion $\lambda = \sum_{i=0}^{r-1} \lambda_i p^i$ with $0 \le \lambda_i < p$. Then $\dim U_r(0) = (2^r - 1)p^r$, while for $\lambda \ne 0$,
$$\dim U_r(\lambda) = 2^n p^r, \quad \text{where } n := \#\{i \mid \lambda_i \ne p - 1\}.$$

  When $p$ is odd, earlier work of Srinivasan [387] determines these dimensions indirectly by comparing ordinary and modular characters (see Chapter 16 below). When $p = 2$, PIMs for the groups $\mathrm{SL}(2, 2^r)$ lend themselves to direct computation, since the only 2-restricted modules are the trivial module and St. In this case Alperin [5] works out a considerable amount of detail about the dimensions and structure of the PIMs.

- When $p = 2$, Chastkofsky and Feit [105] parametrize the simple modules for $\mathrm{SL}(3, 2^r)$ or $\mathrm{SU}(3, 2^r)$ by graphs to keep track of the combinatorics of tensor products. This allows them to compute explicitly the dimensions of PIM's in each case. For example, in both cases the dimension of the projective cover of the trivial module is $8^r(6^r - 5^r)$.

- Independently, Archer [24] studies tensor products for $\mathrm{SL}(3, 2^r)$ in the spirit of Alperin's work on $\mathrm{SL}(2, 2^r)$, thereby obtaining explicit combinatorial formulas for dimensions of PIMs.

- Chastkofsky–Feit [104] use their graphical method in characteristic 2 to study in tandem the PIMs of $\mathrm{Sp}(4, 2^r)$ and the Suzuki groups, obtaining a precise formula for the dimensions along with some other information.

- Cheng [106] uses a similar method to work out the dimensions of certain PIMs for the groups $G_2(2^r)$ and $G_2(3^r)$.

- Holmes [208, 3.5] summarizes known dimensions of PIMs in some low rank cases (with $p = 2$ or 3) including some new ones worked out by his algorithmic method. This involves refinements to the Chastkofsky–Feit approach, allowing him to compute some dimensions in the cases $\mathrm{SL}(3, 3^r)$ and $\mathrm{SL}(4, 2^r)$. In [209, §3] he exhibits detailed tables for the groups $\mathrm{SL}(4, 2^r)$ when $r \le 3$.

# CHAPTER 10

# Comparison with Frobenius Kernels

In the context of simple modules for a group $G$ of Lie type over $\mathbb{F}_p$, the work of Curtis and Steinberg revealed a surprisingly close parallel between the category of $KG$-modules and the category of restricted modules for the Lie algebra $\mathfrak{g}$ of $\mathbf{G}$. Both categories involve the same simple modules $L(\lambda)$ obtained by "restriction" from $\mathbf{G}$ in the applicable sense. Later work, especially by Jantzen, extended this parallel to a group $G$ over a field of order $p^r$ and the $r$th Frobenius kernel $\mathbf{G}_r$ (2.5), within the framework of Steinberg's Tensor Product Theorem.

There is a similar, but more subtle, parallel in the case of projective modules. It is clear by dimension comparison that the parallel cannot be exact. For any finite dimensional algebra, the sum of dimensions of simple modules times dimensions of projective covers must equal the dimension of the algebra (just as in 9.1(1)). But $|G| = \dim KG$ always differs from the dimension $p^{rd}$ of the hyperalgebra $\mathrm{Dist}(\mathbf{G}_r)$, where $d = \dim \mathfrak{g}$. For example, $|\mathrm{SL}(2,p)| = p^3 - p$.

The Steinberg module provides the main steppingstone from the setting of Frobenius kernels to the setting of finite groups. We already saw in 9.3 that $\mathrm{St}_r$ is projective for $KG$ when $G$ is defined over a field of $p^r$ elements. This module also turns out to be projective for the hyperalgebra of $\mathbf{G}_r$, while all other PIMs are obtained as direct summands of various $\mathbf{G}$-modules $L(\mu) \otimes \mathrm{St}_r$ as in the case of $KG$. This suggests the possibility of comparing PIMs for $\mathbf{G}_r$ and $KG$ in the context of these tensor products.

Our main goal in this chapter is to sort out these relationships, starting with a brief review of known results on projective modules for Frobenius kernels $\mathbf{G}_r$ (10.1). The PIMs for the algebra $\mathrm{Dist}(\mathbf{G}_r)$ are denoted by $Q_r(\lambda)$ for $\lambda \in X_r$. When $p$ is not too small, $Q_r(\lambda)$ lifts naturally to a $\mathbf{G}$-module and then restricts to a projective module for $KG$ having $U_r(\lambda)$ as a summand (10.4). But for arbitrary $p$ we must for the present rely on formal character arguments to carry out a comparison with the PIMs $U_r(\lambda)$. As a special case we look at a class of "small" PIMs which can be extracted easily from a tensor product with the Steinberg module (10.7).

Next we derive our main theoretical result, a comparison formula (10.11) due independently to Chastkofsky [96] and Jantzen [251]. The chapter concludes with a summary of specific calculations of the dimensions of PIMs for some groups of small rank, to illustrate how far one can go in practice with the formula.

## 10.1. Injective Modules for Frobenius Kernels

In 2.5 we summarized the basic results about simple modules for a Frobenius kernel $\mathbf{G}_r$. Following Jantzen [**RAGS**, II.9, II.11], we examine further the category of rational $\mathbf{G}_r$-modules (meaning modules for $\mathrm{Dist}(\mathbf{G}_r)$). In particular we look for projective covers and injective hulls. These always exist, and moreover the notions of "projective" and "injective" coincide: as in the case of the group algebra $KG$, we are dealing here with modules for a finite dimensional Hopf algebra, which is automatically a Frobenius algebra. Like $KG$, the algebra $\mathrm{Dist}(\mathbf{G}_r)$ is even a "symmetric" algebra, which implies that the projective covers of modules coincide with their injective hulls.

By way of contrast, it can be shown that projectives fail to exist in the category of rational **G**-modules, whereas there are enough injectives in this category (these being infinite dimensional). So the term "injective" is preferred in Jantzen's treatment. To maintain the parallel with $KG$, we continue to refer to projective covers and PIMs in the context of $\mathbf{G}_r$-modules. Our notation also differs slightly from that in [**RAGS**]: we write $Q_r(\lambda)$ rather than $Q(r, \lambda)$ for the projective cover (=injective hull) of $L(\lambda)$ when $\lambda \in X_r$. It soon becomes clear that the subscript $r$ is essential: for a fixed $\lambda \in X_r$ and for $s \geq r$, the modules $Q_s(\lambda)$ get larger as $s$ grows.

The category of $\mathbf{G}_r$-modules turns out to be similar in many (but not all) respects to the category of $KG$-modules when $G = \mathbf{G}^F$ is defined over a field of order $p^r$. Here we outline some of the key features, first observed for $r = 1$ in Humphreys [**213**] and then developed more deeply for arbitrary $r$ in various papers of Jantzen [**247, 248, 249**].

To simplify the exposition (and the notation), we look first at the case $r = 1$, which captures the flavor of the general theory. Here $\mathrm{Dist}(\mathbf{G}_1)$ is the same as the restricted enveloping algebra $u(\mathfrak{g})$, so we are studying the category of finite dimensional restricted $\mathfrak{g}$-modules. Call it $\mathcal{M}$.

- The simple modules in $\mathcal{M}$ are precisely the restrictions of the **G**-modules $L(\lambda)$ for $\lambda \in X_1$. However, the "torus" $\mathbf{T}_1$ (corresponding to a Cartan subalgebra of $\mathfrak{g}$) cannot distinguish weights in $X$ which are congruent modulo $p$. So the set $\Lambda := X/pX$ can be regarded as a more intrinsic parameter set for the simple modules. This emphasizes the fact that the partial ordering of weights is invisible in $\mathcal{M}$. (For $\mathbf{G}_r$ the restrictions of the $L(\lambda)$ for $\lambda \in X_r$ may be denoted $L_r(\lambda)$ to emphasize the role of $r$.)
- Each $\lambda \in \Lambda$ defines a 1-dimensional restricted $\mathfrak{b}$-module when $\mathfrak{b} = \mathrm{Lie}(\mathbf{B})$. Tensoring it with $u(\mathfrak{g})$ over $u(\mathfrak{b})$ then yields a module $Z(\lambda)$ of dimension $p^m$ (where $m = |\Phi^+|$), sometimes referred to as a **baby Verma module**. This plays the role of universal highest weight module in $\mathcal{M}$ and has a unique simple quotient isomorphic to $L(\lambda)$. In the special case $\lambda = (p-1)\rho$, we get $Z(\lambda) = L(\lambda) = \mathrm{St}$. The module $Z(\lambda)$ is "coinduced" and can be viewed as analogous to the Weyl module $V(\lambda)$. One can also make a dual construction of "induced" modules analogous to $H^0(\lambda)$. (For $\mathbf{G}_r$ the analogue of $Z(\lambda)$ is denoted $Z_r(\lambda)$.)
- The projective cover $Q(\lambda)$ of $L(\lambda)$ in $\mathcal{M}$ coincides with the injective hull of $L(\lambda)$. In particular, $L(\lambda)$ occurs once as a simple quotient and once as the simple socle of $Q(\lambda)$. The Steinberg module St is itself projective in $\mathcal{M}$, and is the only simple projective module.
- Tensoring a projective module with any other module yields a projective module. When $\lambda \in \Lambda$ and $\mu = (p-1)\rho + w_\circ\lambda$ in $\Lambda$, the module $Q(\lambda)$ occurs precisely once as a summand of $L(\mu) \otimes \mathrm{St}$.
- The module $Q(\lambda)$ has at least one filtration with quotients isomorphic to various $Z(\mu)$, so in particular $\dim Q(\lambda)$ is a multiple of $p^m$. Moreover,

there is a reciprocity law: the number $(Q(\lambda) : Z(\mu))$ of quotients isomorphic to $Z(\mu)$ in such a filtration is equal to $d_{\mu\lambda} := [Z(\mu) : L(\lambda)]$, the multiplicity of $L(\lambda)$ as a composition factor of $Z(\mu)$.

- There is a version of the Linkage Principle in $\mathcal{M}$: If $L(\lambda)$ and $L(\mu)$ are composition factors of an indecomposable module in $\mathcal{M}$ (such as $Z(\lambda)$ or $Q(\lambda)$), then $\mu = w \cdot \lambda$ in $\Lambda$ for some $w \in W$. Here the dot action of the Weyl group is defined as in 3.6. But $W$-linkage in this sense corresponds in $X$ to conjugacy by an element of the **extended affine Weyl group** $\widetilde{W}_p := W \ltimes pX$. It is usually not difficult to pass between $W$-linkage and linkage by $W_p$, but if $p$ divides the *index of connection* (the index of the root lattice in the weight lattice) extra care is needed.
- The blocks in $u(\mathfrak{g})$ are in natural bijection with the $W$-linkage classes in $\Lambda$. (Here the situation is very different from that in $KG$.)
- Study of intertwining maps between baby Verma modules shows that $Z(\lambda)$ and $Z(\mu)$ have precisely the same composition factors whenever $\lambda$ and $\mu$ are $W$-linked. In particular, $d_{\mu\lambda} = d_{\lambda\lambda}$, so this number may be called $d_\lambda$. As a result, $\dim Q(\lambda) = a_\lambda d_\lambda p^m$, where $a_\lambda$ is the size of the $W$-linkage class in $\Lambda$.

Even when generalized to the setting of $\mathbf{G}_r$-modules, this theory cannot readily be compared with the theory of $KG$-modules. Nor can one directly predict the values of the decomposition numbers $d_\lambda$. What one needs is a systematic comparison with $\mathbf{G}$-modules. A step in this direction is the addition of an action of the algebraic torus $\mathbf{T}$, which makes it possible to work more naturally with weights and their partial ordering.

## 10.2. Torus Action on $\mathbf{G}_r$-Modules

Jantzen's study of the category of modules for the "mixed" group scheme $\mathbf{G}_r\mathbf{T}$ reveals a close parallel with the theory for $\mathbf{G}_r$: see [**RAGS**, II.9]. In particular, there are liftings denoted $\widehat{L}_r(\lambda), \widehat{Z}_r(\lambda), \widehat{Q}_r(\lambda)$ of the $\mathbf{G}_r$-modules $L_r(\lambda), Z_r(\lambda), Q_r(\lambda)$ when $\lambda \in X_r$. But there are also versions of these $\mathbf{G}_r\mathbf{T}$-modules defined for arbitrary $\lambda \in X$. This adds flexibility to comparisons with $\mathbf{G}$-modules and facilitates the use of formal characters. (In fact Jantzen goes even further in making $\mathbf{B}$ act on some of the modules, but we will not go into that here.)

Using the normality and natural embeddings of the $\mathbf{G}_r\mathbf{T}$, one can define analogues of Frobenius twisting for the modules $\widehat{Q}_r(\lambda)$ so that $\widehat{Q}_r(\lambda)^{[i]}$ is a module for $\mathbf{G}_{r+i}\mathbf{T}$. This leads in turn to a good analogue of the Tensor Product Theorem, given in part (c) of the following theorem. (See [**RAGS**, II.11] and the references there to work of Donkin, Jantzen, and others.)

THEOREM. *Fix $r$ and let $\lambda \in X$.*

(a) *Write $\lambda = \mu + p^r\nu$ for unique $\mu \in X_r, \nu \in X$. Then $\widehat{L}_r(\lambda) \cong \widehat{L}_r(\mu) \otimes p^r\nu$, where $p^r\nu$ denotes a 1-dimensional $\mathbf{T}$-module affording this character, with trivial action of $\mathbf{G}_r$. Similar isomorphisms hold for $\widehat{Z}_r(\lambda)$ and $\widehat{Q}_r(\lambda)$.*

(b) *If $\lambda \in X_r$, then $\widehat{L}_r(\lambda), \widehat{Z}_r(\lambda), \widehat{Q}_r(\lambda)$ restrict to the respective $\mathbf{G}_r$-modules $L_r(\lambda), Z_r(\lambda), Q_r(\lambda)$. In particular, $\widehat{Z}_r((p^r - 1)\rho)$ restricts to $\mathrm{St}_r$.*

(c) *If $\lambda \in X_r$ is written uniquely as $\lambda = \lambda_0 + p\lambda_1 + \cdots + p^{r-1}\lambda_{r-1}$ with all $\lambda_i \in X_1$, then*

$$\widehat{Q}_r(\lambda) \cong \widehat{Q}_1(\lambda_0) \otimes \widehat{Q}_1(\lambda_1)^{[1]} \otimes \cdots \otimes \widehat{Q}_1(\lambda_{r-1})^{[r-1]}$$

as $\mathbf{G}_r\mathbf{T}$-modules. Thus $\dim \widehat{Q}_r(\lambda)$ and $\operatorname{ch}\widehat{Q}_r(\lambda)$ are products of the corresponding dimensions and characters for $\mathbf{G}_1\mathbf{T}$.

(d) For arbitrary $\lambda \in X$, there exists a filtration of $\widehat{Q}_r(\lambda)$ with quotients isomorphic to various $\widehat{Z}_r(\mu)$. In any such filtration, $\widehat{Z}_r(\mu)$ occurs as a quotient $[\widehat{Z}_r(\mu) : \widehat{L}_r(\lambda)]$ times.

## 10.3. Formal Character of $\widehat{Q}_r(\lambda)$

To make the theory of $\mathbf{G}_r\mathbf{T}$-modules effective one has to see that the formal characters involved are actually computable in terms of standard character data for $\mathbf{G}$. The key point is to describe $\operatorname{ch}\widehat{Q}_r(\lambda)$ when $\lambda \in X_r$. Here the work of Jantzen [248, 249] exploits the reciprocity in part (d) of Theorem 10.2 together with the fact that $\operatorname{ch}\widehat{Z}_r(\lambda) = e(\lambda - (p^r - 1)\rho)\,\operatorname{ch}\operatorname{St}_r$ [**RAGS**, II.9.2(3)].

THEOREM. *Fix $r > 0$.*

(a) *For all $\mu \in X$,*

$$\chi(\mu) = \sum_{\lambda \in X_r} \sum_{\nu \in X} [\widehat{Z}_r(\mu) : \widehat{L}_r(\lambda + p^r\nu)]\, \chi_p(\lambda)\chi(\nu)^{[r]}.$$

(b) *If $\lambda \in X_r$, then*

$$\operatorname{ch}\widehat{Q}_r(\lambda) = \sum_{\mu \in X} [\widehat{Z}_r(\mu) : \widehat{L}_r(\lambda)]\, \chi(\mu).$$

(c) *If $\lambda \in X_r$, then there exists a formal character $q(r, \lambda)$ in $\mathcal{X} = \mathbb{Z}[X]^W$ such that $\operatorname{ch}\widehat{Q}_r(\lambda) = q(r, \lambda)\operatorname{ch}\operatorname{St}_r$ in $\mathcal{X}$.*

*In particular, $\operatorname{ch}\widehat{Q}_r(\lambda)$ lies in $\mathcal{X}$ whenever $\lambda \in X_r$.*

Recall that for any $\mu \in X$, $\chi(\mu)$ is defined formally to be $(-1)^{\ell(w)}\chi(\lambda)$ if for some $\lambda \in X^+$ and $w \in W$ we have $w \cdot \mu = \lambda$, or to be $0$ if no such $\lambda$ and $w$ exist. Moreover, the formal characters $\chi_p(\lambda)\chi(\nu)^{[r]}$ with $\lambda \in X_r$ and $\nu \in X$ form an alternative $\mathbb{Z}$-basis of $\mathcal{X}$, since each $\mu \in X$ can be written uniquely in the form $\lambda + p^r\nu$. So part (a) of the theorem, coupled with standard character data for $\mathbf{G}$, makes it possible in principle to compute the decomposition numbers $[\widehat{Z}_r(\mu) : \widehat{L}_r(\lambda)]$. In turn one can compute $\operatorname{ch}Q_r(\lambda)$ for $\lambda \in X_r$, since it is easy to specify upper and lower bounds on the $\mu$ which occur in part (b) in the context below of tensoring with $\operatorname{St}_r$: namely $2(p^r - 1)\rho + w_\circ\lambda$ and its image under $w_\circ$. None of this easy to carry out, of course.

Within this framework, Jantzen is able to sort out some conjectures of Verma [426] for $r = 1$ on the relationship between the decomposition numbers $d_\lambda$ of 10.1 and the multiplicities of composition factors in Weyl modules. (Another conjecture by Verma asserts that the $d_\lambda$ are essentially independent of $p$ for $\lambda$ inside a fixed $p$-alcove. This is then seen to depend on the parallel phenomenon for Weyl modules which would follow from Lusztig's Conjecture.)

Roughly speaking, the idea behind the formalism for $r = 1$ amounts to this: Lift a module $Z(\lambda)$ with $\lambda \in X_1$ to $\mathbf{G}_1\mathbf{T}$, then translate $\lambda$ by a suitable $p\nu \in X^+$ and show that the composition factors of $\widehat{Z}(\lambda + p\nu)$ are in natural bijection with those of the Weyl module $V(\lambda + p\nu)$. Translating all weights back into $X_1$ then yields the $\mathbf{G}_1$-composition factors of $Z(\lambda)$. Translation functors show that $d_\lambda$ depends only on the alcove to which $\lambda$ belongs, and also allow one to predict results for weights in alcove walls (at least when $p \geq h$). A similar procedure applies to $\mathbf{G}_r$, but with systematic blowing up of the patterns of composition factors as for Weyl modules.

EXAMPLES. In type $A_2$ there are two alcoves for $W_p$ in $X_1$: "top" and "bottom". For a $p$-regular weight in either alcove, the generic Weyl module compared with $Z(\lambda)$ has highest weight in an alcove of the same type and has 9 composition factors with multiplicity 1 (as pictured in 3.10). Three of these lie in alcoves of top type, six in alcoves of bottom type. When translated back into $X_1$, some weights coincide. As a result $Z(\lambda)$ has only 6 distinct composition factors, with $d_\lambda = 1$ for $\lambda$ in the top alcove and $d_\lambda = 2$ in the bottom alcove.

Similarly, in type $B_2$ a generic $V(\lambda)$ with $\lambda$ in the lowest $p^2$-alcove has 20 composition factors with multiplicity 1. For the four alcoves in $X_1$ these yield (from top to bottom) values $d_\lambda = 1, 2, 3, 4$. Multiplicities start to show up in generic Weyl modules for type $G_2$. Here the generic $d_\lambda$ are $1, 2, 3, 4, 5, 6, 6, 12, 18, 17, 16, 29$ (again from top to bottom). For type $A_3$ one gets $d_\lambda = 1, 2, 3, 3, 6, 11$. The calculations for $B_2, G_2, A_3$ were first done by Jantzen, together with results on generic behavior [**245, 246, 247, 249**].

Note that the values of $d_\lambda$ can be "ramified" when $p$ divides $[X : \mathbb{Z}\Phi]$. For example, in type $A_2$, generic values of $d_\lambda$ for $p = 3$ are actually 3 and 6.

## 10.4. Lifting PIMs to the Algebraic Group

The fact that all simple modules for $\mathbf{G}_r$ come by restriction from $\mathbf{G}$-modules, coupled with the $\mathbf{T}$-action found by Jantzen (which extends even to $\mathbf{B}$) on the PIMs $Q_r(\lambda)$, makes it natural to ask:

QUESTION. *Does every PIM $Q_r(\lambda)$ for $\mathbf{G}_r$ have a compatible structure of rational $\mathbf{G}$-module, and if so is this structure unique up to isomorphism?*

When $G = \mathbf{G}^F$ over a field of $p^r$ elements, one also wants such a $\mathbf{G}$-module to restrict to a projective (though not necessarily indecomposable) $KG$-module having $U_r(\lambda)$ as a direct summand. The treatments of $\mathrm{SL}(2, q)$ by Jeyakumar [**261**] and Humphreys [**216**] led Verma and Humphreys to expect a general result of this type, which is formulated in [**221**, 8.1]; but the "proof" there is unfortunately oversimplified. The question remains open at this writing, but for large enough $p$ there is a positive answer:

THEOREM. *Assume $p \geq 2(h - 1)$, where $h$ is the Coxeter number of $\mathbf{G}$. For any fixed $r > 0$ and $\lambda \in X_r$, the PIM $Q_r(\lambda)$ for $\mathbf{G}_r$ lifts to a (rational) $\mathbf{G}$-module, which is uniquely determined up to isomorphism. Moreover, when $G = \mathbf{G}^F$ over a field of $p^r$ elements, this $\mathbf{G}$-module is projective as a $KG$-module and involves $U_r(\lambda)$ exactly once as a summand.*

Ballard [**30**] proved a slightly weaker version of this theorem without the uniqueness, under the assumption $p \geq 3(h - 1)$. Then Jantzen [**249**, §4] (see [**RAGS**, II.11]) proved the theorem as stated. For this he introduced the category of $p^r$-*bounded* modules for $\mathbf{G}$: all weights $\mu$ of such a module must satisfy $\langle \mu, \alpha_0^\vee \rangle < 2p^r(h - 1)$. This bound is designed so that the category contains all tensor products $L(\mu) \otimes \mathrm{St}_r$ with $\mu \in X_r$. (If $L(\lambda)$ is $p^r$-bounded, one may also say that $\lambda$ is $p^r$-bounded.)

By considering $\mathrm{Hom}_{\mathbf{G}}(L(\mu) \otimes \mathrm{St}_r, L(\lambda)) \cong \mathrm{Hom}_{\mathbf{G}}(\mathrm{St}_r, L(\lambda) \otimes L(\mu)^*)$ as in 9.4, one sees that $L(\lambda)$ occurs once as a quotient of the tensor product when $\mu = \lambda^\circ = (p^r - 1)\rho + w_\circ\lambda$. (Thus the PIM $Q_r(\lambda)$ occurs as a $\mathbf{G}_r$-summand.)

The more subtle step in the proof, requiring the bounded category and the lower bound on $p$, shows that the block component of the tensor product involving the highest weight has a **G**-summand which restricts to $Q_r(\lambda)$. In particular, it follows that this lifted **G**-module restricts to a projective $KG$-module.

Jantzen's characterization of the lifted module as injective hull in the bounded category shows that the lifting is unique up to isomorphism. (See also Donkin [**139**].) In fact the category has enough injectives for arbitrary $p$, but unless $p \geq 2(h-1)$ the injective hull of $L(\lambda)$ may not be the desired lifting of $Q_r(\lambda)$.

REMARKS. (1) Andersen–Polo–Wen [**22**, 5.9] treat parallel questions for quantum groups at a root of unity, making it plausible that the lifting of PIMs to **G** will be possible at least for primes between $h$ and $2(h-1)$: this would follow from Lusztig's conjectures about the comparison of simple modules for quantum groups and algebraic groups when the highest weights involved are suitably limited.

(2) Pillen [**333**] and Donkin [**142**, §2] reinterpret the lifting problem in the language of tilting modules, where the basic theory is valid for all $p$. This also adds to the hope that the lifting will exist in general.

(3) Answering a question of Parshall [**331**, §5.3], Lin and Nakano [**301**, 3.5] prove: *Let* **G** *be split over* $\mathbb{F}_p$, *with* $G = \mathbf{G}(\mathbb{F}_p)$. *If* $M$ *is a finite dimensional rational* **G***-module which restricts to a projective* $\mathbf{G}_1$*-module, then* $M$ *is a projective* $KG$*-module.* This is a byproduct of homological arguments inspired by work of Quillen, related to the study of "complexity" (see 15.4 below), and should carry over to twisted groups as well.

Other indications that the theorem above may be valid for all $p$ include computations by Donkin and others for small $p$. But for the moment we substitute for it some arguments based just on formal character manipulation, thereby avoiding any condition on $p$.

## 10.5. Some Consequences

Apart from its value for the study of projective modules for finite groups of Lie type, the existence of **G**-structure on the PIMs of $\mathbf{G}_r$ has important implications for the category of **G**-modules. Here are some further consequences developed in [**RAGS**, II.11] and the papers cited there:

- If all $Q_1(\lambda)$ have compatible **G**-structure for $\lambda \in X_1$, then one can derive a twisted tensor product theorem for PIMs closely analogous to the Tensor Product Theorem for simple modules; this strengthens the result for $\mathbf{G}_r\mathbf{T}$-modules in Theorem 10.2(c) above. Define the Frobenius twist $Q_1(\lambda)^{[i]}$ of the **G**-module $Q_1(\lambda)$ in the obvious way. Then if $\lambda \in X_r$ is written uniquely as $\lambda = \lambda_0 + p\lambda_1 + \cdots + p^{r-1}\lambda_{r-1}$ with all $\lambda_i \in X_1$,

$$Q_r(\lambda) \cong Q_1(\lambda_0) \otimes Q_1(\lambda_1)^{[1]} \otimes \cdots \otimes Q_1(\lambda_{r-1})^{[r-1]}$$

  as $\mathbf{G}_r$-modules. In particular, all $Q_r(\lambda)$ lift to **G**-modules. This result is found in Ballard [**30**], Jantzen [**249**], Donkin [**138**].

- If the modules $Q_r(\lambda)$ lift to **G**-modules for arbitrary $r$ and $\lambda \in X_r$, then the injective hull of $L(\lambda)$ ($\lambda \in X_r$) in the category of rational **G**-modules is the direct limit of modules $Q_s(\lambda)$ for $s \geq r$ ([**RAGS**, 11.17]). To embed $Q_s(\lambda)$ naturally in $Q_{s+1}(\lambda)$, observe that the tensor product theorem expresses the latter as the tensor product of the former and the module $Q_1(0)^{[s]}$ which contains a unique trivial **G**-submodule.

- Jantzen's translation functors in the category of rational **G**-modules can be applied directly to the study of PIMs with **G**-structure.
- When they are **G**-modules, the $Q_r(\lambda)$ have good filtrations and Weyl filtrations as defined in 6.1.

## 10.6. Tensoring with the Steinberg Module

After this digression we return to the comparison of PIMs for Frobenius kernels and finite groups of Lie type. Even though this comparison cannot yet be made directly for arbitrary $p$, the technique of tensoring with $\mathrm{St}_r$ provides a bridge between the subjects. In this and the following section we retrace some of the ideas which were explored in the finite group context in Chapter 9. Since $\mathrm{St}_r$ is itself a PIM for $\mathbf{G}_r$, tensoring it with another module yields a direct sum of various PIMs. As discussed in 10.4, $L(\lambda^\circ) \otimes \mathrm{St}_r$ always has $Q_r(\lambda)$ as a distinguished summand.

When $\mu$ is "small", something precise can be said about the block summand of the $\mathbf{G}_r$-module $L(\mu) \otimes \mathrm{St}_r$ containing the 1-dimensional highest weight space for the weight $\mu + (p^r - 1)\rho$. The natural setting for this is found in Jantzen [**RAGS**, II.11.10]. His translation functor from the special weight $(p^r - 1)\rho$ to a weight $\lambda$ in the top $p$-alcove of $X_r$ takes the PIM $\mathrm{St}_r$ to the PIM $Q_r(\lambda)$. In the process he finds $\dim Q_r(\lambda)$ and also shows that $Q_r(\lambda)$ has a natural **G**-structure (for arbitrary $p$).

For our comparison with PIMs of $KG$ in the next section, we outline instead a somewhat more self-contained approach due to Ballard [**29**, §7] and Humphreys [**220**, §9]. (Here we limit ourselves to Chevalley groups, with $r = 1$.) If $\mu \in X^+$, let $W_\mu$ be the isotropy group of $\mu$ in $W$ and fix some set $W^\mu$ of coset representatives for $W/W_\mu$. As in 5.6, write $s(\mu) = \sum_{w \in W^\mu} e(w\mu)$. We need an easy calculation using Weyl's character formula (see [**215**, 24.3]): for any $\lambda \in X^+$,

$$\chi(\lambda) = \frac{\sum_{w \in W} \varepsilon(w) e(w(\lambda + \rho))}{\sum_{w \in W} \varepsilon(w) e(w\rho)}, \quad \text{where } \varepsilon(w) = (-1)^{\ell(w)}.$$

LEMMA. *If $\mu \in X^+$ and all $w\mu + (p^r - 1)\rho \in X^+$ for $w \in W$, then*

$$\sum_{w \in W^\mu} \chi(w\mu + (p^r - 1)\rho) = s(\mu) \operatorname{ch} \mathrm{St}_r.$$

PROOF. Abbreviate the denominator in Weyl's formula by $D$.

$$
\begin{aligned}
\sum_{w \in W^\mu} \chi(w\mu + (p^r - 1)\rho) &= \sum_{w \in W^\mu} \sum_{w' \in W} \varepsilon(w') e(w'(w\mu + p^r\rho))/D \\
&= \sum_{w \in W^\mu} \sum_{w' \in W} \varepsilon(w') e(w'w\mu) e(p^r w'\rho)/D \\
&= \Big( \sum_{w \in W^\mu} e(w\mu) \Big) \Big( \sum_{w' \in W} \varepsilon(w') e(p^r w'\rho)/D \Big)
\end{aligned}
$$

The last term is just $s(\mu) \operatorname{ch} \mathrm{St}_r$. $\qquad\square$

This suggests a parallel with the role of the Brauer characters $s_\mu \varphi$ in 9.6.

Recall from 10.1 that the blocks of $\mathbf{G}_1$ (that is, the blocks of the algebra $\mathrm{Dist}(\mathbf{G}_1)$) correspond naturally to $W$-linkage classes in $X/pX$. Fix $\mu \in X_1$ and assume that $\mu$ lies in the lowest $p$-alcove for the affine Weyl group $W_p$:

$$C_\circ = \{\lambda \in X \mid 0 < (\lambda + \rho, \alpha^\vee) < p \text{ for all } \alpha > 0\}.$$

The Linkage Principle shows that for such "small" weights $\mu$, we have $V(\mu) = L(\mu)$. Moreover, an easy calculation shows that all weights $w\mu + (p-1)\rho$ are *dominant*. As a special case of Theorem 6.2(b), each $V(w\mu + (p-1)\rho)$ therefore contributes once to the list of **G**-composition factors of $V(\mu) \otimes \text{St}$. All of the weights $w\mu + (p-1)\rho$ are clearly $W$-linked to $\mu + (p-1)\rho$, so the block containing this weight involves at least these composition factors. Since $\mu \in C_\circ$, a further calculation shows that no other Weyl character occurring in the tensor product has a highest weight linked to $\mu + (p-1)\rho$.

Thanks to the above lemma, the dimension of this block summand of $L(\mu) \otimes \text{St}$ is therefore $|W^\mu| \, p^m$. This provides an upper bound for $\dim Q(\lambda)$, where $\lambda = \mu^\circ = (p-1)\rho + w_\circ \mu$. On the other hand, the filtration described in Theorem 10.2(d) yields the same lower bound on dimension. So $\dim Q(\lambda) = |W^\mu| \, p^m$. To summarize:

PROPOSITION. *Let $r = 1$ and $\lambda \in X_1$. Assume that $\mu = \lambda^\circ$ lies in $C_\circ$. Then $Q(\lambda)$ is a block summand of the $\mathbf{G}_1$-module $L(\mu) \otimes \text{St}$, with $\dim Q(\lambda) = |W^\mu| \, p^m$.*

## 10.7. Small PIMs

It is easy to adapt the results of the preceding section to a Chevalley group $G$ over $\mathbb{F}_p$ (or, with extra care, to its twisted analogue in types $A, D, E_6$). Recall from Theorem 9.7 that the smallest "generic" dimension of a PIM for $KG$ should be $|W| \, p^m$. Here we consider just the case when $\mu$ is regular, i.e., $W^\mu = W$. (The general case is a little more delicate.)

THEOREM. *Let $G$ be a Chevalley group over $\mathbb{F}_p$, with $\lambda \in X_1$. If $\mu = \lambda^\circ$ is regular and lies in the lowest alcove $C_\circ$ for $W_p$, then the $\mathbf{G}$-summand of $L(\mu) \otimes \text{St}$ involving the linkage class of the highest weight $\mu + (p-1)\rho$ restricts to the PIM $U(\lambda)$ for $KG$. Thus $\dim U(\lambda) = |W| \, p^m$.*

PROOF. Under our regularity assumption on $\mu$, no coordinate of $\lambda$ equals $p-1$. So the remarks at the end of 9.7 force $\dim U(\lambda) = |W| \, p^m$. In general, we know that $U(\lambda)$ occurs as a $KG$-summand of the **G**-summand of $L(\mu) \otimes \text{St}$ involving the highest weight. But 10.6 shows that the latter also has dimension $|W| \, p^m$. $\square$

EXAMPLE. In very special cases the entire **G**-module $L(\mu) \otimes \text{St}$ is actually indecomposable and restricts to a PIM for $KG$: take $\mu$ to be *minimal* in the partial ordering of dominant weights, so that $L(\mu)$ has as weights just the $w\mu$ for $w \in W$ and thus $\dim L(\mu) = |W\mu|$. When $\mu \neq 0$, Bourbaki calls it "minuscule": see [**56**, VI, §1, ex. 24 and §4, ex. 15]. All such weights are fundamental, which insures that Ballard's lower bound for the dimension of a PIM in 9.7 is $|W\mu|$. It follows by dimension comparison that $U(\lambda) = L(\mu) \otimes \text{St}$.

Essentially the same argument works over a field of order $p^r$. Tsushima [**422**], generalizing work of Lusztig [**307**] on the general linear group over a finite field, derives this directly for each of the classical groups. He also treats a few other small cases in which $L(\mu)$ involves a single $W$-orbit of nonzero weights together with the weight 0. (For example, take $\mu$ to be the highest short root.) Then the tensor product is the direct sum of a PIM and $\text{St}_r$.

## 10.8. The Category of $\mathbf{G}_r\mathbf{T}$-Modules

Now we consider what can be said in general about the computation of characters or dimensions of PIMs for $KG$. For this we have to work first in the category of $\mathbf{G}_r\mathbf{T}$-modules. Denote by $\mathcal{C}_r$ and $\mathcal{P}_r$ respectively free abelian groups having as bases the formal characters of the $\widehat{L}_r(\lambda)$ and the $\widetilde{Q}_r(\lambda)$. Note that $\mathcal{X} \subset \mathcal{C}_r$. (The inclusion is proper because the characters need not be $W$-invariant.) In particular, when $\chi \in \mathcal{X}$ is written uniquely as a $\mathbb{Z}$-linear combination of various $\operatorname{ch}\widehat{L}_r(\lambda)$, the coefficient of $\widehat{L}_r(\lambda)$ may be written as $[\chi : \widehat{L}_r(\lambda)]$. In this setting one sees readily (see Jantzen [251, 2.2]) that for all $\chi \in \mathcal{X}, \lambda \in X_r, \mu \in X$:

$$
(1) \qquad [\chi : \widehat{L}_r(\lambda + p^r\mu)] = \sum_{\nu \in X^+} [\chi : L(\lambda + p^r\nu)]_\chi \dim L(\nu)_\mu.
$$

If $L$ and $M$ are modules for $\mathbf{G}_r\mathbf{T}$, the space $\operatorname{Hom}_{\mathbf{G}_r}(L, M)$ has a natural $\mathbf{T}$-module structure, whose character lies in $\mathbb{Z}[p^r X]$. Following [251, §2], we can define a bilinear pairing $\mathcal{P}_r \times \mathcal{C}_r \to \mathbb{Z}[p^r X]$, by setting

$$
\langle \operatorname{ch} Q, \operatorname{ch} M \rangle := \operatorname{ch}\operatorname{Hom}_{\mathbf{G}_r}(Q, M).
$$

One feature of the pairing is reminiscent of the usual pairing on the space of class functions of a finite group. There is an involutive automorphism $\chi \mapsto \overline{\chi}$ of the group ring $\mathbb{Z}[X]$, defined by sending $e(\mu) \mapsto e(-\mu)$. For a $\mathbf{T}$-module $M$, with contragredient $M^*$, this gives $\operatorname{ch} M^* = \overline{\operatorname{ch} M}$. Now we get for any projective $\mathbf{G}_r\mathbf{T}$-module $Q$ and arbitrary $L, M$:

$$
\langle \operatorname{ch} Q \operatorname{ch} L, \operatorname{ch} M \rangle = \langle \operatorname{ch} Q, \overline{\operatorname{ch} L} \operatorname{ch} M \rangle,
$$

using the fact that $Q \otimes L$ is projective while $\operatorname{ch}(Q \otimes L) = \operatorname{ch} Q \operatorname{ch} L$. In turn, for any $\chi \in \mathcal{P}_r$ and $\phi, \psi \in \mathcal{C}_r$:

$$
\langle \chi\phi, \psi \rangle = \langle \chi, \overline{\phi}\psi \rangle.
$$

Jantzen shows in [251, 2.4] that for all $\lambda \in X_r$ and all $\chi \in \mathcal{C}_r$:

$$
(2) \qquad \langle \operatorname{ch}\widehat{Q}_r(\lambda), \chi \rangle = \sum_{\mu \in X} [\chi : \widehat{L}_r(\lambda + p^r\mu)]\, e(p^r\mu).
$$

Now combine equations (1) and (2) to get:

LEMMA. *For all $\lambda \in X_r$ and all $\chi \in \mathcal{X}$,*

$$
\langle \operatorname{ch}\widehat{Q}_r(\lambda), \chi \rangle = \sum_{\nu \in X^+} [\chi : L(\lambda + p^r\nu)]_\chi \chi_p(\nu)^{[r]}.
$$

## 10.9. Rewriting Multiplicities

Equipped with this formalism for $\mathbf{G}_r\mathbf{T}$-modules, we can now rewrite a complicated multiplicity formula for characters in $\mathcal{X}$ in terms of something much simpler. This will be a key ingredient in the proof of the main theorem. Recall from Theorem 10.3 the formula $\operatorname{ch}\widehat{Q}_r(\lambda) = q(r, \lambda)\operatorname{ch}\operatorname{St}_r$, where $q(r, \lambda) \in \mathcal{X}$.

LEMMA. *If $\lambda \in X_r, \nu \in X^+, \chi \in \mathcal{X}$, then*

$$
[\chi : L(\lambda + p^r\nu)]_\chi = [\chi\,\overline{q(r, \lambda)} : \operatorname{St}_r \otimes L(p^r\nu)]_\chi.
$$

PROOF. Using the associative property of the pairing in 10.8, we have:

$$
\begin{aligned}
\langle \operatorname{ch}\widehat{Q}_r(\lambda), \chi \rangle &= \langle q(r, \lambda)\operatorname{ch}\operatorname{St}_r, \chi \rangle \\
&= \langle \operatorname{ch}\operatorname{St}_r, \chi\overline{q(r, \lambda)} \rangle.
\end{aligned}
$$

By applying Lemma 10.8 to the right side, one gets

$$\sum_{\nu \in X^+} [\chi \overline{q(r, \lambda)} : \mathrm{St}_r \otimes L(p^r \nu)]_\chi \, \chi_p(\nu)^{[r]}.$$

Now compare the coefficients for the left side given by Lemma 10.8.          □

## 10.10.  Brauer Characters

To reach our main goal, we still have to relate the formal characters of the $\mathbf{G}_r \mathbf{T}$-modules $\widehat{Q}_r(\lambda)$ to the Brauer characters of PIMs $U_r(\lambda)$ for $KG$ when $\lambda \in X_r$. For large enough $p$, this might be done in a concrete way, by lifting the $\mathbf{G}_r \mathbf{T}$-module $\widehat{Q}_r(\lambda)$ to $\mathbf{G}$ and then restricting to $G$ (10.4). But even without such a $\mathbf{G}$-structure, we can relate the two settings formally.

Recall from 5.6 that we can define br $\chi$ for any $\chi \in \mathcal{X}$, agreeing with br $V$ when $\chi$ is the formal character of a $\mathbf{G}$-module $V$. This satisfies

$$\mathrm{br}\,\overline{\chi} = \overline{\mathrm{br}\,\chi},$$

where $\overline{\chi}$ is the duality involution on $\mathcal{X}$.

We know (9.3) that tensoring an arbitrary $\mathbf{G}$-module with $\mathrm{St}_r$ and restricting to $KG$ yields a projective module. This implies that for any $\chi \in \mathcal{X}$, $\mathrm{br}(\chi \,\mathrm{ch}\, \mathrm{St}_r)$ is a $\mathbb{Z}$-linear combination of various $\mathrm{br}\, U_r(\mu)$. Since $\mathrm{ch}\,\widehat{Q}_r(\lambda) = q(r, \lambda)\,\mathrm{ch}\, \mathrm{St}_r$ with $\chi = q(r, \lambda) \in \mathcal{X}$, we obtain for all $\lambda, \mu \in X_r$ integers $m_{\lambda\mu} := [\widehat{Q}_r(\lambda) : U_r(\mu)]$ for which

$$(1) \qquad \mathrm{br}\,\mathrm{ch}\,\widehat{Q}_r(\lambda) = \sum_{\mu \in X_r} m_{\lambda\mu}\,\mathrm{br}\, U_r(\mu).$$

When $p$ is large enough, this formula has a simple interpretation: lift the $\mathbf{G}_r \mathbf{T}$-module $\widehat{Q}_r(\lambda)$ to $\mathbf{G}$, then restrict to $G$ and decompose into a direct sum of various PIMs $U_r(\mu)$. In particular, all $m_{\lambda\mu} \geq 0$. This last statement actually turns out to be true for all $p$, as a consequence of the theorem in the following section.

## 10.11.  Statement of the Main Theorem

Finally we can formulate a result, discovered independently by Chastkofsky [**96**] and Jantzen [**251**], which reduces the determination of the Brauer characters of all PIMs for $KG$ to the parallel problem for the $r$th Frobenius kernel $\mathbf{G}_r$. There the answer depends only on standard character data for $\mathbf{G}$.

THEOREM.  *Let $G = \mathbf{G}^F$ over a field of $p^r$ elements. Fix $\lambda \in X_r$ and $\chi \in \mathcal{X}$, and define $m_{\lambda\mu} := [\widehat{Q}_r(\lambda) : U_r(\mu)]$ as in 10.10(1). Then:*

$$(1) \qquad \sum_{\mu \in X_r} m_{\lambda\mu}[\chi : L(\mu)]_G = \sum_{\nu \in X^+} [\chi \chi_p(\widetilde{\nu}) : L(\lambda + p^r \nu)]_\chi.$$

*In particular, setting $\chi = \chi_p(\mu)$ for $\mu \in X_r$:*

$$(2) \qquad m_{\lambda\mu} = \sum_{\nu \in X^+} [L(\mu) \otimes L(\widetilde{\nu}) : L(\lambda + p^r \nu)]_{\mathbf{G}}.$$

*Therefore $m_{\lambda\mu}$ is nonnegative and depends just on standard character data for $\mathbf{G}$.*

The proof will be given in the following section; it makes clear that the sums on the right side of (1) and (2) are actually finite. First we want to make explicit several consequences which are easily checked:

COROLLARY. *Let $\lambda, \mu \in X_r$.*

(a) $m_{\lambda\mu} \geq 0$ *and is equal to 1 when $\lambda = \mu$.*

(b) *If $m_{\lambda\mu} \neq 0$ for $\mu \neq \lambda$, then $\langle \lambda, \alpha_0^\vee \rangle < \langle \mu, \alpha_0^\vee \rangle$.*

(c) *All $\nu$ contributing nonzero terms to the right side of formula (2) satisfy $\langle \nu, \alpha_0^\vee \rangle \leq \langle \rho, \alpha_0^\vee \rangle$.*

Note that in (b) and (c), the inequality in the conclusion yields a similar inequality for the weights involved if $G$ is a Chevalley group.

The inequalities in the corollary show that one can compute $\dim U_r(\lambda)$ recursively once all $\dim Q_r(\lambda)$ are known. This is of course not so easy to carry out explicitly. See 10.14 below for some rank 2 calculations.

EXAMPLE. It is easy to apply the theorem in the case of $\mathrm{SL}(2, p)$, where as usual we identify weights with integers. Given $0 \leq \lambda, \mu < p^r$ with $\lambda \neq \mu$, we ask which $\nu > 0$ can contribute to the sum on the right side of (2). It is necessary that $\lambda + p^r \nu \leq \mu + \nu$, or $\lambda + (p^r - 1)\nu \leq \mu$. Since $\nu > 0$, the only possibility is $\lambda = 0, \nu = 1, \mu = p^r - 1$. This agrees with the earlier finding (9.8) that $\dim U_r(0) = (2^r - 1)p^r$, whereas $\dim Q_r(0) = 2^r p^r$.

## 10.12. Proof of the Main Theorem

Roughly speaking, the idea of the proof is to use the associativity properties of pairings in our module categories to separate the factor $q(r, \lambda)$ from $\mathrm{ch}\,\widehat{Q}_r(\lambda)$ and then exploit the pivotal role of $\mathrm{St}_r$ as both a simple and a projective module.

An essential ingredient will be Theorem 6.9: For any $\chi \in \mathcal{X}$, we have

$$(1) \qquad [\chi : \mathrm{St}_r]_G = \sum_{\nu \in X^+} [\chi \chi_p(\widetilde{\nu}) : \chi_p(\sigma_r + p^r \nu)]_\chi.$$

To detect the value $m_{\lambda\mu}$ as $\mu$ varies, we take any $\chi \in \mathcal{X}$ and rewrite the left side of 10.11(1) as follows (taking into account the Orthogonality Relations 9.5):

$$\begin{aligned}
\sum_{\mu \in X_r} m_{\lambda\mu}[\chi : L(\mu)]_G &= \sum_{\mu \in X_r} m_{\lambda\mu}(\mathrm{br}\, U_r(\mu), \chi)_G \\
&= (\mathrm{br}\,\widehat{Q}_r(\lambda), \mathrm{br}\,\chi)_G \\
&= (\mathrm{br}\, q(r, \lambda)\,\mathrm{br}\,\mathrm{St}_r, \mathrm{br}\,\chi)_G \\
&= (\mathrm{br}\,\mathrm{St}_r, \mathrm{br}\,\chi\,\overline{\mathrm{br}\, q(r, \lambda)})_G \\
&= [\chi\,\overline{q(r, \lambda)} : \mathrm{St}_r]_G.
\end{aligned}$$

In turn, (1) above (with $\chi\,\overline{q(r, \lambda)}$ in place of $\chi$) allows us to rewrite this in terms of **G**-data:

$$\sum_{\nu \in X^+} [\chi \chi_p(\widetilde{\nu})\overline{q(r, \lambda)} : \chi_p(\sigma_r + p^r \nu)]_\chi.$$

Since $\mathrm{St}_r \otimes L(p^r \nu) \cong L(\sigma_r + p^r \nu)$, Lemma 10.9 yields the right side of 10.11(1), completing the proof. □

## 10.13. Letting $r$ Grow

Here we examine further the dependence of the multiplicity formula in 10.11 on the power of $p$ involved. This contributes to the explicit calculation of dimensions of PIMs as $r$ grows, as well as related calculation of Cartan invariants (Chapter

11 below). The ideas here have been developed by several authors, starting with Chastkofsky [**97**] (see Jantzen [**257**, App. 2]). We follow the notational scheme of Hu and Ye [**210**]. The idea is to reduce essential calculations to the case $r = 1$, using the Tensor Product Theorem and its analogue for PIMs.

To simplify, let $G$ be a Chevalley group; only minor modifications are needed for twisted groups. To each $p^r$-restricted weight $\lambda$ and $\mathbf{G}$-module $V$ we can associate an infinite array $M_r^\lambda(V)$ with entries indexed by $X^+ \times X^+$ and $(\nu, \nu')$ entry equal to $[V \otimes L(\nu) : L(\lambda + p^r \nu')]_{\mathbf{G}}$. (We omit the subscript $\mathbf{G}$ in the following discussion, when no other type of multiplicity is considered.)

If we totally order $X^+$ in some way we can regard $M_r^\lambda(V)$ as a matrix. Notice that each row has only finitely many nonzero entries, so it will make sense to multiply such matrices. We are mainly interested in the special case $V = L(\mu)$, with $\mu \in X_r$. Here the matrix has only finitely many nonzero diagonal entries. Thanks to Theorem 10.11, the resulting trace is given by the formula

$$\operatorname{Tr} M_r^\lambda(L(\mu)) = [\widehat{Q}_r(\lambda) : U_r(\mu)].$$

We can rewrite the matrix in terms of $p$-adic expansions of weights:

PROPOSITION. *Let* $\lambda, \mu \in X^+$ *have* $p$-*adic expansions* $\lambda = \lambda_0 + p\lambda_1 + \cdots + p^{r-1}\lambda_{r-1}$ *and* $\mu = \mu_0 + p\mu_1 + \cdots + p^{r-1}\mu_{r-1}$. *Then*

$$M_r^\lambda(L(\mu)) = M_1^{\lambda_0}(L(\mu_0)) \ldots M_1^{\lambda_{r-1}}(L(\mu_{r-1})).$$

PROOF. Proceed by induction on $r$. The idea is to rewrite multiplicities in terms of the $p$-adic expansions of $\lambda$ and $\mu$. For the induction step, we abbreviate $\lambda = \lambda_0 + p\lambda'$ and $\mu = \mu_0 + p\mu'$. Writing $M := M_r^\lambda(L(\mu))$, we have:

$$
\begin{aligned}
M_{\nu,\nu'} &= [L(\mu) \otimes L(\nu) : L(\lambda + p^r \nu')] \\
&= [L(\mu_0) \otimes L(\mu')^{[1]} \otimes L(\nu) : L(\lambda_0 + p\lambda' + p^r \nu')],
\end{aligned}
$$

which can be expressed in terms of multiplicities of various $L(\nu'')$ in $L(\mu_0) \otimes L(\nu)$:

$$
M_{\nu,\nu'} = \sum_{\nu'' \in X^+} [L(\mu_0) \otimes L(\nu) : L(\nu'')] \; [L(\nu'') \otimes L(\mu')^{[1]} : L(\lambda_0 + p\lambda' + p^r \nu')]
$$

It is clear that the second factor here can be nonzero only if the $p$-restricted part of $\nu''$ is $\lambda_0$, so we can write $\nu'' = \lambda_0 + p\nu_1$ ($\nu_1 \in X^+$):

$$
\begin{aligned}
M_{\nu,\nu'} &= \sum_{\nu_1 \in X^+} [L(\mu_0) \otimes L(\nu) : L(\lambda_0 + p\nu_1)] \; [L(\mu') \otimes L(\nu_1) : L(\lambda' + p^{r-1}\nu')] \\
&= M_1^{\lambda_0}(L(\mu_0))_{\nu,\nu_1} \; M_{r-1}^{\lambda'}(L(\mu'))_{\nu_1,\nu'}
\end{aligned}
$$

Now the induction hypothesis takes over.                                              □

As observed by Hu and Ye, there is a parallel formula in which $L(\mu), L(\mu_0), \ldots$ are replaced by $\widehat{Q}_r(\mu), \widehat{Q}_1(\mu_0), \ldots$.

In a related direction, Chastkofsky [**96, 97**] studies the dimensions of PIMs for a fixed $\lambda \in X_1$ as $r$ increases. For this he defines a "zeta function"

$$f(\lambda, t) := \exp\left(\sum_{r=1}^{\infty} \dim U_r(\lambda(r)) t^r\right),$$

where $\lambda(r) := \sum_{i=0}^{r-1} p^i \lambda$, together with a similar expression for Cartan invariants (see 11.10 below).

## 10.14. Some Comparisons in Low Ranks

To illustrate how the method of Chastkofsky and Jantzen is applied in practice, we summarize calculations done for several families of groups in rank 2. In these cases $Q_r(\lambda)$ is known to lift to a **G**-module for all $p$. The results here are still somewhat limited, but go well beyond the dimension computations cited in 9.11.

To compute $\dim U_r(\lambda)$ explicitly one has to combine the information of the type given in the tables below with computations of $\dim Q_r(\lambda)$ for $\lambda \in X_r$ based on a knowledge of standard character data for **G** (see 10.3 above). The reader should be cautioned that the results here are still relatively uncomplicated compared with the intricate results of Mertens [**319**] for $G_2(p)$ (see Chapter 18). In any case, one sees that "most" PIMs for the finite group are of the same dimension as the corresponding PIMs for $\mathbf{G}_r$ when $p$ is sufficiently large.

(A) Let $G = \mathrm{SL}(3, p^r)$. Here some calculations (mainly when $r = 1$) have been done by Andersen [**14**], Humphreys [**221, 226, 233**]. For $p = 2$ some examples are worked out by Andersen–Jorgensen–Landrock [**19**]. Koshitani [**281**] looks more closely at the structure of PIMs for $\mathrm{SL}(3,3)$.

For $\mathrm{SL}(3,p)$, the only cases in which the dimensions of $Q_1(\lambda)$ and $U_1(\lambda)$ differ are those listed in Table 1, when $p > 2$. Here and in the following tables, the

| | |
|---|---|
| $(0,0)$ | $(0,0) + (p-1,0) + (0,p-1) + (p-1,p-1)$ |
| $(a,0)$ | $(a,0) + (a,p-1)$ if $1 \leq a \leq p-1$ |
| $(0,b)$ | $(0,b) + (p-1,b)$ if $1 \leq b \leq p-1$ |

TABLE 1. Decomposable $Q_1(\lambda)$ for $\mathrm{SL}(3,p)$, $p$ odd

weight $\lambda$ is labelled by its coordinates $(a,b)$ relative to the fundamental dominant weights in their usual ordering. The first column gives the weight $\lambda$, followed by an abbreviated list of summands $U_1(\lambda)$ in $Q_1(\lambda)$ when lifted to **G** and then restricted to $KG$. Recall from 10.2 that PIMs $Q_1(\lambda)$ for $p$-regular weights $\lambda$ have dimension either $6p^3$ or $12p^3$, with degeneracies to $3p^3$ or (in the case of St) $p^3$. This makes it easy to compute $\dim U_1(\lambda)$ in each case. For example, $\dim U_1(0) = 7p^3$ if $p$ is odd.

In his 1988 Diplomarbeit, Dordowsky [**144**] follows the method of Chastkofsky and Jantzen to obtain by an induction on $r$ dimension formulas for $\mathrm{SL}(3,p^r)$ and $\mathrm{SU}(3,p^r)$, including explicit formulas for $\dim U_r(\lambda)$ when $r = 1, 2$. In the process he recovers results of Chastkofsky–Feit in the special case $p = 2$. When $p$ is odd and $r \geq 2$, he can write [**144**, 4.3.11]:

$$\dim U_r(\lambda) = \dim Q_r(\lambda) - 2^{r_1(\lambda)} 3^{r_2(\lambda)} 5^{r_3(\lambda)} p^{3r}.$$

Here the exponents are defined explicitly in terms of $\lambda$, but differ slightly in the case of $\mathrm{SU}(3,p^r)$.

While it is hard to get completely explicit results for $\mathrm{SL}(3,p^r)$, Ye [**441**] is able to compute $\dim U_r(0)$ for powers of an odd prime: $(12^r - 6^r + 1^r)p^{3r}$.

(B) Now let $G = \mathrm{SU}(3,p^r)$. Here the computations for $r = 1$ were done by Chastkofsky (unpublished), correcting earlier estimates by the author in [**221**]. (Andersen's table in [**14**] has a few errors.) The only cases in which the dimensions

| | |
|---|---|
| $(0,0)$ | $(0,0) + (p-1,1) + (1,p-1) + (p-1,p-1)$ |
| $(a,0)$ | $(a,0) + (a+1,p-1)$ if $1 \leq a \leq p-3$ |
| $(0,b)$ | $(0,b) + (p-1,b+1)$ if $1 \leq b \leq p-3$ |
| $(p-2,0)$ | $(p-2,0) + 2(p-1,p-1)$ |
| $(0,p-2)$ | $(0,p-2) + 2(p-1,p-1)$ |

TABLE 2. Decomposable $Q_1(\lambda)$ for SU$(3,p)$, $p$ odd

of $Q_1(\lambda)$ and $U_1(\lambda)$ differ are those listed Table 2, when $p > 2$. Note here that the weights $(0, p-1)$ and $(p-1, 0)$ yield no decomposition, in contrast to the case of SL$(3, p)$. As in the case of SL$(3, p)$, it is easy to derive from the table the dimensions of all $U_1(\lambda)$. For example, $\dim U_1(0) = 5p^3$ when $p$ is odd.

As noted above, Dordowsky [144] studies dimensions for SU$(3, p^r)$ (but his notational convention replaces $r$ here by $2r$) in tandem with SL$(3, p^r)$. He also explores Chastkofsky's zeta functions (10.13) in this setting, showing that these are not always rational functions in the case of SU$(3, p^r)$ if $p$ is odd. For SU$(3, p^r)$ and $p$ odd, Ye [441] earlier computed $\dim U_r(0) = (12^r - 6^r - 1^r)p^{3r}$. Note how close this is to the result above for SL$(3, p^r)$.

(C) Next let $G = \mathrm{Sp}(4, p^r)$, viewed as a group of type C$_2$, with short simple root $\alpha$ and long simple root $\beta$. Calculations when $r = 1$ can be found (sometimes with the opposite convention on root lengths) in Andersen [14], Chastkofsky [95, 96], Humphreys [223, 230], Jantzen [257]. The only cases in which the dimensions of $Q_1(\lambda)$ and $U_1(\lambda)$ differ involve weights having at least one coordinate equal to 0. For $p > 3$ they are listed in Table 3, with $\lambda = (0,0)$ followed by weights $\lambda = (a, 0)$ with $1 \leq a \leq p - 1$ and then weights $\lambda = (0, b)$ with $1 \leq b \leq p - 1$. In this case the generic dimensions of the $Q_1(\lambda)$ are $8p^4, 16p^4, 24p^4, 32p^4$: see 10.2.

For arbitrary $r$ and $p \geq 7$, Ye [442] computes $\dim U_r(0) = (32^r - a^r - b^r + 1)p^{4r}$, where $a$ and $b$ are the roots of $x^2 - 24x + 64$.

| | |
|---|---|
| $(0,0)$ | $(0,0) + (p-1,0) + (p-1,1) + (0,p-1) + (2,p-1) + (p-1,p-1)$ |
| $(a,0)$ | $(a,0) + (a,p-1) + (a+2,p-1)$ if $1 \leq a \leq p-4$ |
| $(p-3,0)$ | $(p-3,0) + (p-3,p-1) + 3(p-1,p-1)$ |
| $(p-2,0)$ | $(p-2,0) + (p-2,p-1)$ |
| $(p-1,0)$ | $(p-1,0) + (p-1,p-1)$ |
| $(0,b)$ | $(0,b) + (p-1,b) + (p-1,b+1)$ if $1 \leq b \leq p-3$, $b \neq \frac{1}{2}(p-3)$ |
| $(0,\frac{1}{2}(p-3))$ | $(0,\frac{1}{2}(p-3)) + (p-1,\frac{1}{2}(p-3))$ |
| $(0,p-2)$ | $(0,p-2) + (p-1,p-2) + 3(p-1,p-1)$ |
| $(0,p-1)$ | $(0,p-1) + (p-1,p-1)$ |

TABLE 3. Decomposable $Q_1(\lambda)$ for Sp$(4,p)$, $p > 3$

CHAPTER 11

# Cartan Invariants

The **Cartan invariants** of a group algebra $KG$ (or other finite dimensional algebra) are the multiplicities of simple modules as composition factors of the principal indecomposable modules. If a set $\Lambda$ (ordered in some way) indexes the isomorphism classes of simple modules $L_\lambda$ and their projective covers $U_\lambda$, the resulting $|\Lambda| \times |\Lambda|$ **Cartan matrix** $C$ has entries $c_{\lambda\mu} := [U_\lambda : L_\mu]$. (There should be no confusion with the use of "Cartan matrix" in the context of root systems.)

For a group of Lie type there are two overlapping questions: Can one actually compute the matrix $C$ for a specific group or family of groups? Can one find meaningful patterns in the entries of $C$, such as uniformities for a Lie type when $p$ is sufficiently large?

After reviewing some general facts about Cartan invariants for a group algebra (11.1–11.3), we shall discuss what is known in the case of finite groups of Lie type, with emphasis on "generic" behavior for large $p$ (11.9–11.11). We then survey the known computations and exhibit some small explicit examples, updating the 1985 survey [**228**]. The chapter concludes with a look at some recent conjectures involving block invariants.

## 11.1. Cartan Invariants for Finite Groups

If $G$ is any finite group, consider the simple $KG$-modules $L_\lambda$ (indexed by some set $\Lambda$) and their projective covers $U_\lambda$. Denote the composition factor multiplicities by $c_{\lambda\mu} = [U_\lambda : L_\mu]$. These are the Cartan invariants, forming a $|\Lambda| \times |\Lambda|$ matrix $C$. The "symmetry" of the group algebra $KG$ implies that $C$ is a symmetric matrix. Here is a summary of standard facts:

THEOREM. *Let $G$ be any finite group, with Cartan matrix $C$ relative to $p$.*

(a) *$C$ is a symmetric, positive-definite $t \times t$ matrix, where $t$ is the number of $p$-regular classes of $G$.*

(b) *The determinant of $C$ is a power of $p$.*

(c) *If $x_1, \ldots, x_t$ are representatives of the $p$-regular conjugacy classes of $G$, the elementary divisors of $C$ are $p^{e_1}, \ldots, p^{e_t}$, where $p^{e_i}$ is the exact power of $p$ dividing $|C_G(x_i)|$.*

(d) *If $\Lambda$ is partitioned and ordered according to the blocks of $KG$, then $C$ has a corresponding block diagonal form.*

For the proof, see Curtis–Reiner [**125**, §84]. (Note however that the determinant calculation there only shows that $\det C$ is $\pm$ a power of $p$.) See also [**126**, §18] and Serre [**366**, 16.1]. When we list the elementary divisors of $C$ below in special cases, we omit those equal to 1.

Part (d) of the theorem allows us to speak of the Cartan matrix of a block of $G$. For a block of defect 0, the matrix is just a $1 \times 1$ identity matrix. For other blocks, the diagonal entries must be at least 2, since the projective cover of a nonprojective simple module is also its injective hull. Moreover, the positive definiteness of $C$ implies that the largest entry in $C$ can be found on the diagonal.

It is not clear whether one can say anything else in general about the size of Cartan invariants. At one time it was expected that all Cartan invariants of a $p$-block of defect $d$ would be bounded above by $p^d$. This is however false, and there appears to be no other natural candidate for such an upper bound in terms of $p$ and $d$. In particular, Landrock [295] shows that the principal 2-block of the Suzuki group Sz(8) has defect 6 but involves a Cartan invariant at least 160: the multiplicity of the trivial module in its PIM. (In fact, this Cartan invariant is exactly 160, as shown explicitly by Zaslawsky [460] and others.) The more general results of Chastkofsky and Feit [104] on Suzuki groups in characteristic 2 show that Landrock's observation is not isolated. See 11.17 below for further discussion of such numerical questions.

## 11.2. Brauer Characters and Cartan Invariants

For an arbitrary finite group $G$ and prime $p$, it is actually possible in principle to compute the Cartan matrix (or rather its inverse) from a knowledge of the irreducible Brauer characters. Let $C_1, \ldots, C_t$ be the $p$-regular conjugacy classes of $G$, of respective sizes $h_1, \ldots, h_t$. Denote the irreducible Brauer characters of $G$ by $\varphi_1, \ldots, \varphi_t$, and abbreviate the value of $\varphi_i$ at an element of $C_j$ by $\varphi_{ij}$. Replacing the index $j$ by $j^*$ here indicates that $\varphi_i$ is instead evaluated at an element of the class $C_{j^*}$ consisting of inverses of the elements of $C_j$. Further, denote by $\gamma_{ij}$ the $(i, j)$-entry of the inverse matrix $C^{-1}$. With this notation, the relevant version of one of the classical orthogonality relations reads as follows. (See Curtis–Reiner [125, (84.12)] or [126, (18.23)]. The classical case corresponds to $C = I$.)

THEOREM. *If $C$ is the matrix of Cartan invariants of a finite group $G$, relative to a prime $p$, the entries of the inverse matrix $C^{-1} = [\gamma_{ij}]$ are given in terms of Brauer character values by:*

$$\frac{1}{|G|} \sum_{k=1}^{t} h_k \varphi_{ik} \varphi_{jk^*} = \gamma_{ij}$$

In practice it is very difficult to carry out such calculations when $t$ is at all large. But the theorem does emphasize how tightly the data about simple modules and projective modules fit together.

## 11.3. Decomposition Numbers

When we look at the reduction modulo $p$ of ordinary characters of a finite group $G$ in Chapter 16, we will encounter another important integer matrix $D$: the **decomposition matrix**. This is a $t \times s$ matrix, where $s$ is the total number of conjugacy classes of $G$. (There is a minor notational conflict with the use of $D$ to denote a defect group of a block, which should cause no confusion.) It turns out that knowledge of $D$ is sufficient to compute $C$. This goes back to early work of Brauer–Nesbitt [61] (see Curtis–Reiner [125, 83.9] or [126, 18.10]).

THEOREM. *Let $C$ be the Cartan matrix and $D$ the decomposition matrix of an arbitrary finite group $G$, with transpose $D^t$. Then $C = D^t D$.*

Note that this exhibits in a new way the symmetry of the matrix $C$ already observed in 11.1 above. Some of the computations surveyed later in this chapter were in fact done by first determining $D$.

## 11.4. Groups of Lie Type

When $G$ is a finite group of Lie type over a field of order $p^r$, we write the entries of the Cartan matrix of $KG$ as $c_{\lambda\mu}$, or $c_{\lambda\mu}^{(r)}$ if we need to emphasize the dependence on $r$. Here the index set is $X_r$. In particular, we denote the **first Cartan invariant** by $c_{00}$ or $c_{00}^{(r)}$. (In the literature, the first Cartan invariant for an arbitrary group is often written $c_{11}$.)

For a single group $G$ we might ask for an explicit computation of $C$, though this becomes impractical for very large values of $p$ or $r$ once the Lie rank exceeds 1. Some examples are given at the end of the chapter, including the relatively small case of SL(5, 2), which already takes a lot of work to determine.

It is more interesting to study simultaneously a family of groups of a given Lie type. Then we may consider how the entries of $C$ depend on $p$ or $r$. We look in particular for "generic" behavior based on the proven or expected generic behavior of formal character data for **G**. This may for example involve, for fixed $r$, finding an explicit lower bound on $p$ above which all values of the Cartan invariant $c_{00}^{(r)}$ remain constant. A weaker notion of "generic" (as in the work of Andersen–Jantzen–Soergel [**18**]) would rely on the existence of an unknown lower bound on $p$. In any case, precise lower bounds would not be the main issue here.

Here are typical questions, some of which can be answered at least partially:

- For fixed $r$, are there "generic" values of Cartan invariants for large $p$ and weights in sufficiently general position?
- For fixed $r$, is there a "generic" value of the first Cartan invariant, for large enough $p$?
- For fixed weights $\lambda$ and $\mu$ (both in $X_1$, for example), how does the corresponding Cartan invariant $c_{\lambda\mu}^{(r)}$ vary as $r$ increases?
- How does the first Cartan invariant vary as $r$ grows (when $p$ is fixed but sufficiently large)?
- How are the Cartan invariants of a Chevalley group of type $A, D,$ or $E_6$ related to the Cartan invariants of the corresponding twisted group?
- Is there a uniform upper bound in terms of block invariants for the size of Cartan invariants?

As discussed below, the known results indicate that Cartan invariants for groups of Lie type behave quite reasonably. Moreover, the generic values of the invariants coincide for Chevalley groups and their twisted analogues.

Reasonable algorithmic approaches have been developed for the computation of Cartan invariants in special cases. But the combinatorics involved can be heavy even in rank 1, so it remains a challenging problem to get insight into the more conceptual questions just raised. A closer look at the groups SL(2, $q$) will be instructive.

## 11.5. Example: SL(2, $p$)

As noted in 9.8, the PIMs for SL(2, $p$) are easy to describe from several viewpoints, including the theory of blocks with a cyclic defect group. The resulting matrices $D$ and $C$ were first written down by Brauer–Nesbitt [61] in the case of $PSL(2,p)$, which inherits the 1-block of $SL(2,p)$ as well as St; the other block of highest defect behaves similarly. (See Humphreys [219] for an elementary exposition emphasizing the connection with ordinary character theory.)

When $p = 2$ or $3$, the Cartan matrix of the 1-block is $1 \times 1$ with entry respectively 2 or 3. For $p \geq 5$ both matrices $C$ and $D$ have a uniform structure independent of what $p$ is, as shown in Tables 1 and 2. A typical row (= column) has only three nonzero entries $1, 2, 1$. (The matrices for the other block of defect 1 look quite similar to these.)

$$\begin{bmatrix}
2 & 1 & & & & & & \\
1 & 2 & 1 & & & & & \\
  & 1 & 2 & 1 & & & & \\
  &   &   & \cdot & & & & \\
  &   &   &   & \cdot & & & \\
  &   &   &   &   & \cdot & & \\
  &   &   &   & 1 & 2 & 1 & \\
  &   &   &   &   & 1 & 2 & 1 \\
  &   &   &   &   &   & 1 & 3
\end{bmatrix}$$

TABLE 1. Cartan invariants of 1-block of $SL(2,p)$, $p \geq 5$

$$\begin{bmatrix}
1 & & & & & & & \\
1 & 1 & & & & & & \\
  & 1 & 1 & & & & & \\
  &   & 1 & \cdot & & & & \\
  &   &   & \cdot & \cdot & & & \\
  &   &   & \cdot & 1 & & & \\
  &   &   &   & 1 & 1 & & \\
  &   &   &   &   & 1 & & \\
  &   &   &   &   & 1 & &
\end{bmatrix}$$

TABLE 2. Decomposition numbers of 1-block of $SL(2,p)$, $p \geq 5$

## 11.6. Example: $SL(2,q)$

For the groups $SL(2,q)$ when $q = p^r$, it takes considerably more work to exhibit the Cartan invariants explicitly. But this can be done. The special case $p = 2$ is treated combinatorially by Alperin [2, 5]. He finds that all Cartan invariants are powers of 2. In particular, $c_{00}^{(r)} = 2^r$.

When $p$ is odd, the computation has been organized in several different ways. In each case there are explicit combinatorial formulas for the Cartan invariants. These are usually powers of 2 bounded by $2^r$, with $c_{00}^{(r)} = 2^r$. But there are also some exceptional ones of the form $2^i + 2^{r-i}$, a phenomenon already seen for $SL(2,p)$.

(1) Following Jeyakumar's construction of PIMs inside twisted tensor products, Uphadhaya [423] arrives at explicit but rather complicated combinatorial formulas for the Cartan invariants.

(2) Elmer [**161**] also works in the characteristic $p$ setting, but streamlines the combinatorics by taking advantage of the lifting of PIMs for $\mathbf{G}_r$ to $SL(2, K)$. The affine Weyl group helps to organize the multiplicity data for $\mathbf{G}_r$ (where for $\lambda$ in general position the total number of composition factors of $Q_r(\lambda)$ is $4^r$). This data is then adapted to $SL(2, q)$ by a sort of deformation process involving the way twisted tensor products restrict to ordinary tensor products.

(3) The approach of van Ham, Springer, and van der Wel [**202**] is completely different. They first use the known ordinary and Brauer characters to work out the decomposition numbers (which are all 0 or 1), thereby recovering the results of Srinivasan [**387**]). Then an efficient algorithm for the Cartan invariants follows from $C = D^t D$.

## 11.7. Using Standard Character Data to Compute Cartan Invariants

In order to work systematically with groups of higher rank, one usually needs to invoke some knowledge of standard character data for the ambient algebraic group $\mathbf{G}$. Since this is not yet available in most cases, what can be said about the behavior of Cartan invariants for a family of groups of Lie type is often formulated modulo Lusztig's Conjecture (supplemented by treatment of primes $< h$).

There are several computational approaches worked out in the literature: see especially the papers by Chastkofsky, Hu, Hu–Ye, and Pillen cited below. In order to work with a group of Lie type over a field of $p^r$ elements, these authors rely on various mixtures of the following character data, which are typically needed just for a limited family of weights. All of these are computable in principle from standard character data for $\mathbf{G}$.

- Formal characters $\chi_p(\lambda)$ for $\lambda \in X_1$, coupled with the Tensor Product Theorem (Chapter 2).
- The analogue of the Tensor Product Theorem expressing $\operatorname{ch} \widehat{Q}_r(\lambda)$ in terms of Frobenius twists $\operatorname{ch} \widehat{Q}_1(\lambda_i)^{[i]}$ (10.2).
- Formal multiplicities $[\widehat{Q}_r(\lambda) : U_r(\mu)]$ for $\lambda, \mu \in X_r$ (10.11).
- Multiplicities $[L(\mu) \otimes L(\nu_1) : L(\lambda + p^r \nu_2)]_{\mathbf{G}}$ when $\nu_1, \nu_2 \leq_{\mathbb{Q}} \rho$.
- Analogous multiplicities when $Q_r(\mu)$ has compatible $\mathbf{G}$-module structure: $[Q_r(\mu) \otimes L(\nu_1) : L(\lambda + p^r \nu_2)]_{\mathbf{G}}$ when $\nu_1, \nu_2 \leq_{\mathbb{Q}} 2\rho$.

In the spirit of 10.13, the Cartan invariants are expressed combinatorially as signed sums of traces of products of integer matrices. The example $SL(2, p^r)$ already shows that the computations can be expected to get overwhelming as the rank increases. Moreover, there is no straightforward way to use the matrix $C$ for $r = 1$ to generate the matrix for larger values of $r$. But there is hope of finding interesting patterns, including "generic" ones.

## 11.8. Conditions for Genericity

For finite groups of Lie type, several approaches have been developed to study the generic behavior of Cartan invariants when $r$ is fixed but $p$ is allowed to grow: see Chastkofsky [**98**], Ye [**443, 444**], Humphreys [**232**]. These all have in common some reliance on the data just summarized in 11.7, but differ in the emphasis placed on actual computation of generic invariants. In [**443**], for example, it is shown that knowledge of the generic Cartan invariants for $r = 1$ will enable one to compute the generic invariants for all $r$.

When $r$ is fixed but $p$ varies, what does it mean for generic Cartan invariants to exist? Say $p \geq h$, so $p$-regular weights exist. Such weights in $X_r$ lie in certain $p$-alcoves (as indicated in 3.5). For weights in such an alcove $A$, one wants to impose conditions which imply that the Cartan invariants attached to $U_r(\lambda)$ are the same for all $\lambda \in A$ meeting the conditions. Moreover, one wants the conditions to hold for an arbitrarily large fraction $< 1$ of weights in $A$ when $p$ is sufficiently large.

Here we recall the main points of the discussion in [232], which focuses on a comparison with the parallel theory of $\mathbf{G}_r\mathbf{T}$-modules. Keep $r$ fixed throughout.

To insure that all $Q_r(\lambda)$ have compatible $\mathbf{G}$-module structure (10.4), we always assume $p \geq 2(h-1)$. For a fixed $\lambda \in X_r$ we then impose four genericity conditions which will imply a bijection (respecting multiplicities) between the composition factors of $U_r(\lambda)$ and those of the $\mathbf{G}_r\mathbf{T}$-module $\widehat{Q}_r(\lambda)$. The latter are already known to depend generically just on the $p$-alcove to which $\lambda$ belongs. For sufficiently large $p$ it will be clear that "most" $\lambda \in X_r$ satisfy these conditions, though we do not attempt to work out specific lower bounds on $p$ for each type of group.

(A) The weight $\lambda$ is $p$-regular, i.e., lies in the interior of some alcove for $W_p$.

(B) The restriction of the $\mathbf{G}$-module $Q_r(\lambda)$ to $G$ is indecomposable, hence is isomorphic to $U_r(\lambda)$.

(C) Take any composition factor of the $\mathbf{G}$-module $Q_r(\lambda)$, say $L(\mu_0) \otimes L(\mu_1)^{[r]}$ (with $\mu_0 \in X_r$). If $A$ is the alcove containing $\mu_0$, then $\mu_0$ lies at a distance from all walls of $A$ at least half the diameter of the weight diagram of $L(\mu_1)$. (Here diameter is measured in the euclidean space $\mathbb{R} \otimes X$.)

(D) Suppose $L(\mu)$ and $L(\nu)$ are composition factors of the $\mathbf{G}$-module $Q_r(\lambda)$. Write $\mu = \mu_0 + p^r\mu_1$ and $\nu = \nu_0 + p^r\nu_1$, with $\mu_0, \nu_0 \in X_r$. If $\mu_0$ and $\nu_0$ lie in the same $p$-alcove but are distinct, then $\mu_0 + \sigma \neq \nu_0 + \tau$ for all weights $\sigma, \tau$ of $L(\mu_1), L(\nu_1)$ respectively. (If $G$ is a twisted group, replace $\mu_1$ and $\nu_1$ here by their twists $\widetilde{\mu}_1$ and $\widetilde{\nu}_1$ using the graph automorphism $\pi$.)

The role of (C) is to insure that the $\mathbf{G}$-module $L(\mu_0) \otimes L(\mu_1)$ (or $L(\widetilde{\mu}_1)$ in the twisted case) splits into a direct sum of simple modules whose highest weights lie in the alcove $A$.

## 11.9. Generic Cartan Invariants

The main theorem in [232] compares composition factors of certain $\mathbf{G}_r\mathbf{T}$-modules with those of corresponding modules for a finite group of Lie type, showing that the multiplicities coincide. In this way the expected generic behavior of PIMs for $\mathbf{G}_r\mathbf{T}$ will imply similar behavior for $G$.

THEOREM. *Let $G = \mathbf{G}^F$ over a field of $p^r$ elements, with $p \geq 2(h-1)$. Suppose $\lambda \in X_r$ lies inside a $p$-alcove $A$ and satisfies conditions (A)–(D) above. Then there is a natural 1–1 correspondence between the distinct composition factors of the $\mathbf{G}_r\mathbf{T}$-module $\widehat{Q}_r(\lambda)$ and the distinct composition factors of the $G$-module $U_r(\lambda) = Q_r(\lambda)$. The correspondence sends $\widehat{L}_r(\mu + p^r\nu)$, with $\mu \in X_r$ and $\nu \in X^+$, to $L(\mu + \nu)$. Moreover, the composition factor multiplicities agree.*

The proof follows easily from the genericity conditions. In the case $r = 1$, $SL(3, p)$ is used in [224] as an example to show how Cartan invariants for $\mathbf{G}_1\mathbf{T}$ "deform" to give those for the finite group.

COROLLARY. *Let $G$ be a Chevalley group of type $A_\ell$ (with $\ell > 1$), $D_\ell$, or $E_6$. If $\lambda \in X_r$ satisfies conditions (A)–(D) for both $G$ and its twisted analogue, then the*

*Cartan invariants attached to the PIM for $L(\lambda)$ of $KG$ coincide with those for the twisted group.*

PROOF. The theorem applies to either group, insuring a multiplicity-preserving bijection between composition factors of each projective cover of $L(\lambda)$ and those of $Q_r(\lambda)$ □

This shows that generic behavior for PIMs of $\mathbf{G}_r\mathbf{T}$ will imply identical generic behavior for both of the finite groups: "most" weights in a fixed alcove will satisfy the genericity conditions (A)–(D) for both groups when $p$ is sufficiently large.

REMARKS. (1) Even when one knows the existence of generic Cartan invariants for sufficiently large $p$, it is a challenging problem to compute them explicitly when the standard character data for $\mathbf{G}$ are available. We saw already in 11.5 that most PIMs for $SL(2,p)$ are of dimension $2p$, with Cartan invariants $2, 1, 1$. In the case of $SL(3,p)$, the generic Cartan invariants for large enough $p$ distribute as follows: Approximately half of the PIMs $U(\lambda)$ have dimension $6p^3$, with generic invariants

6 (twice), 4 (six times), 2 (six times), 1 (six times),

while PIMs of dimension $12p^3$ have generic invariants

12, 6 (seven times), 4 (six times), 2 (twelve times), 1 (six times).

This is stated in Humphreys [221, pp. 57–58] and confirmed by Chastkofsky [98], who observes also that the same generic invariants occur for $SU(3,p)$.

(2) The existence of generic Cartan invariants shows that the matrix $C$ is "generically sparse" for large $p$. To make this more precise one would need to find a fixed upper bound (independent of $p$) for the number of nonzero entries in every row of the matrix.

(3) Ye [443] explains how to compute generic Cartan invariants for arbitrary $r$ when these are known for $r = 1$.

## 11.10. Growth of Cartan Invariants

If we fix $\lambda, \mu \in X_1$, the Cartan invariants $c_{\lambda\mu}^{(r)}$ are defined for all $r$. So one might ask how these numbers behave as $r$ increases. They are of course very isolated entries in a typical matrix $C$ for large $r$, but the case $\lambda = \mu = 0$ has special interest (and will be explored further in the next two sections).

Chastkofsky [97] (see also [100]) develops the computational formalism far enough to extract a conceptual formulation:

THEOREM. *Let $G = \mathbf{G}^F$ over a field of $p^r$ elements, with Cartan invariants $c_{\lambda\mu}^{(r)}$. Given $\lambda, \mu \in X_1$, there exist algebraic integers $\theta_1, \ldots, \theta_t$ such that $c_{\lambda\mu}^{(r)}$ is a sum of terms $\pm\theta_i^r$ for all $r \geq 1$. A similar statement holds for the numbers $\dim U_r(\lambda)$. Moreover, the $\theta_i$ are the same for a Chevalley group of type $A, D, E_6$ and its twisted analogue, though some of the signs may differ.*

The statement about dimensions of PIMs is already reflected in several low rank examples discussed in 9.11 and 10.14 above, while examples involving Cartan invariants $c_{00}^{(r)}$ will be given below.

Chastkofsky's arguments are outlined carefully in Jantzen [257, App. 2]. The algebraic integers $\theta_i$ are eigenvalues of certain integer matrices. These enter naturally into the computation of Cartan invariants, which are expressed as signed sums of traces of products of $r$ integer matrices.

## 11.11. The First Cartan Invariant

When $G$ is a Chevalley group or twisted analogue over a field of order $p^r$, we are denoting by $c_{00}^{(r)}$ the multiplicity of the trivial module in its projective cover. Here we fix $r$ but allow $p$ to increase. Based on low rank examples, one might expect $c_{00}^{(r)}$ to have a value independent of $p$ for all sufficiently large $p$. In order to prove a statement of this type, it seems essential to know that the relevant character data for $\mathbf{G}$ are independent of $p$ when $p$ is large enough. As in other "generic" questions, one gets stronger or weaker versions depending on whether a definite lower bound for $p$ can be given.

Ye [441, 442] and Hu and Ye [210, 212] emphasize computational approaches to the first Cartan invariant. We follow Pillen [335]; he works with a modified version of Conjecture V in Verma [426], which is implied by Lusztig's Conjecture:

(∗) If $x, y \in W_p$ with $x \cdot 0, y \cdot 0 \in X^+$ and $\langle x \cdot 0, \alpha_0^\vee \rangle \leq p(p-h-2)$, then $[\chi(x \cdot 0) : \chi_p(y \cdot 0)]_\chi$ is independent of $p$.

THEOREM. *Let* $G = \mathbf{G}^F$ *over a field of order* $p^r$. *Assume* (∗) *above. If* $p > 4h - 3$, *the first Cartan invariant* $c_{00}^{(r)}$ *assumes a constant value independent of* $p$.

Examples show that the bound $4h - 3$ will not be optimal, but for a generic assertion this makes little difference.

Pillen's proof involves fairly heavy combinatorics coming from the methods of Chastkofsky and Jantzen, thus requires a lot of notation. But his main idea can be formulated briefly. He relies especially on character data (encoded in matrices) of the last two types listed in 11.7 above, involving tensor product multiplicities in the $\mathbf{G}$-module setting. Under his special assumptions he can show that the only weights required here form a finite collection which can be labelled independently of $p$ in terms of affine Weyl group data.

REMARKS. (1) As suggested in [228, §7], the known examples suggest that for large enough $p$ the entire row (= column) of $C$ indexed by the weight 0 should have precisely the same nonzero entries. For $SL(2, p)$ these entries are 2 and 1, while for $SL(3, p)$ they should be $8, 4$ ( seven times), 2 ( three times), 1 ( five times). The heuristic discussion in [228, 7.1] suggests that such "generic" behavior may be best understood in the wider context of the reduction modulo $p$ of ordinary characters.

(2) It is tempting to conjecture that the generic values of the first Cartan invariant for a Chevalley group $G$ of type $A, D, E_6$ and its twisted analogue will coincide (as seen experimentally in type $A_2$). But this does not follow immediately from Pillen's set-up.

## 11.12. Special Cases

To compute the generic value of the first Cartan invariant in any particular case may be quite challenging, as examples in ranks 1 and 2 show. We summarize here some systematic calculations by type; see also 11.13 below for a survey of isolated groups whose full Cartan matrix is known.

$A_1$: As we remarked in 11.6, $c_{00}^{(r)} = 2^r$ for all primes.

$A_2$: Using the method of Chastkofsky–Feit [105] when $p = 2$, Cheng [108] works out $c_{00}^{(r)}$ for $SL(3, 2^r)$ and $SU(3, 2^r)$.

For $G = \mathrm{SL}(3, p)$, an algorithmic approach in Humphreys [226] shows that $c_{00}^{(1)} = 8$ for $p \geq 7$. (This holds also for $p = 5$ according to [217].)

More generally, Ye [441, 4.6] shows for $p \geq 7$ that the first Cartan invariant of $\mathrm{SL}(3, p^r)$ equals that of $\mathrm{SU}(3, p^r)$. It is given explicitly as follows. Set $a := 9 + \sqrt{33}$ and $b := 9 - \sqrt{33}$ (these are the roots of $x^2 - 18x + 48$). Then

$$c_{00}^{(r)} = a^r + b^r + 6^r - 2 \cdot 8^r.$$

As a result,

$$\lim_{r \to \infty} c_{00}^{(r)} / (9 + \sqrt{33})^r = 1.$$

Using a somewhat different algorithm, Chastkofsky [100] recomputes these values and also treats the case $p = 5$, where the result is slightly different for the two families of groups:

$$c_{00}^{(r)} = \begin{cases} a^r + b^r + 6^r - 2 \cdot 8^r + 2^r + (-2)^r & \text{if } G = \mathrm{SL}(3, p^r) \\ a^r + b^r + 6^r - 2 \cdot 8^r + 2^r - (-2)^r & \text{if } G = \mathrm{SU}(3, p^r) \end{cases}$$

$B_2$: Chastkofsky–Feit [104] study the groups $\mathrm{Sp}(4, 2^r)$ along with the Suzuki groups (Chapter 20), arriving inductively at formulas for the first Cartan invariant.

When $G = \mathrm{Sp}(4, p^r)$ with $p > 7$, Ye [442] finds a uniform value for $c_{00}^{(r)}$. For the special case $\mathrm{Sp}(4, 7)$, he computes $c_{00}^{(1)} = 16$, in contrast to the generic value $c_{00}^{(1)} = 14$ for $p > 7$. Hu and Ye [210] go on to compute all $c_{00}^{(r)}$ in the case $p = 7$.

$G_2$: Computation of $C$ by Mertens [319, 3.5, 3.6] shows that $c_{00}^{(1)} = 32$ when $p = 2$ and $c_{00}^{(1)} = 78$ when $p = 3$. (See Chapter 18 below.)

Hu and Ye [212] calculate $c_{00}^{(r)}$ for $G_2(p^r)$ when $p \geq 17$, confirming that $c_{00}^{(1)} = 168$ (this corrects their remark at the end of [210]). When $p = 5$ they find that $\dim U_1(1, 0) = 13 \cdot 5^6$, while $c_{00}^{(1)} = 20$.

## 11.13. Computations of Cartan Matrices

Here is a brief summary of cases in which the Cartan matrix $C$ of a Chevalley group or twisted analogue has been fully computed (sometimes via the decomposition matrix $D$). This updates the list given in the survey by Humphreys [228]. (The groups of Ree and Suzuki are discussed separately in Chapter 20.) Several examples will be exhibited below. As usual $q = p^r$ and $\ell$ is the Lie rank.

$A_1$: When $r = 1$, the matrix $C$ appears above in 11.5. For arbitrary $r$ see the discussion in 11.6.

$A_2$: Zaslawsky [460] computes $D$ and $C$ for $\mathrm{PSL}(3, q)$ when $q = 2, 3, 4, 5, 7, 8, 9$. In particular, this yields respective values for the first Cartan invariant: $2, 10, 34, 8, 8, 182, 212$.

A direct approach to $C$ for $\mathrm{SL}(3, p)$ is given by Humphreys [217] when $p = 3, 5$ and by Ye [439] when $p = 7$.

Andikfar [23] works out some of the entries of $C$ for $\mathrm{SL}(3, 2^r)$ as well as $\mathrm{SU}(3, 2^r)$.

$^2A_2$: Zaslawsky [460] computes $C$ for $\mathrm{PSU}(3, q)$ (in our notation) when $q = 2, 3, 4, 5, 7, 8$. This yields respective values for the first Cartan invariant: $8, 10, 16, 12, 8, 440$.

The matrices $C$ and $D$ for $\mathrm{SU}(3, 4)$ are exhibited by Benson and Martin [51], while the matrix $C$ for $\mathrm{SU}(3, 5)$ appears in Humphreys [221, 15.3].

$B_2$: Humphreys [217] works out $C$ for $\mathrm{Spin}(5, 3) \cong \mathrm{Sp}(4, 3)$, finding in particular $c_{00} = 7$. (His later work also produces $D$.) Similar computations for $\mathrm{PSp}(4, 5)$, where $c_{00} = 21$, are summarized in [223], but $C$ is not exhibited. In a preliminary version of [257], Jantzen computes the decomposition matrix $D$ for both blocks of highest defect in $\mathrm{Sp}(4, 5)$.

Thackray [410] determines $C$ and $D$ for $\mathrm{Sp}(4, 3)$. (These data play a further role in Benson [46].)

Ye [442] computes $C$ for $\mathrm{Sp}(4, 7)$.

$G_2$: When $q = 3$, Cheng [106] computes $C$ (but with one incorrect entry). Mertens [319] studies $G_2(p)$ for all $p$ and computes the matrices $C$ and $D$ in the cases $p = 2, 3$. The matrix $C$ is exhibited for $G_2(5)$ by Hu and Ye [211]. See Chapter 18 for some of the details.

$A_3$: Benson [44] studies $\mathrm{SL}(4, 2) \cong A_8$ using a mixture of ordinary and 2-modular character data; he computes both $D$ and $C$. Here $C$ is $8 \times 8$ of determinant $2^9$, involving the 1-block and St. Independently, Du [158] outlines the way in which the Chastkofsky–Jantzen method can be used for the groups $\mathrm{SL}(4, 2^r)$ when $p = 2$, and then exhibits the matrix $C$ for $\mathrm{SL}(4, 2)$ as well as the $64 \times 64$ matrix $C$ for $\mathrm{SL}(4, 4)$ (or rather the $63 \times 63$ matrix coming from the 1-block).

$B_3$: Ye and Zhou [451] compute the matrix $C$ for $\mathrm{SO}(7, 3)$, based on their computations for algebraic groups of types $B_3$ and $C_3$ when $p = 3$. The 1-block is $17 \times 17$ (with $c_{00} = 741$) and has determinant $3^{29}$. The other block of highest defect is $9 \times 9$ and has determinant $3^{12} = 3^9 \cdot 3^2 \cdot 3$.

$C_3$: Unpublished work of Y. Cheng uses character data to compute $C$ for $\mathrm{Sp}(6, 2)$. Liu and Ye [302] work out the $27 \times 27$ matrix $C$ for $\mathrm{Sp}(6, 3)$. Here the first Cartan invariant is 974. The matrix $C$ involves two blocks of highest defect ($= 9$), of sizes $14 \times 14$ and $12 \times 12$, having respective determinants $3^{28}$ and $3^{23}$.

$A_4$: The matrix $C$ for $\mathrm{SL}(5, 2)$ has been computed independently by Jantzen and by Ye: see 11.16 below.

### 11.14. Example: SL(3, 3)

Humphreys [217, Table 4] works out the Cartan 3-invariants of $\mathrm{SL}(3, 3)$: see Table 3. Here there are just two blocks, one containing St, and the determinant is $3^4$ as predicted by Brauer theory.

| dim $L(\lambda)$ | $\lambda\backslash\mu$ | $(0,0)$ | $(1,1)$ | $(1,0)$ | $(0,2)$ | $(2,1)$ | $(0,1)$ | $(2,0)$ | $(1,2)$ |
|---|---|---|---|---|---|---|---|---|---|
| 1 | $(0,0)$ | 10 | 5 | 2 | 1 | 4 | 2 | 1 | 4 |
| 7 | $(1,1)$ | 5 | 7 | 4 | 2 | 2 | 4 | 2 | 2 |
| 3 | $(1,0)$ | 2 | 4 | 4 | 2 | 0 | 2 | 1 | 1 |
| 6 | $(0,2)$ | 1 | 2 | 2 | 3 | 0 | 1 | 2 | 0 |
| 15 | $(2,1)$ | 4 | 2 | 0 | 0 | 3 | 1 | 0 | 1 |
| 3 | $(0,1)$ | 2 | 4 | 2 | 1 | 1 | 4 | 2 | 0 |
| 6 | $(2,0)$ | 1 | 2 | 1 | 2 | 0 | 2 | 3 | 0 |
| 15 | $(1,2)$ | 4 | 2 | 1 | 0 | 1 | 0 | 0 | 3 |
| dim $U(\lambda)/3^3$ | | 7 | 6 | 3 | 2 | 3 | 3 | 2 | 3 |

TABLE 3. Cartan invariants of 1-block of SL(3, 3), $p = 3$

## 11.15. Example: Sp(4, 3) and Related Groups

Humphreys [**217**, Table 8] exhibits the Cartan invariants of Spin(5, 3) $\cong$ Sp(4, 3). Here the algebraic group $\mathbf{G}$ is Spin(5, $K$) $\cong$ Sp(4, $K$), of type $B_2 = C_2$. The first simple root is taken to be short.

When $p = 3$ and $G = $ Sp(4, 3), there are two 3-blocks of highest defect and a single block of defect 0 containing St: see Table 4. One can check that the Cartan matrix of the principal block has determinant $3^8 = 3^4 \cdot 3^2 \cdot 3 \cdot 3$, while the other block yields the determinant $3^5 = 3^4 \cdot 3$.

| dim | $\lambda\backslash\mu$ | $(0,0)$ | $(0,1)$ | $(2,0)$ | $(0,2)$ | $(2,1)$ | $(1,0)$ | $(1,1)$ | $(1,2)$ | $(2,2)$ |
|---|---|---|---|---|---|---|---|---|---|---|
| 1 | $(0,0)$ | 7 | 6 | 3 | 5 | 1 | | | | |
| 5 | $(0,1)$ | 6 | 15 | 9 | 6 | 6 | | | | |
| 10 | $(2,0)$ | 3 | 9 | 21 | 6 | 9 | | | | |
| 14 | $(0,2)$ | 5 | 6 | 6 | 7 | 2 | | | | |
| 25 | $(2,1)$ | 1 | 6 | 9 | 2 | 7 | | | | |
| 4 | $(1,0)$ | | | | | | 20 | 18 | 7 | |
| 16 | $(1,1)$ | | | | | | 18 | 21 | 6 | |
| 40 | $(1,2)$ | | | | | | 7 | 6 | 5 | |
| 81 | $(2,2)$ | | | | | | | | | 1 |

TABLE 4. Cartan invariants of Sp(4, 3), $p = 3$

In his 1980 Cambridge thesis [**410**], Thackray studies these and related groups from the viewpoint of modular characters. He computes $D$ and $C$ for CSp(4, 3) and its quotient SO(5, 3). These are reproduced in the Appendix of Benson [**46**], where the groups are denoted respectively by $\widehat{G}$ and $G$. To pass from Thackray's Cartan matrix to the one exhibited here, one has to combine his entries coming from simple modules of the same dimension with superscripts $\pm$. See 13.15 below for further discussion of the PIMs associated with these groups.

## 11.16. Example: SL(5, 2)

In unpublished work, Jantzen computes the matrix $D$ for SL(5, 2) when $p = 2$, which leads to the Cartan matrix $C$ later computed independently by Ye [**448**]. (However, Ye's matrix is arranged differently, with the first Cartan invariant in the lower right corner.) Table 5 shows Jantzen's initial data, which he works out in an ad hoc way via his sum formula.

| $\lambda$ | $\dim L(\lambda)$ |
|---|---|
| $(0,0,0,0)$ | 1 |
| $(1,0,0,0)$ | 5 |
| $(0,0,0,1)$ | 5 |
| $(0,1,0,0)$ | 10 |
| $(0,0,1,0)$ | 10 |
| $(1,0,0,1)$ | 24 |
| $(1,1,0,0)$ | 40 |
| $(0,0,1,1)$ | 40 |
| $(1,0,1,0)$ | 40 |
| $(0,1,0,1)$ | 40 |
| $(0,1,1,0)$ | 74 |
| $(1,1,0,1)$ | 160 |
| $(1,0,1,1)$ | 160 |
| $(1,1,1,0)$ | 280 |
| $(0,1,1,1)$ | 280 |

TABLE 5. Data for $SL(5,2), p = 2$

In Jantzen's matrix $D$, the rows correspond to the ordinary irreducible characters labelled by their dimensions: see Table 6. The matrix could instead be computed by comparing ordinary and Brauer character tables, using the data in [243]. (The resulting matrix $D$ is currently available at the Modular Atlas homepage [243].)

Table 7 displays the Cartan invariants of $SL(5,2)$ which result from either Ye's direct calculation or Jantzen's indirect calculation via $D$. It can be checked that $\det C = 2^{10} \cdot 2^3 \cdot 2^2 \cdot 2 \cdot 2$, as predicted by Brauer's theory.

## 11.17. Conjectures on Block Invariants

Let $G$ be an arbitrary finite group and $p$ a prime dividing its order. By now we have encountered most of the main numerical invariants attached to the category of $KG$-modules or to the module category for an individual block $\mathcal{B}$. These include $\dim KG = |G|$ (or the $p'$-part $|G|_{p'}$ of the order) and $\dim \mathcal{B}$, as well as the order $|D|$ of a defect group of $\mathcal{B}$; for a group of Lie type this is either 1 or the full power of $p$ dividing $|G|$. Another interesting invariant is the number $l(G)$ of nonisomorphic simple $KG$-modules (equal for a group of Lie type to the number of $p'$-classes), or the number $l(\mathcal{B})$ of simple modules belonging to $\mathcal{B}$. We can also consider the various degrees $\varphi(1)$ of the corresponding Brauer characters; call the set of characters $\mathrm{Irr}_p(G)$ or $\mathrm{Irr}_p(\mathcal{B})$. Further invariants are the trace and determinant of the Cartan matrix of $KG$ or $\mathcal{B}$.

These invariants of the group algebra or its blocks figure in a number of intriguing conjectures. Sometimes analogous invariants from the characteristic 0 representation theory of $G$ are also mixed in. Conjectures for $G$ are regarded as "global", while conjectures for a single block are regarded as "local". It is always a high

| | | | | | | | | | | | | | | | |
|---|---|---|---|---|---|---|---|---|---|---|---|---|---|---|---|
| 1 | 1 | 0 | 0 | 0 | 0 | 0 | 0 | 0 | 0 | 0 | 0 | 0 | 0 | 0 | 0 |
| 30 | 0 | 1 | 1 | 1 | 1 | 0 | 0 | 0 | 0 | 0 | 0 | 0 | 0 | 0 | 0 |
| 124 | 0 | 0 | 0 | 1 | 1 | 1 | 0 | 0 | 1 | 1 | 0 | 0 | 0 | 0 | 0 |
| 155 | 1 | 1 | 1 | 2 | 2 | 1 | 0 | 0 | 1 | 1 | 0 | 0 | 0 | 0 | 0 |
| 217 | 3 | 2 | 2 | 2 | 2 | 0 | 0 | 0 | 1 | 1 | 1 | 0 | 0 | 0 | 0 |
| 280 | 2 | 2 | 2 | 0 | 0 | 1 | 1 | 1 | 1 | 1 | 1 | 0 | 0 | 0 | 0 |
| 315 | 0 | 1 | 0 | 2 | 1 | 0 | 0 | 0 | 0 | 0 | 0 | 0 | 0 | 1 | 0 |
| 315 | 0 | 0 | 1 | 1 | 2 | 0 | 0 | 0 | 0 | 0 | 0 | 0 | 0 | 0 | 1 |
| 315 | 1 | 2 | 2 | 1 | 1 | 0 | 0 | 0 | 1 | 0 | 1 | 1 | 0 | 0 | 0 |
| 315 | 1 | 2 | 2 | 1 | 1 | 0 | 0 | 0 | 0 | 1 | 1 | 0 | 1 | 0 | 0 |
| 315 | 1 | 0 | 2 | 0 | 0 | 1 | 1 | 0 | 1 | 1 | 0 | 1 | 0 | 0 | 0 |
| 315 | 1 | 2 | 0 | 0 | 0 | 1 | 0 | 1 | 1 | 1 | 0 | 0 | 1 | 0 | 0 |
| 465 | 2 | 2 | 3 | 1 | 1 | 1 | 1 | 0 | 2 | 1 | 1 | 1 | 0 | 0 | 0 |
| 465 | 2 | 3 | 2 | 1 | 1 | 1 | 0 | 1 | 1 | 2 | 1 | 0 | 1 | 0 | 0 |
| 465 | 2 | 2 | 3 | 1 | 1 | 1 | 1 | 0 | 2 | 1 | 1 | 1 | 0 | 0 | 0 |
| 465 | 2 | 3 | 2 | 1 | 1 | 1 | 0 | 1 | 1 | 2 | 1 | 0 | 1 | 0 | 0 |
| 496 | 2 | 2 | 2 | 0 | 0 | 0 | 0 | 0 | 1 | 1 | 1 | 1 | 1 | 0 | 0 |
| 651 | 2 | 2 | 3 | 2 | 1 | 0 | 1 | 0 | 1 | 0 | 1 | 1 | 0 | 1 | 0 |
| 651 | 2 | 3 | 2 | 1 | 2 | 0 | 0 | 1 | 0 | 1 | 1 | 0 | 1 | 0 | 1 |
| 651 | 3 | 3 | 3 | 2 | 2 | 1 | 0 | 0 | 2 | 2 | 1 | 1 | 1 | 0 | 0 |
| 868 | 2 | 4 | 4 | 2 | 2 | 3 | 1 | 1 | 3 | 3 | 1 | 1 | 1 | 0 | 0 |
| 930 | 2 | 3 | 3 | 3 | 1 | 1 | 1 | 0 | 2 | 1 | 1 | 1 | 1 | 1 | 0 |
| 930 | 2 | 3 | 3 | 1 | 3 | 1 | 0 | 1 | 1 | 2 | 1 | 1 | 1 | 0 | 1 |
| 930 | 4 | 5 | 5 | 2 | 2 | 2 | 1 | 1 | 3 | 3 | 2 | 1 | 1 | 0 | 0 |
| 960 | 0 | 2 | 2 | 3 | 3 | 0 | 0 | 0 | 0 | 0 | 0 | 1 | 1 | 1 | 1 |
| 1240 | 2 | 4 | 4 | 3 | 3 | 1 | 1 | 1 | 1 | 1 | 1 | 1 | 1 | 1 | 1 |

TABLE 6. Decomposition numbers of 1-block of $SL(5,2)$, $p = 2$

priority to check such conjectures for groups of Lie type, both in the defining characteristic $p$ and in other characteristics.

Here we outline some recent formulations for arbitrary groups by Willems [**434**] and Holm–Willems [**207**]. The first paper states a global conjecture and verifies it in a number of cases including the finite groups of Lie type in the defining characteristic (where the squared degree of the Steinberg character already dominates the right side).

$$\textit{Global Conjecture:} \quad |G|_{p'} \leq \sum_{\varphi \in \mathrm{Irr}_p(G)} \varphi(1)^2,$$

with equality iff the Sylow $p$-subgroup of $G$ is normal. In the second paper a weaker version is proposed:

$$\textit{Weak Global Conjecture:} \quad \frac{|G|_{p'}}{l(G)} \leq \max_{\varphi \in \mathrm{Irr}_p(G)} \{\varphi(1)^2\},$$

with an explicit condition for equality to hold.

The authors go on to compare these global conjectures with a local version (independent of the nature of $G$) for a single $p$-block $\mathcal{B}$ having defect group $D$.

$$\begin{bmatrix}
88 & 108 & 108 & 56 & 56 & 36 & 19 & 19 & 57 & 57 & 40 & 25 & 25 & 6 & 6 \\
108 & 154 & 146 & 84 & 84 & 50 & 24 & 29 & 73 & 76 & 52 & 34 & 39 & 12 & 12 \\
108 & 146 & 154 & 84 & 84 & 50 & 29 & 24 & 76 & 73 & 52 & 39 & 34 & 12 & 12 \\
56 & 84 & 84 & 66 & 60 & 26 & 14 & 11 & 40 & 37 & 26 & 21 & 20 & 13 & 9 \\
56 & 84 & 84 & 60 & 66 & 26 & 11 & 14 & 37 & 40 & 26 & 20 & 21 & 9 & 13 \\
36 & 50 & 50 & 26 & 26 & 26 & 11 & 11 & 32 & 32 & 16 & 12 & 12 & 2 & 2 \\
19 & 24 & 29 & 14 & 11 & 11 & 9 & 4 & 16 & 12 & 9 & 8 & 4 & 3 & 1 \\
19 & 29 & 24 & 11 & 14 & 11 & 4 & 9 & 12 & 16 & 9 & 4 & 8 & 1 & 3 \\
57 & 73 & 76 & 40 & 37 & 32 & 16 & 12 & 47 & 42 & 26 & 20 & 16 & 4 & 2 \\
57 & 76 & 73 & 37 & 40 & 32 & 12 & 16 & 42 & 47 & 26 & 16 & 20 & 2 & 4 \\
40 & 52 & 52 & 26 & 26 & 16 & 9 & 9 & 26 & 26 & 20 & 12 & 12 & 3 & 3 \\
25 & 34 & 39 & 21 & 20 & 12 & 8 & 4 & 20 & 16 & 12 & 13 & 8 & 4 & 3 \\
25 & 39 & 34 & 20 & 21 & 12 & 4 & 8 & 16 & 20 & 12 & 8 & 13 & 3 & 4 \\
6 & 12 & 12 & 13 & 9 & 2 & 3 & 1 & 4 & 2 & 3 & 4 & 3 & 5 & 2 \\
6 & 12 & 12 & 9 & 13 & 2 & 1 & 3 & 2 & 4 & 3 & 3 & 4 & 2 & 5
\end{bmatrix}$$

TABLE 7. Cartan invariants of 1-block of $SL(5,2)$, $p = 2$

Here their conjecture states:

$$\textit{Local Conjecture:} \quad \frac{\dim \mathcal{B}}{l(\mathcal{B})\,|D|} \le \sum_{\varphi \in \mathrm{Irr}_p(\mathcal{B})} \varphi(1)^2.$$

While the local analogue of the Weak Global Conjecture holds in many special cases, it can fail for symmetric groups in characteristic 2. Substantial evidence for the Local Conjecture is presented, based in part on a comparison with the Cartan matrix of the block:

PROPOSITION. *Let $\mathcal{B}$ be a p-block of a finite group with defect group $D$, and let $C$ be the Cartan matrix of $\mathcal{B}$. Then:*

(a) *In all cases,*

$$\frac{\dim \mathcal{B}}{\operatorname{Tr} C} \le \sum_{\varphi \in \mathrm{Irr}_p(\mathcal{B})} \varphi(1)^2, \quad \textit{with equality iff } l(\mathcal{B}) = 1.$$

(b) *If $\operatorname{Tr} C \le l(\mathcal{B})\,|D|$, then the Local Conjecture holds for $\mathcal{B}$.*

Using this, it is shown that the Local Conjecture holds for blocks of tame representation type. The authors also suggest that the hypothesis in (b) is always true, though it is not at all obvious in view of the example in Landrock [295] where the first Cartan invariant exceeds $|D|$.

For groups of Lie type in the defining characteristic, the conjectures are plausible in view of the known generic behavior. Blocks of defect $> 0$ have a Sylow $p$-subgroup as defect group, and much is known (modulo Lusztig's Conjecture) about the generic behavior or independence of $p$ of various data involved in the conjectures. For small $p$ it is instructive to check directly cases such as $SL(5,2)$.

# CHAPTER 12

# Extensions of Simple Modules

In this and the next chapter we take a closer look at the structure of projective modules. This is much harder to get at than the numerical information about Cartan invariants and dimensions. But it plays a central role in sorting out the structure of arbitrary modules in the category of $KG$-modules. What we do here is a starting point for the broader study of cohomology. Initially we focus on $\mathrm{Ext}^1$ for pairs of simple modules (abbreviated by Ext), which in a sense is just the tip of the iceberg. But this aspect of cohomology admits a concrete interpretation as well as a fairly direct connection with the algebraic group theory. It includes as a special case the study of $H^1(G, L(\mu)) = \mathrm{Ext}_G^1(K, L(\mu))$, which will be placed in the more general context of cohomology in Chapter 14.

In the case of an arbitrary finite group $G$, there is no guaranteed method to determine Ext groups. For groups of Lie type we formulate in 12.1 the problem of comparison with algebraic groups, where more systematic methods are available. This is illustrated for $\mathrm{SL}(2, p)$ in 12.2. Then we examine what is currently known (perhaps modulo the determination of standard character data) about Ext for the group $\mathbf{G}$ (12.4).

One decisive comparison can be worked out for all $p$ by elementary means (12.5), but in order to go further it seems necessary to impose some lower bound on $p$. Early work of Andersen is outlined in 12.7, followed by a discussion of the more comprehensive recent work by Bendel, Nakano, and Pillen (12.8–12.10).

The chapter concludes with a look at special cases treated in the literature (12.11) and a discussion of criteria for semisimplicity of $KG$-modules (12.12).

## 12.1. The Extension Problem

For a finite group $G = \mathbf{G}^F$ over a field of $p^r$ elements, a basic problem involving the Ext functor can be formulated as follows. Given two simple modules $L(\lambda)$ and $L(\mu)$ with $\lambda, \mu \in X_r$, what are the possible $KG$-modules $E$ which fit into a short exact sequence

(1) $$0 \to L(\mu) \to E \to L(\lambda) \to 0?$$

There is a natural equivalence relation on such sequences, yielding the collection $\mathrm{Ext}_G(L(\lambda), L(\mu))$ of equivalence classes. (Here we abbreviate $\mathrm{Ext}_{KG}$ by $\mathrm{Ext}_G$.) This set has a standard group structure, written additively, with the class of the split extension playing the role of zero. It even has the structure of a vector space over $K$, whose dimension is finite but usually quite difficult to compute using just finite group methods.

In principle one can study nonsplit extensions of type (1) by embedding the module $E$ into the injective hull $U_r(\mu)$ of $L(\mu)$ or by realizing $E$ as a quotient of the projective cover $U_r(\lambda)$ of $L(\lambda)$. Denoting by $\mathrm{Rad}\, U_r(\lambda)$ the unique maximal submodule of $U_r(\lambda)$, one has for example the standard isomorphism

$$\mathrm{Ext}_G(L(\lambda), L(\mu)) \cong \mathrm{Hom}_G(\mathrm{Rad}\, U_r(\lambda), L(\mu)).$$

But it is extremely difficult to determine the precise structure of PIMs in our situation. (This question will be explored further in Chapter 13.)

For the algebraic group $\mathbf{G}$, there are closely parallel extension problems in the categories of rational $\mathbf{G}$-modules and $\mathbf{G}_r$-modules. Here some systematic methods are available to study Ext groups: see 12.4 below. As in previous chapters, a fundamental question then arises: Can $\mathrm{Ext}_G(L(\lambda), L(\mu))$ be determined from a knowledge of various Ext groups for $\mathbf{G}$ and $\mathbf{G}_r$ together with standard character data? This is of course part of a much broader investigation of the relationship between cohomology of $G$ over $K$ and (rational) cohomology of $\mathbf{G}$, to be discussed further in Chapter 14.

One preliminary observation can be made for both finite and algebraic groups:

PROPOSITION. *If $\lambda, \mu \in X^+$, then*

$$\mathrm{Ext}_{\mathbf{G}}(L(\lambda), L(\mu)) \cong \mathrm{Ext}_{\mathbf{G}}(L(\mu^*), L(\lambda^*)) \cong \mathrm{Ext}_{\mathbf{G}}(L(\mu), L(\lambda)).$$

*The same isomorphisms hold for $\lambda, \mu \in X_r$ when $\mathbf{G}$ is replaced by a finite subgroup $G = \mathbf{G}^F$ over a field of $p^r$ elements.*

PROOF. Recall from 2.2 that $L(\lambda)^* \cong L(\lambda^*)$, where $\lambda^* := -w_\circ\lambda$. The first isomorphism is then a standard fact about the Ext functor, obtained by dualizing the short exact sequence (1) above for either $\mathbf{G}$ or $G$. The argument in [**RAGS**, II.2.12] giving the second isomorphism adapts immediately to $G$, which is stable under the anti-automorphism $\tau$ (2.4) used to define contravariant forms. $\qquad\square$

It should be emphasized that Ext (or higher cohomology) for finite groups of Lie type has been studied in a variety of ways, often using ideas from group cohomology or topology (see Adem–Milgram [**1**]). For the most part our discussion is limited to situations in which algebraic group techniques play a major role. But for cases like $p = 2$ a broader spectrum of methods is likely to be needed.

## 12.2. Example: SL$(2, p)$

The rank 1 case already shows the need for an indirect approach to comparison of Ext groups for $\mathbf{G}$ and $G$. When $\mathbf{G} = \mathrm{SL}(2, K)$, the fact that all $p$-restricted weights lie in the closure of a single $p$-alcove shows that there are no nonsplit extensions involving pairs of such weights (see Proposition 12.4 below). So we have to look further for possible $G$-extensions.

The discussion in 9.8 makes it easy to read off the precise structure of PIMs (hence Ext groups) for $G = \mathrm{SL}(2, p)$. We interpret this in the $\mathbf{G}$-module setting of Chapter 10, to see what might be attempted in higher ranks. Note first that in the degenerate case $p = 2$ (when $|G| = 6$), the projective cover of the trivial module just gives a nonsplit extension of this module by itself.

Now assume $p$ is odd and denote weights as usual by integers. The Steinberg module $L(p - 1) = U(p - 1)$ admits no nonsplit extensions. If $0 \le \lambda < p - 1$, we set $\lambda^\circ = (p - 1) - \lambda > 0$ and extract $U(\lambda)$ as a summand of $L(\lambda^\circ) \otimes \mathrm{St}$. This coincides with the corresponding PIM $Q(\lambda)$ for $\mathbf{G}_1\mathbf{T}$ (lifted to an indecomposable $\mathbf{G}$-module). Here $Q(\lambda)$ is uniserial of length 3, with composition factors

$$L(\lambda), \quad L(p - 2 - \lambda) \otimes L(1)^{[1]}, \quad L(\lambda).$$

In particular, $\mathrm{Ext}_{\mathbf{G}}(L(\lambda), L(2p-2-\lambda)) \neq 0$.

On restriction to $G$ the middle factor just becomes the ordinary tensor product $L(p-2-\lambda) \otimes L(1)$. By the Linkage Principle this splits for $\mathbf{G}$ and hence for $G$ into a direct sum $L(p-1-\lambda) \oplus L(p-3-\lambda)$ if $\lambda < p-2$. In this way we find various nonzero Ext groups. There are a couple of degenerate cases. When $\lambda = 0$, a copy of St splits off from $Q(0)$; thus $U(0)$ is uniserial with composition factors $L(0), L(p-3), L(0)$. This yields $\dim \mathrm{Ext}_G(L(0), L(p-3)) = 1$. When $p = 3$ this produces a "self-extension" of the trivial module by itself, as occurred for $p = 2$. At the other extreme, when $\lambda = p-2$, $Q(\lambda) = U(\lambda)$ is uniserial with composition factors $L(\lambda), L(p-1-\lambda), L(\lambda)$. This yields $\dim \mathrm{Ext}_G(L(p-2), L(1)) = 1$. Again when $p = 3$, we get a self-extension. To summarize:

PROPOSITION. *Let* $\mathbf{G} = \mathrm{SL}(2, K)$ *and* $G = \mathrm{SL}(2, p)$. *Then all cases in which* $\mathrm{Ext}_G(L(\lambda), L(\mu)) \neq 0$ *are given as follows.*

   (a) *If* $p = 2$, *then* $\dim \mathrm{Ext}_G(L(0), L(0)) = 1$.

   (b) *If* $p$ *is odd, then* $\mathrm{Ext}_G(L(\lambda), L(p-1-\lambda))$ *and* $\mathrm{Ext}_G(L(\lambda), L(p-3-\lambda))$ *are 1-dimensional for* $0 < \lambda < p-2$, *while* $\dim \mathrm{Ext}_G(L(0), L(p-3)) = 1$ *and* $\dim \mathrm{Ext}_G(L(p-2), L(1)) = 1$. $\quad\square$

## 12.3. The Optimal Situation

The example of $\mathrm{SL}(2, p)$ illustrates one way in which we might expect knowledge of $\mathrm{Ext}_{\mathbf{G}}$ to contribute systematically to the study of $\mathrm{Ext}_G$. An obvious question is whether anything more complicated can occur, apart from this indirect procedure or (in higher ranks) the more direct restriction to $G$ of nonsplit $\mathbf{G}$-extensions involving two restricted weights.

Before we attempt to formulate in precise language what is currently known about the extension problem, it may be useful to sketch the best possible scenario and indicate potential difficulties. Suppose the PIMs $Q_r(\lambda)$ for $\mathbf{G}_r$ lift to $\mathbf{G}$-modules (known when $p \geq 2(h-1)$ and perhaps true for all $p$). The analysis in Chapter 10 shows that "most" of these modules restrict to PIMs for $G$. Now given any nonsplit $G$-extension $0 \to L(\mu) \to E \to L(\lambda) \to 0$, we can embed $E$ in $U_r(\mu)$. Suppose moreover $U_r(\mu) = Q_r(\mu)$.

Possibly $E$ is a $\mathbf{G}$-submodule of $Q_r(\mu)$, in which case we get a direct connection with $\mathrm{Ext}_{\mathbf{G}}(L(\lambda), L(\mu))$. Otherwise we might hope to embed $E$ into a $\mathbf{G}$-submodule of $Q_r(\mu)$ having just two composition factors $L(\mu)$ and $L(\nu) = L(\nu_0 + p^r \nu_1)$ with $\nu_1 \neq 0$. As in the rank 1 case, $L(\lambda)$ would be recovered here as a $G$-submodule of the tensor product $L(\nu_0) \otimes L(\nu_1)$ (with $\tilde{\nu}_1$ replacing $\nu_1$ if $G$ is twisted). So the given $G$-extension would be indirectly recovered from the study of $\mathrm{Ext}_{\mathbf{G}}(L(\nu), L(\mu))$. Necessarily $\nu \leq \mu + (p^r - 1)\rho$, which limits the range of weights for $\mathbf{G}$ one needs to consider to the $p^r$-bounded ones.

There are some obvious pitfalls in this scenario. Sometimes $U_r(\mu)$ is smaller than $Q_r(\mu)$, making comparisons more subtle. More seriously, there is no obvious reason why the $KG$-module $E$ should live in a $\mathbf{G}$-submodule of $Q_r(\mu)$ with just two composition factors as above. Moreover, if one starts with a $\mathbf{G}$-extension and then restricts to $G$, some of the ordinary tensor products encountered might not be semisimple. Only in "generic" cases can one expect things to go as smoothly as they do for $\mathrm{SL}(2, p)$.

## 12.4. Extensions for Algebraic Groups

Here we outline briefly what is known about the extension problem for **G**, referring to [**RAGS**] for details and references. Modulo the determination of standard character data, there turns out to be considerable uniformity in the behavior of Ext**G** (at least in "generic" cases, for $p$ not too small).

For arbitrary $\lambda, \mu \in X^+$, how can Ext**G**$(L(\lambda), L(\mu))$ be determined? This involves finding all rational **G**-modules $E$ belonging to nonsplit short exact sequences $0 \to L(\mu) \to E \to L(\lambda) \to 0$. As in the case of finite groups, an indecomposable module $E$ of this type must embed into the injective hull of $L(\mu)$. But this injective hull is infinite dimensional. Even when it can be realized as the direct limit of (lifted) PIMs for Frobenius kernels **G**$_r$ (as in 10.5), its internal structure will be hard to study directly.

To bypass this difficulty we can exploit some special features of the algebraic group situation. For example, the Linkage Principle (3.6) implies that nonsplit extensions cannot exist unless $\lambda$ and $\mu$ are linked by the affine Weyl group $W_p$.

Given a nonsplit extension as above, it is easy to see that either $\lambda > \mu$ or $\lambda < \mu$. Indeed, if $\lambda = \mu$ then linearly independent vectors of weight $\lambda$ generate submodules isomorphic to $L(\lambda)$ having $E$ as direct sum. If the two weights are noncomparable, then each must be maximal among the weights of $E$. Therefore a nonzero vector of weight $\lambda$ is a maximal vector. It generates a submodule of $E$ having a composition factor $L(\lambda)$, with all its weights $\leq \lambda$. This excludes $\mu$, so we just get a copy of $L(\lambda)$ as a submodule, splitting the exact sequence.

By invoking some standard but nontrivial theory, we can go further. Given a nonsplit extension with $\lambda > \mu$, the universal highest weight property of Weyl modules in 3.3 (which depends on Kempf's Vanishing Theorem) implies that $V(\lambda)$ must have $E$ as a quotient. On the other hand, if $\lambda < \mu$ then we can argue that $E \hookrightarrow H^0(\mu)$ and $\lambda < \mu$. For this we invoke Frobenius reciprocity in the category of rational **G**-modules: [**RAGS**, I.3.4], which insures that a **B**-homomorphism from the restriction of a **G**-module $N$ into a **B**-module $M$ induces naturally a **G**-homomorphism $N \to \text{Ind}_{\mathbf{B}}^{\mathbf{G}} M$. In particular, the hypothesis $\lambda < \mu$ allows the 1-dimensional **B**-module afforded by $\mu$ to play the role of $M$, while $E$ plays the role of $N$. This yields an embedding $E \hookrightarrow H^0(\mu)$.

We can summarize this discussion as follows, denoting by $\text{Rad}\, V(\lambda)$ the unique maximal submodule of $V(\lambda)$. (See [**RAGS**, II.2.14, II.5.6–5.7, II.6.17].)

PROPOSITION. *Let* $\lambda, \mu \in X^+$.

(a) Ext**G**$(L(\lambda), L(\mu)) = 0$ *unless* $\mu = w \cdot \lambda$ *for some* $w \neq 1$ *in the affine Weyl group* $W_p$.

(b) *If* Ext**G**$(L(\lambda), L(\mu)) \neq 0$, *then either* $\lambda > \mu$ *or* $\lambda < \mu$. *In particular,* Ext**G**$(L(\lambda), L(\lambda)) = 0$.

(c) *If* $\lambda > \mu$, *then every nonsplit extension of* $L(\lambda)$ *by* $L(\mu)$ *can be realized as a quotient of* $V(\lambda)$, *with*

$$\text{Ext}_{\mathbf{G}}(L(\lambda), L(\mu)) \cong \text{Hom}_{\mathbf{G}}(\text{Rad}\, V(\lambda), L(\mu)).$$

(d) *If* $\lambda < \mu$, *then any nonsplit extension can be embedded in* $H^0(\mu)$.  □

It is also known that the validity of Lusztig's Conjecture would carry with it a combinatorial description of the dimension of Ext, for pairs of weights in the alcoves covered by the conjecture, in terms of coefficients of Kazhdan–Lusztig polynomials for $W_p$. (See the discussion in 12.6 below.) So one can hope for effective computations based just on the study of Weyl modules when $p$ is not too small.

Another crucial feature of the algebraic group setting (not available for the finite groups) is the fact that each Frobenius kernel $\mathbf{G}_r$ is a *normal* subgroup scheme of $\mathbf{G}$. Moreover, $\mathbf{G}$ acts naturally on the cohomology spaces for $\mathbf{G}_r$ and $\mathbf{G}/\mathbf{G}_r$ is scheme-theoretically isomorphic to $\mathbf{G}$. In this setting it is natural to use spectral sequence techniques. A number of authors have studied extensions in this spirit, leading in particular to inductive computational methods: see [**RAGS**, II.10.17] and Andersen [**13**], Cline–Parshall–Scott [**114**], Cline–Parshall–Scott–van der Kallen [**116**], Donkin [**140**], Bendel–Nakano–Pillen [**40**].

## 12.5. Injectivity Theorem

We already saw in the case $G = \mathrm{SL}(2,p)$ that $\mathrm{Ext}_G(L(\lambda), L(\mu))$ can be nonzero when the corresponding Ext group for $\mathbf{G}$ is zero. But in all cases there is a natural inclusion of the latter group into $\mathrm{Ext}_G$.

THEOREM. *Let $G = \mathbf{G}^F$ over a field of $p^r$ elements. If $\lambda, \mu \in X_r$, then the restriction map $\mathrm{Ext}_\mathbf{G}(L(\lambda), L(\mu)) \to \mathrm{Ext}_G(L(\lambda), L(\mu))$ is injective.*

PROOF. This is a consequence of more general theorems in Cline–Parshall–Scott–van der Kallen [**116**, 7.4], where use is made of earlier work by Wong [**437**, 2D]. A short version suggested by Humphreys [**229**] goes as follows: Given a nonsplit $\mathbf{G}$-extension $0 \to L(\mu) \to E \to L(\lambda) \to 0$, we have to show that $E$ fails to split as a $KG$-module. Thanks to Proposition 12.4, we may assume that $\lambda > \mu$ and thus that $E$ is a quotient of $V(\lambda)$. Theorem 5.9 shows that $L(\lambda)$ is the only simple quotient of the $KG$-module $V(\lambda)$. Thus $E$ fails to split on restriction to $G$, as required.

Implicit in this method of argument is the construction of Weyl modules (beginning in characteristic 0), combined with Kempf's Vanishing Theorem to insure that $V(\lambda)$ is a universal highest weight module for $\mathbf{G}$. A proof in the characteristic $p$ framework has been given by Andersen [**14**, 2.7]. He assumes at first that $G$ is a Chevalley group, then adapts everything to twisted groups. (Later work of Bendel–Nakano–Pillen [**38**] also recovers this result.)

Since $\mathrm{Ext}_\mathbf{G}(L(\lambda), L(\mu)) \cong \mathrm{Ext}_\mathbf{G}(L(\mu), L(\lambda))$ and similarly for $G$, while moreover $\mathrm{Ext}_\mathbf{G}(L(\lambda), L(\lambda)) = 0$, we may as well assume that $\lambda < \mu$. (This uses elementary parts of Propositions 12.1 and 12.4.)

Given a nonsplit $\mathbf{G}$-extension $0 \to L(\mu) \to E \to L(\lambda) \to 0$, we have to show that $\mathrm{Hom}_G(L(\lambda), E) = 0$. This is difficult to see directly, so first we embed the $\mathbf{G}$-module $E$ into $H^0(\mu)$ as in 12.4 above. In turn, we can exploit the Steinberg module $\mathrm{St}_r = L(\sigma_r)$ (where $\sigma_r = (p^r - 1)\rho$): Embed $H^0(\mu)$ into the larger $\mathbf{G}$-module $\mathrm{St}_r \otimes H^0(\sigma_r - \mu^*)$, using Frobenius reciprocity and the "tensor identity" in conjunction with the fact that the 1-dimensional $\mathbf{B}^-$-module $\mu$ occurs as a $\mathbf{B}^-$-submodule of $\sigma_r \otimes H^0(\sigma_r - \mu^*)$. (The weight $\mu$ is the minimal weight of this tensor product.)

To show that $\mathrm{Hom}_G(L(\lambda), E) = 0$, it therefore suffices to prove that

$$0 = \mathrm{Hom}_G(L(\lambda), \mathrm{St}_r \otimes H^0(\sigma_r - \mu^*)) \cong \mathrm{Hom}_G(\mathrm{St}_r, L(\lambda^*) \otimes H^0(\sigma_r - \mu^*)).$$

For this, look at an arbitrary $\mathbf{G}$-composition factor $L(\nu)$ of this latter tensor product. Obviously

$$(*) \quad \nu \le \lambda^* + \sigma_r - \mu^* < \sigma_r,$$

since $\lambda^* < \mu^*$. In particular, $\nu \neq \sigma_r$. But if $\nu$ fails to be $p^r$-restricted, we have to argue further that $\mathrm{St}_r$ cannot occur in the restriction to $G$ of $L(\nu) \cong L(\nu_0) \otimes L(\nu_1)^{[r]}$. Here $\nu_0 \in X_r$ and $0 \neq \nu_1 \in X^+$. In particular, $\langle \nu_1, \alpha_0^\vee \rangle > 0$ (this integer being a "weighted" sum of the coordinates of $\nu_1$ relative to fundamental weights).

Assume first that $G$ is a Chevalley group. We know that as a $KG$-module, $L(\nu) \cong L(\nu_0) \otimes L(\nu_1)$. Thanks to (∗), any $\mathbf{G}$-composition factor $L(\theta)$ of this tensor product satisfies

$$\langle \theta, \alpha_0^\vee \rangle \leq \langle \nu_0 + \nu_1, \alpha_0^\vee \rangle < \langle \nu, \alpha_0^\vee \rangle \leq \langle \sigma_r, \alpha_0^\vee \rangle.$$

If $\theta \in X_r$, it follows that $L(\theta) \not\cong \mathrm{St}_r$ and we are done. Otherwise repeat the argument, which eventually terminates since the successive inner products with $\alpha_0^\vee$ decrease.

If $G$ is a twisted group, the argument is almost identical: the $KG$-module $L(\nu)$ becomes $L(\nu_0) \otimes L(\widetilde{\nu}_1)$, but the chain of inequalities is unchanged because $\langle \widetilde{\nu}_1, \alpha_0^\vee \rangle = \langle \nu_1, \alpha_0^\vee \rangle$. □

Andersen goes on in [14, 2.8] to derive sufficient conditions for the restriction map $\mathrm{Ext}_{\mathbf{G}}(L(\lambda), L(\mu)) \to \mathrm{Ext}_G(L(\lambda), L(\mu))$ to be an isomorphism. For this he has to assume that $p \geq 3(h-1)$, while excluding the special case $G = \mathrm{SL}(2, p)$. Then he gets an isomorphism whenever $\lambda + \mu$ satisfies the condition: $\langle \lambda + \mu, \alpha_0^\vee \rangle < p^r - p^{r-1} - 1$. More refined estimates occur in later literature, based for example on work of Cline–Parshall–Scott–van der Kallen [116] and Avrunin [27] to be discussed in Chapter 14. See especially Jantzen [258], McNinch [317], and the series of papers by Bendel–Nakano–Pillen discussed in 12.8 below.

REMARK. Note that for types $A_\ell$ $(\ell \geq 2), D_\ell, E_6$, the theorem provides a common contribution to Ext for both a Chevalley group of this type and its twisted analogue. Even so, Andersen's computation of Ext groups for $\mathrm{SL}(3, p)$ and $\mathrm{SU}(3, p)$ in [14] shows that these are not always isomorphic for $\lambda, \mu \in X_1$. The differences arise from more subtle contributions involving $\mathrm{Ext}_{\mathbf{G}}$ for not necessarily restricted weights.

## 12.6. Dimensions of Ext Groups

Theorem 12.5 provides a lower bound for the dimension of $\mathrm{Ext}_G(L(\lambda), L(\mu))$ which is computable in special cases. For this one needs a formal consequence of the validity of Lusztig's Conjecture (3.11), mentioned above in 12.4: While the values of Kazhdan–Lusztig polynomials at 1 predict formal characters of simple modules, the individual coefficients of the polynomials predict (in a somewhat complicated way described in [RAGS, II.C]) dimensions of related $\mathrm{Ext}^i$ groups. Consider the case of $\mathrm{Ext}_{\mathbf{G}}(L(\lambda), L(\mu))$, when $\lambda$ and $\mu$ are linked $p$-regular weights lying in $p$-alcoves labelled by two elements $w, y$ of $W_p$. If $\lambda > \mu$, the degree of the associated polynomial $P_{y,w}$ is at most $(\ell(w) - \ell(y) - 1)/2$, and the coefficient of this power of the indeterminate then measures the dimension of the Ext group. (See Scott [362].)

Since the Lusztig Conjecture is known to hold whenever $p$ is sufficiently large (but with an unknown bound on $p$), one can draw some conclusions about the size of Ext groups for $G$. This is done by Scott [363], using computations by him and his students. Specifically, they study the groups $G = \mathrm{PSL}(6, q)$. Provided Lusztig's Conjecture is true for $p$, it follows that $\dim H^1(G, L(\mu)) \geq 3$ for a weight

$\mu$ which they specify explicitly. Similarly, a pair of weights is specified for which $\dim \operatorname{Ext}_G(L(\lambda), L(\mu)) \geq 4$. (It is hoped that computer calculations may soon be able to verify Lusztig's Conjecture in this case for a prime as small as 7, thereby providing more explicit examples.)

These arguments give the first known examples where $H^1$ for a finite group has dimension $> 2$, showing that earlier expectations of Guralnick for a bound as small as 2 must be revised. At this point it is unclear whether any universal upper bound on the dimension exists.

## 12.7. The Generic Case

Andersen [**14**, §3] goes on to study more general situations. If $G$ is a Chevalley group and $\lambda, \mu \in X_r$, there is a natural map

$$(1) \quad \bigoplus_{\nu} \operatorname{Hom}_{\mathbf{G}}(L(\lambda), L(\nu_0) \otimes L(\nu_1)) \otimes \operatorname{Ext}_{\mathbf{G}}(L(\nu), L(\mu)) \to \operatorname{Ext}_G(L(\lambda), L(\mu)),$$

with the sum taken over all $p^r$-bounded weights $\nu = \nu_0 + p^r \nu_1$. Recall from 10.4 that a weight $\lambda \in X^+$ is called $p^r$-*bounded* if all weights $\mu$ of $L(\lambda)$ satisfy $\langle \mu, \alpha_0^\vee \rangle < 2p^r(h-1)$. (If $G$ is a twisted group, one just replaces $\nu_1$ by $\widetilde{\nu}_1$ in (1).)

In a "generic" situation, where $\lambda$ and $\mu$ are $p$-regular weights not too close to the walls of their $p$-alcoves, Andersen shows:

THEOREM. *Let $G = \mathbf{G}^F$ over a field of order $p^r$. Assume $\lambda, \mu \in X_r$ satisfy $\langle \lambda, \alpha_0^\vee \rangle \leq \langle \mu, \alpha_0^\vee \rangle$. In addition, suppose $\lambda$ (resp. $\mu$) has distance at least $h - 1$ (resp. $2(h-1)$) to any wall of the $p$-alcove it lies in. Then the map in (1) is an isomorphism.*

Here the requirement that $\lambda$ have distance at least $h - 1$ to an $\alpha$-wall of its alcove can be expressed combinatorially: if $p$ divides $\langle \lambda + \rho, \alpha^\vee \rangle + n$ for some $n \in \mathbb{Z}$, then $|n| \geq h - 1$.

The hypotheses of the theorem insure that $p$ is not too small. The proof of the theorem then makes effective use of the $\mathbf{G}$-structure on various PIMs $Q_r(\nu)$ for $\mathbf{G}_r$. Andersen remarks that if one assumes $p \geq 3(h-1)$, then the theorem actually holds for all $\lambda \in X_r$. (A closely related formulation for Chevalley groups, in the case $r \geq 2$, is given by Bendel–Nakano–Pillen [**38**, 3.3].)

## 12.8. Truncated Module Categories and Cohomology

The results of Andersen just sketched give a good sense of how the Ext functors for $\mathbf{G}$ and $G$ interact when $p$ is not too small. But it remains difficult to make detailed comparisons when $p$ is small. Moreover, there is little direct impact on the study of cohomology in higher degrees.

In an ongoing series of papers [**37**]–[**43**], Bendel, Nakano, and Pillen have refined these Ext results in a unified setting which encompasses higher cohomology and permits further explicit computations. We outline first their basic philosophy, then discuss some specific results in the following two sections.

Their most computable comparisons of $\operatorname{Ext}_G$ with $\operatorname{Ext}_{\mathbf{G}}$ still require that $p \geq 3(h-1)$. But some of their formulations are valid for all $p$. While they initially treat just Chevalley groups in [**39**, **38**], the results and proofs carry over to twisted groups as well with only minor modifications [**41**]. In [**43**] they go further by allowing $p$ to be arbitrary; here the results depend instead on bounding the power $p^r$ below (by something on the order of $h^2$).

As in earlier work of Andersen, Donkin, and Jantzen, essential use is made of various truncated categories of **G**-modules. These are thought of as approximations to categories of modules for $\mathbf{G}_1$ and $G$. In a truncated category the weights of all **G**-modules are suitably bounded above, which results in "highest weight categories" having enough projectives. This permits the use of Grothendieck spectral sequences to connect such a category with the related module categories.

There are still significant homological differences between these module categories for **G** and $G$, since the former has finite global dimension and the latter does not. So one gets reasonable comparisons of cohomology only in a fixed range of degrees (and for $p$ large enough): the precise upper bound imposed on weights increases as the degree increases. Here we just look at $H^1$.

Their work depends heavily on previous comparisons of cohomology for $\mathbf{G}, \mathbf{G}_r$, and $G$ in their paper [**37**] (which are carried further in [**40**]). This involves at times the full category of rational **G**-modules, including the infinite dimensional injectives.

Consider the category $\mathcal{C}$ consisting of (necessarily finite dimensional) **G**-modules all of whose composition factors have highest weights $\lambda$ satisfying $\langle \lambda + \rho, \alpha_0^\vee \rangle < 2p^r(h-1)$. (Note that this is slightly different from the definition above of $p^r$-bounded weight.) Denote by $\mathcal{F}$ the truncation functor which assigns to a **G**-module $M$ its largest submodule belonging to $\mathcal{C}$. Then define the functor $\mathcal{G} = \mathcal{F} \circ \mathrm{Ind}_G^{\mathbf{G}}$ from $KG$-modules to **G**-modules. While the induction functor usually produces infinite dimensional **G**-modules, the composite functor leads back to the finite dimensional setting.

## 12.9. Comparison Theorems

Now we can formulate some of the specific results obtained by Bendel–Nakano–Pillen in the setting just outlined. It turns out that the value of the functor $\mathcal{G}$ at the trivial module $K$ is crucial. Using a considerable amount of spectral sequence technology and related homological algebra, the following theorems are obtained in [**38**, 2.2, 2.5].

THEOREM. *Let $G$ be a Chevalley group over a field of order $p^r$. Define $\Gamma := \{\nu \in X^+ \mid \langle \nu, \alpha_0^\vee \rangle \leq h-1\}$. Then for all $\lambda, \mu \in X_r$:*

(a) $\mathrm{Ext}_G(L(\lambda), L(\mu)) \cong \mathrm{Ext}_{\mathbf{G}}(L(\lambda), L(\mu) \otimes \mathcal{G}(K))$.

(b) *Assume $p \geq 3(h-1)$ and set $\Gamma := \{\nu \in X^+ \mid \langle \nu, \alpha_0^\vee \rangle < h\}$. Then*

$$\mathrm{Ext}_G(L(\lambda), L(\mu)) \cong \bigoplus_{\nu \in \Gamma} \mathrm{Ext}_{\mathbf{G}}(L(\lambda) \otimes L(\nu)^{[r]}, L(\mu) \otimes L(\nu)).$$

The more explicit formula in part (b) results from the fact that $\mathcal{G}(K)$ is semisimple when $p \geq 3(h-1)$ and can then be explicitly described [**37**, 7.4]. But for smaller $p$ the situation is sometimes more complicated. Note that it is easy to recover from (b) the earlier results for $G = \mathrm{SL}(2,p)$ when $p$ is odd (12.2). In this case $\Gamma = \{0,1\}$.

The isomorphism in (b) implies that Ext for $G$ can be computed from a knowledge of certain Ext groups for **G**. In [**39**, §3] a more precise version of this comparison is worked out in case $r \geq 2$: for $\lambda, \mu \in X_r$, there is an isomorphism between $\mathrm{Ext}_G(L(\lambda), L(\mu))$ and $\mathrm{Ext}_{\mathbf{G}}(L(\lambda'), L(\mu'))$ for a pair of weights $\lambda', \mu' \in X_r$. This gives, for $p \geq 3(h-1)$ and $r \geq 2$, a satisfying way to construct all $G$-extensions from a knowledge of **G**-extensions. (Here and elsewhere, the theory becomes somewhat smoother when $r \geq 2$.)

REMARK. As the authors point out in [**38**, 2.5], there are intriguing similarities between their Ext formula and the formula of Chastkofsky and Jantzen (see 10.11 above) relating PIMs for $\mathbf{G}_r$ and $G$.

## 12.10. Self-Extensions

At another extreme from the study of generic cases, one might ask:

QUESTION. *For a finite group $G = \mathbf{G}^F$ over a field of $p^r$ elements, which simple modules $L(\lambda)$ can extend themselves nontrivially?*

When $\operatorname{Ext}_G(L(\lambda), L(\lambda)) \neq 0$, we say that $L(\lambda)$ has a **self-extension**. Self-extensions do not exist for the algebraic group $\mathbf{G}$. But they do sometimes occur for the associated Frobenius kernel $\mathbf{G}_1$, where Andersen [**13**, §4] showed that self-extensions exist just when $\mathbf{G}$ (assumed simple) is of type $C_\ell$ and $p = 2$. (See also [**RAGS**, II.12.9] and Bendel–Nakano–Pillen [**38**, §4]). Here we make the natural convention that the group of rank 1 has type $C_1$ rather than $A_1$. The unusual feature of the root system $C_\ell$ which leads to the existence of self-extensions is that its long simple root is twice a weight.

For the finite group $G$, self-extensions have already been seen in 12.2 to exist for $G = \mathrm{SL}(2, p)$ when $\lambda = p - 2$. With the example of Frobenius kernels in mind, Humphreys [**229**] constructed explicit examples of self-extensions (for $p$ odd) when $G = \mathrm{Sp}(4, p)$ and suggested that here too the groups of type $C_\ell$ might be special. This has since been confirmed for large enough $p$ in the work of Bendel–Nakano–Pillen [**38**, **41**], with additional constructions of self-extensions then given by Pillen [**338**].

There is also an interesting related result in [**39**, 3.3]: If $p \geq 3(h - 1)$, while $r \geq 2$, the maximum dimension of $\operatorname{Ext}_G(L(\lambda), L(\mu))$ for $\lambda, \mu \in X_r$ agrees with the similar maximum dimension for $\mathbf{G}$ *unless* $r = 2$ and $\mathbf{G}$ is of type $C_\ell$; then $\operatorname{Ext}_G$ can be twice as large as $\operatorname{Ext}_\mathbf{G}$. Moreover, the $p$-adic coefficients of weights involved in the latter case are just the candidates for self-extensions over $\mathbb{F}_p$.

While the results on self-extensions are not yet definitive, the picture is becoming reasonably clear. The examples constructed in [**229**] and [**338**] follow roughly the pattern for $\mathrm{SL}(2, p)$ discussed in 12.2 above: Find a nonsplit $\mathbf{G}$-extension involving a restricted weight and a nonrestricted $p^r$-bounded weight, then extract a nonsplit self-extension for $G$.

Here we just formulate the general result obtained from the combined study of Chevalley groups and twisted groups. For full details see [**38**, 3.4, 4.2] and [**41**, 4.1, 4.2]. (Earlier, Cabanes [**76**] and Völklein [**429**] proved weaker results in this direction.) At this point the lower bound on $p$ is essential for the proofs, but the known examples and the parallel case of Frobenius kernels encourage the expectation that self-extensions will exist only in type C and when $r = 1$.

THEOREM. *Let $G = \mathbf{G}^F$ over a field of $p^r$ elements. Assume that $p \geq 3(h - 1)$.*

(a) *If $r \geq 2$, $G$ admits no self-extensions.*

(b) *If $r = 1$, the only possible self-extensions occur when $G$ is a Chevalley group of type $C_\ell$ with $\ell \geq 1$ and the weight $\lambda \in X_1$ satisfies the following condition relative to the unique long simple root $\alpha_\ell$:*

$$\langle \lambda, \alpha_\ell^\vee \rangle = \frac{p - 2 - c}{2}, \text{ with } c \text{ odd and } 0 < |c| \leq h - 1 = 2\ell - 1.$$

The method of Bendel–Nakano–Pillen involves a transition from finite groups to Frobenius kernels. First the finite group cohomology is related to algebraic group cohomology. Then a Lyndon–Hochschild–Serre spectral sequence permits passage to $\mathbf{G}_r$, where they use and extend Andersen's work on cohomology of line bundles on the flag variety to prove their results on self-extensions.

REMARK. The self-extension question comes up not just in the modular setting, but in the comparison of ordinary and modular characters. Here are two instances:

(1) As explained by Landrock [**296**, I, Cor. 16.12], Brandt [**59**] exhibits a lower bound for the number of "ordinary" irreducible characters in a block of a finite group $G$. We describe this in ad hoc terms, since some of the standard notation conflicts with ours. It will be recalled in Chapter 16 below that the irreducible characters of $G$ over $\mathbb{C}$ distribute naturally into the $p$-blocks. If more than two of them lie in a given block containing $t$ simple modules $L_1, \ldots, L_t$, then Brandt shows that the number of these characters must in fact be at least

$$1 + t + \sum_{i=1}^{t} \dim \operatorname{Ext}_G(L_i, L_i).$$

(2) For groups of Lie type, Tiep and Zalesskii [**412**] show that certain simple $KG$-modules can be lifted to $\mathbb{C}G$-modules only if they admit self-extensions. They construct for groups $\operatorname{Sp}(2\ell, p)$ with $\ell > 2$ examples of the type found independently by Pillen. (See 16.14 below.)

## 12.11. Special Cases

Here we survey briefly a variety of papers which determine Ext groups in special cases for algebraic groups and/or finite groups of Lie type. We focus mainly on recent work using algebraic group methods, but a few older papers are cited as well. (See Chapter 14 below for related work on group cohomology.) Much of this literature deals with small values of $p$ not covered directly by the general theory, while the groups involved are usually of small rank. Often the simple modules can be managed efficiently, thanks to the Tensor Product Theorem. For the connections with algebraic groups, the "generic" comparisons of Cline–Parshall–Scott–van der Kallen [**116**] are frequently invoked along with the work of Donkin and Andersen.

In most of the computations below, dimensions of Ext groups turn out to be $\leq 1$, though dimension 2 can also occur. But as pointed out in 12.6, it is hard to make any general dimension estimates based just on low rank examples.

Although there are practical limits to computing all extensions between simple modules in higher ranks, the mixture of techniques used in these papers may have further application to small primes and special weights.

- Pollatsek [**341**, **342**] works out 1-cohomology for various linear groups with coefficients in the standard module $V$, over rather general fields $k$ of characteristic $p$. In [**341**], $H^1(G, V)$ is computed when $G = \operatorname{Sp}(2n, k)$ with $n \geq 2$ and $p$ odd, or else $p = 2$ and $k$ perfect, as well as when $G = \operatorname{O}(2n, k)$ with $n \geq 2$ and $p = 2$ ($k$ perfect). In each case $H^1(G, V)$ has $k$-dimension at most 1. The methods are concrete and specific to the geometry of the group action on $V$. Further orthogonal groups are treated similarly in [**342**].
- Bell [**35**] studies the cohomology of finite special linear groups.

- Jones and Parshall [264] summarize in a table their detailed computations of $H^1(G, V)$, for all finite (universal) groups $G = \mathbf{G}^F$ when $V$ is a "minimal" $KG$-module. This depends on the work of Cline–Parshall–Scott [112, 113] as well as the 1975 thesis of Jones at U. Minnesota. Here a simple $KG$-module $L(\lambda)$ is called *minimal* if $\lambda$ is minimal among the nonzero dominant weights, relative to the usual partial order. Usually such a weight is fundamental or (in a few cases) a sum of two fundamental weights.

- As a byproduct of his study of PIMs for $\mathrm{SL}(2, 2^r)$, Alperin [5] determines all Ext groups between simple modules. In the process he develops inductive methods for computation which figure in other papers cited below, especially those of Sin and Dowd.

- Cline [111] applies methods of the joint paper [116] to the case $\mathbf{G} = \mathrm{SL}(2, K)$ in order to work out dimensions of all $\mathrm{Ext}_{\mathbf{G}}(L(\lambda), L(\mu))$: these are always $\leq 2$.

- In his 1982 thesis, Yehia [453] shows how to determine inductively which Ext groups for $\mathrm{SL}(3, K)$ are nonzero, using the methods of Donkin and others. In particular, he finds that $\dim \mathrm{Ext}_{\mathbf{G}}(L(\lambda), L(\mu)) \leq 1$ in all cases.

- Andersen–Jørgensen–Landrock [19] obtain detailed descriptions of Loewy series of PIMs (see Chapter 13 below) for the groups $\mathrm{SL}(2, p^r)$. In particular they determine the Ext groups, which are sometimes 2-dimensional when $r = 2$. (Carlson [83, Thm. 4.1] recovers their Ext results as a byproduct of his more comprehensive cohomology calculations.) They also describe extensions for the groups $\mathrm{SL}(3, 2^r)$.

- Working in the context of his general theorems, Andersen [14] computes the Ext data for $\mathrm{SL}(3, p)$ and $\mathrm{SU}(3, p)$. (But there are a few errors in 4.3 and Table III.) He also treats $\mathrm{Sp}(4, p)$ in detail when $p > 5$.

- In several papers, Völklein [427, 428, 429] explores the 1-cohomology of the adjoint module (getting an explicit description in most cases for Chevalley groups), more general cases of 1-cohomology of "small" modules, and further features of Ext groups in the spirit of Smith [379] and Cabanes [76]. He extends or improves a number of previous results, for example by treating very small prime powers.

- Benson and Martin [51] determine Ext for certain pairs of "small" modules (some simple) when $G = \mathrm{SU}(3, 2^r)$, using a mixture of finite group and algebraic group methods. For $G = \mathrm{PSU}(3, 4)$ they exhibit in a table all $\dim \mathrm{Ext}_G(L(\lambda), L(\mu))$; these are 0 or 1. (There is some overlap with independent work of Sin [371].)

- Kleshchev [276] finds conditions for the vanishing of 1-cohomology of $\mathrm{SL}_n$ with coefficients in the module of truncated polynomials (polynomials in $n$ variables, modulo $p$th powers). In [277] he uses this to compute $H^1$ for $\mathrm{SL}(n, p^r)$ (with $n \geq 3$) with coefficients in twisted tensor products coming from modules of truncated polynomials or exterior powers of the natural module for smaller rank groups.

  In [278] Kleshchev studies the vanishing of $H^1$ (and $H^2$) for Chevalley groups with coefficients in restricted modules $L(\lambda)$ whose weight spaces are all 1-dimensional; he lists the exceptional cases where nonvanishing occurs. (The $\mathbf{G}$-modules in question are determined by Seitz [364, §6].)

- Lin [**299**] studies Ext groups in the setting of $\mathbf{G}_r\mathbf{T}$-modules, deriving socle series of induced modules with $p$-singular highest weights when $r = 1$ and $p \geq 15$ in the case of type $G_2$. Applying translation functors and known results, he shows how to compute all $\mathrm{Ext}_{\mathbf{G}}(L(\lambda), L(\mu))$.

- Ye [**445**, **446**] applies inductive methods to the study of Ext for the algebraic groups $\mathrm{Sp}(4, K)$, treating odd $p$ and $p = 2$ separately. Here all Ext groups are of dimension $\leq 1$.

  In collaboration with Liu [**303**] (but not using the approach of Lin [**299**]), Ye determines where Ext is nonzero for the much more complicated algebraic group of type $G_2$ when $p \geq 13$ in a similar way. Here all Ext groups have dimensions $\leq 3$; the dimension is 2 in some explicit examples. Then in [**447**] he goes on to apply these results to the finite group $G_2(p)$, also when $p \geq 13$. A considerable amount of computation is required to find $\mathbf{G}$-socles of certain tensor products of simple modules.

- In an ambitious series of papers, Sin and Dowd explore Ext and $H^1$ thoroughly for algebraic and finite groups having rank as high as 4, most often when $p$ is 2 or 3. For such small primes the simple modules are usually manageable. But the determination of Ext groups requires the full range of available techniques.

  In [**370**] Sin describes all Ext groups for $\mathrm{Sp}(4, 2^r)$ in tandem with the Suzuki groups, in the spirit of earlier work by Chastkofsky and Feit [**104**]. The groups $\mathrm{SL}(3, 2^r)$ and $\mathrm{SU}(3, 2^r)$ are treated in [**371**], using different methods from those of Benson and Martin [**51**], while the Ext groups for $G_2(2^r)$ are treated in [**372**]. In [**373**] $H^1$ for the groups of type $G_2$ is computed in characteristic 3, along with the Ree groups. A similar pairing occurs in [**374**], where $H^1$ for groups of type $F_4$ in characteristic 2 and related Ree groups is computed. The special isogenies leading to Suzuki and Ree groups are the focus of [**375**], where computations of Ext groups are completed for algebraic groups of types $C_2$ and $F_4$ (when $p = 2$), $G_2$ (when $p = 3$).

  Dowd [**153**] works out all $H^1(G, L(\lambda))$ for $\mathrm{SL}(3, 3^r)$ and $\mathrm{SU}(3, 3^r)$ when $r > 4$. In [**154**] he does similar computations for the finite (universal) groups of types $A_3, {}^2A_3, D_3$ over fields of order $2^r$.

  Dowd and Sin [**156**] (following Sin [**375**] and Dowd [**154**]) describe simple modules for the algebraic groups of rank up to 4 in the case $p = 2$, then complete the computation of Ext groups between simple modules in all cases. Their idea is to use the special isogeny relating types B and C in characteristic 2 (as in [**154**]), while relating these groups to groups of type D. They also exploit a special relationship between types $D_4$ and $F_4$ in characteristic 2. All of this involves a considerable amount of case-by-case work, including the determination of simple submodules in many tensor products.

  In [**155**] Dowd considers more broadly what is involved in determining all $H^1$ groups for a Chevalley group $G = \mathbf{G}(\mathbb{F}_{p^r})$. The context here is the rational and "generic" cohomology comparison of Cline–Parshall–Scott–van der Kallen [**116**]. Dowd shows that for fixed $\mathbf{G}$ and $p$, all $H^1(G, L(\lambda))$ with $\lambda \in X_r$ and $r \geq 1$ can be determined from a finite number of cases. In a later Addendum, he gives a corrected restatement of his Corollary 2, making explicit the role of Galois conjugation over a field of order $p^r$: there is an integer $N(\mathbf{G}, p)$ such that $r > N(\mathbf{G}, p)$ implies that every $\lambda \in X_r$ has a Galois conjugate $\lambda'$ with $H^1(G, L(\lambda)) \cong H^1(\mathbf{G}, L(\lambda'))$. (The statement is slightly different for type C when $p = 2$.)

## 12.12. Semisimplicity Criteria

The study of extensions makes it clear that one cannot expect to get a straight-forward classification of all possible indecomposable modules for finite groups of Lie type (other than $\mathrm{SL}(2, p)$), even those coming by restriction from the ambient algebraic group. On the other hand, it is reasonable to look for criteria which guarantee that certain modules are *semisimple*. For example, the above results on self-extensions provide sufficient conditions for a module whose composition factors are all isomorphic to be semisimple. (For earlier investigations of semisimplicity, see Smith [379] and Cabanes [76].)

One may also ask for a bound on dimensions below which every $KG$-module is semisimple. This bound might for example depend on the rank $\ell$ and on the prime $p$ or the size $q$ of the finite field involved.

McNinch [317] refines a considerable amount of earlier information (notably the work of Jantzen [258]) to get such a bound, arriving at the following near-definitive result:

THEOREM. *Let* $G = \mathbf{G}^F$ *over a field of* $p^r$ *elements, where* $\mathbf{G}$ *has rank* $\ell > 1$. *Then every* $KG$-*module of dimension* $\leq \ell p$ *is semisimple, with the possible exception of the cases:* $A_2$ $(q = 2)$; $A_\ell$ $(q = 3, 9)$; $C_\ell$ $(q = 5)$.

The rank 1 group $\mathrm{SL}(2, q)$ narrowly misses inclusion in the theorem, since the study of PIMs for $\mathrm{SL}(2, p)$ in 9.8 shows the existence of nonsimple indecomposable modules of dimension $p - 1$. (These come by reduction modulo $p$ from certain "discrete series" characters as discussed in Chapter 17 below).

The theorem reduces quickly to the study of extensions between two simple modules, where previous work of Jantzen and McNinch can be applied. An essential ingredient in the proof is the study of rational and generic cohomology by Cline–Parshall–Scott–van der Kallen [116] (see 14.5 below), as adapted by Jantzen to the case of Ext. This provides general conditions under which the natural restriction map $H^1(\mathbf{G}, M) \to H^1(G, M)$ is an isomorphism, for a rational $\mathbf{G}$-module $M$. A number of interesting configurations have to be looked at separately. For example, one has to rule out a potential self-extension for $G_2(7)$ involving two copies of the natural 7-dimensional module.

By combining the work of Jantzen and McNinch with general facts about finite groups (and an appeal to the classification of finite simple groups), Guralnick [198] is then able to reach the following conclusions:

> Let $G$ be a finite group containing no nontrivial normal $p$-subgroup, and let $M$ be a faithful $KG$-module.
> (a) If $\dim M \leq p - 2$, then $M$ is semisimple.
> (b) If $\dim M \leq p - 3$, then $H^1(G, M) = 0$.

CHAPTER 13

# Loewy Series

The internal structure of indecomposable projective modules encodes, among other things, the possible extensions between simple modules (12.1). In Chapter 12 we looked closely at the Ext functor for finite groups of Lie type. Here we consider the broader question of describing the Loewy series of PIMs (13.1), which in principle can yield minimal projective resolutions of all simple modules (13.4).

Even though Loewy series for groups of Lie type have been worked out in detail only for $SL(2, q)$ (see 13.5) and a few isolated groups of higher rank surveyed at the end of the chapter, the study of closely parallel theories in characteristic 0 (13.6) and characteristic $p$ (13.7) suggests a natural candidate for the "generic" Loewy length of PIMs for $KG$ when $G$ is a finite group of Lie type over a field of $p^r$ elements. This should be $2rm + 1$, where $m = |\Phi^+|$.

Assuming the truth of Lusztig's Conjecture (3.11), Pillen exploits for finite Chevalley groups when $r = 1$ an embedding of principal series modules into Weyl modules (13.10) to prove that the generic Loewy length of PIMs is indeed uniform: $2m + 1$ (13.11). This is in striking contrast to the fact that generic dimensions of PIMs typically vary from alcove to alcove for restricted weights. (Similar results are expected for $r > 1$ and for twisted groups, with generic Loewy length $2rm+1$.)

## 13.1. Loewy Series

Given any module $M$ satisfying chain conditions, we can consider various chains of submodules

$$M = M^0 \supset M^1 \supset \cdots \supset M^n = 0,$$

where each quotient $M^{k-1}/M^k$ is *semisimple* and nonzero. A composition series is the longest such chain. But here we look for a shortest such chain, called a **Loewy series**. Some authors reserve the term "Loewy series" for the special case of the radical series, defined below. Notation and terminology vary quite a bit in the literature. For background, see for example Landrock [**296**, I §8], who discusses "Loewy" (=radical) and socle series in detail, and Benson [**47**, 1.2].

Two extreme types of Loewy series are singled out for special attention. Recall that the **radical** $\mathrm{Rad}\, M$ is defined to be the smallest submodule $N$ of $M$ for which $M/N$ is semisimple, while the **socle** $\mathrm{Soc}\, M$ is defined to be the largest semisimple submodule of $M$. Setting $\mathrm{Rad}^0 M = M$ and $\mathrm{Rad}^1 M = \mathrm{Rad}\, M$, we then define inductively $\mathrm{Rad}^k M := \mathrm{Rad}(\mathrm{Rad}^{k-1} M)$. Similarly, we let $\mathrm{Soc}^0 M = 0$ and $\mathrm{Soc}^1 M = \mathrm{Soc}\, M$, then define $\mathrm{Soc}^k M$ to be the unique submodule for which $\mathrm{Soc}(M/\mathrm{Soc}^{k-1} M) = \mathrm{Soc}^k M/\mathrm{Soc}^{k-1} M$. The **radical series** and **socle series** of $M$ are then defined by:

$$M = \mathrm{Rad}^0 M \supset \mathrm{Rad}^1 M \supset \cdots \supset \mathrm{Rad}^r M = 0,$$
$$0 = \mathrm{Soc}^0 M \subset \mathrm{Soc}^1 M \subset \cdots \subset \mathrm{Soc}^s M = M.$$

Here $r$ and $s$ are the least integers for which (respectively) $\operatorname{Rad}^r M = 0$ and $\operatorname{Soc}^s M = M$. It is well-known that $r = s$. This common value $r$ is called the **Loewy length** of $M$, sometimes denoted $\ell\ell(M)$. If the radical series and socle series of $M$ happen to coincide, we call $M$ **rigid**.

Define the $k$th radical layer to be $\operatorname{Rad}_k M := \operatorname{Rad}^{k-1} M / \operatorname{Rad}^k M$, for $k = 1, \ldots, r$. The **head** $\operatorname{Hd} M$ of $M$ is defined to be the first radical layer $M / \operatorname{Rad} M$. This is the largest semisimple quotient of $M$. (It is also known as the "cap" or "cosocle".) Similarly, we define the $k$th socle layer: $\operatorname{Soc}_k M := \operatorname{Soc}^k M / \operatorname{Soc}^{k-1} M$, for $k = 1, \ldots, r$. So $\operatorname{Soc} M = \operatorname{Soc}_1 M$.

In general, $M$ has many Loewy series. These all have length $r$ and occupy intermediate positions between the radical series and the socle series:

$$\operatorname{Rad}^{r-k} M \subseteq M^{r-k} \subseteq \operatorname{Soc}^k M.$$

In this setting, there are three natural questions about a given module $M$:

- What is the Loewy length of $M$?
- Is $M$ rigid?
- What are the composition factors in each socle or radical layer of $M$?

While these questions are in principle logically independent, the answers are often interrelated in practice. Moreover, if we look at a family of modules in a category (say projectives), it may be possible to answer the questions uniformly—at least under some "genericity" conditions.

## 13.2. Loewy Series for Finite Groups

In the case of a group algebra $KG$, there is no guaranteed method for determining Loewy series of PIMs—or even the extensions between simple modules. But a mixture of ad hoc and general techniques is sometimes successful for particular groups. Even so, it is usually impossible to be absolutely confident about the correctness of published computations unless one repeats every step independently.

Landrock [**296**, Lemma 9.10] discovered a very useful reciprocity principle, which reveals some extra symmetry and also cuts down on the amount of work needed to determine the radical series of all PIMs for $KG$. This result is usually referred to as **Landrock's Lemma** (see [**47**, Thm. 1.7.8]):

PROPOSITION. *Let $G$ be a finite group. If $M$ and $N$ are simple $KG$-modules with respective projective covers $P_M$ and $P_N$, the number of times $M$ appears in the $k$th radical layer of $P_N$ is equal to the number of times the dual module $N^*$ appears in the $k$th radical layer of $P_{M^*}$.*

The reader can observe how this works in the examples presented later.

## 13.3. Example: $\operatorname{SL}(2,p)$

What we already know about extensions when $G = \operatorname{SL}(2,p)$ makes it easy to determine the Loewy series of all PIMs. Leaving aside the trivial case $p = 2$, assume that $p$ is odd. According to Theorem 8.5, $KG$ has two blocks of highest defect involving all simple modules except the Steinberg module: the "even" block (principal block) involves $L(0), L(2), \ldots, L(p-3)$, while the "odd" block involves $L(1), L(3), \ldots, L(p-2)$.

We showed in 12.2 that when $0 < \lambda < p - 2$ and $\mu = p - 2 - \lambda$ is the linked weight, $L(\lambda)$ extends nontrivially both $L(\mu - 1)$ and $L(\mu + 1)$. On the other hand, the latter two modules fail to extend each other. Since the head and the socle of the PIM $U(\lambda)$ are isomorphic to $L(\lambda)$, while $\dim U(\lambda) = 2p$, both the radical series and the socle series must therefore have length 3: in each case, the middle layer has the form $L(\mu - 1) \oplus L(\mu + 1)$. Only small adjustments are needed in the extreme cases: If $\lambda = 0$ and $\mu = p - 2$, we know that $\dim U(0) = p$. Here the radical and socle series must both have just $L(p - 3)$ in the middle layer. Similarly, if $\lambda = p - 2$ and $\mu = 0$, the middle Loewy layer of $U(p - 2)$ involves only $L(1)$.

To summarize:

PROPOSITION. *Let $G = \mathrm{SL}(2, p)$ with $p$ odd. Then all PIMs in a block of highest defect are rigid and have Loewy length 3.*

### 13.4. Minimal Projective Resolutions

In principle, knowledge of the radical series of all PIMs for a finite group $G$ would lead to explicit computations of **minimal projective resolutions**. Recall that each $KG$-module $M$ has a projective resolution $\cdots P_2 \to P_1 \to P_0 \to M \to 0$ with all $\dim P_n$ as small as possible; then the $P_n$ are unique up to isomorphism. Each $P_n$ is of course a direct sum of PIMs, but the dimensions of the $P_n$ may well increase dramatically with $n$. The "complexity" of resolutions will be discussed more carefully in Chapter 15 below.

Here we just take a closer look at the PIMs for $\mathrm{SL}(2, p)$, with $p$ odd. A little experimentation reveals that minimal projective resolutions of non-projective simple modules $L(\lambda)$ are always **periodic** in this case: $P_{n+p-1} = P_n$ for all $n$. (This phenomenon will be discussed further in 15.7.) Moreover, the indecomposable summands of $P_n$ are seen to be distinct and to number at most $(p-1)/2$.

In a minimal resolution of $L(0)$, all $P_n$ are actually indecomposable. The periodic sequence of weights here is

$$0, \quad p-3, \quad 2, \quad p-5, \quad 4, \quad p-7, \ldots, p-7, \quad 4, \quad p-5, \quad 2, \quad p-3, \quad 0.$$

In the middle of this sequence one sees a repeated occurrence of $\lambda = (p-1)/2$ if $p \equiv 1 \pmod 4$ or $\lambda = (p-3)/2$ if $p \equiv 3 \pmod 4$, consistent with the discussion of self-extensions in 12.2.

It is useful to picture the sequence in terms of the graphs in Figure 1, whose vertices are labelled with weights. This is a reversed form of the Brauer tree, which

$$
\begin{array}{ccccccccc}
\circ & \!\!-\!\!-\!\! & \circ & \!\!-\!\!-\!\! & \circ & \!\!-\! & \circ & \!\!-\!\!-\!\! & \circ \cdots \cdots \circ \!\!-\!\!-\!\! \circ \!\!-\!\!-\!\! \circ \!\!-\!\!-\!\! \circ \!\!-\!\!-\!\! \circ \\
0 & & p-3 & & 2 & & p-5 & & 4 \qquad\quad 4 \quad p-5 \quad 2 \quad p-3 \quad 0
\end{array}
$$

$$
\begin{array}{ccccccccc}
\circ & \!\!-\!\!-\!\! & \circ & \!\!-\!\!-\!\! & \circ & \!\!-\! & \circ & \!\!-\!\!-\!\! & \circ \cdots \cdots \circ \!\!-\!\!-\!\! \circ \!\!-\!\!-\!\! \circ \!\!-\!\!-\!\! \circ \!\!-\!\!-\!\! \circ \\
p-2 & & 1 & & p-4 & & 3 & & p-6 \quad\quad p-6 \quad 3 \quad p-4 \quad 1 \quad p-2
\end{array}
$$

FIGURE 1. Graphs for $\mathrm{SL}(2, p)$, with $p$ odd

was introduced in 9.9. Each block of highest defect yields such a graph with vertices labelled by highest weights of simple modules in the block. For the "odd" block, the second graph encodes the minimal projective resolution of $L(p - 2)$, again with slight variations in the middle based on whether $p \equiv 1 \pmod 4$ or $p \equiv 3 \pmod 4$.

REMARK. To put the example of $SL(2, p)$ in perspective, it may be useful to comment briefly on the general theory of blocks with cyclic defect group mentioned in 9.9. In [8], Alperin and Janusz prove:

- If $G$ is a finite group with a cyclic Sylow $p$-subgroup and $k$ is a splitting field of characteristic $p$ for $G$, then each term of the minimal projective resolution of the trivial $kG$-module is indecomposable.
- With $G$ and $k$ as above, let $M$ be any indecomposable $kG$-module. Then each term in its minimal projective resolution has no repeated indecomposable summand. Thus each term is the sum of at most $e$ indecomposables, where $e$ is the number of simple $kG$-modules in the block containing $M$. (The worst case is in fact realized by $SL(2, p)$ for $p > 3$, with $M$ the reduction mod $p$ of an ordinary representation. Here $e = (p-1)/2$.)

Their Theorem 1 shows how to build a minimal projective resolution of the trivial module by following a "canonical" sequence of edges of the Brauer tree for the 1-block. This is identified by Green [195] with his "walk around the Brauer tree", constructed in a somewhat different way.

## 13.5. Example: $SL(2, q)$

In case $\mathbf{G} = SL(2, K)$ and $G = SL(2, p^r)$, Andersen–Jørgensen–Landrock [19] work out completely the Loewy structure of PIMs $U_r(\lambda)$. In this case, all PIMs $Q_r(\lambda)$ for $\mathbf{G}_r$ have compatible $\mathbf{G}$-module structure. Moreover, on restriction to $G$ one has $Q_r(\lambda) = U_r(\lambda)$ unless $\lambda = 0$, when $Q_r(0) = U_r(0) \oplus St_r$. Here all PIMs turn out to be rigid, with precisely described Loewy layers. Moreover:

THEOREM. Let $\mathbf{G} = SL(2, K)$ and $G = SL(2, p^r)$. Fix $0 \leq \lambda < p^r$. Then:

(a) The $\mathbf{G}$-module $Q_r(\lambda)$ is rigid. Its Loewy length is $2(r - \nu_p(\lambda + 1)) + 1$, where $\nu_p(\lambda + 1) = \max\{i \mid p^i \text{ divides } \lambda + 1\}$.

(b) The $KG$-module $U_r(\lambda)$ is rigid. Except in the case $r = 1, p = 2, \lambda = 0$, its Loewy length is $2r + 1$.

Their proof is concrete but complicated, involving induction on $r$ and careful bookkeeping with weights which cannot be imitated in higher ranks. The ideas are worked out first in the setting of algebraic groups and Frobenius kernels, then adapted systematically to the finite groups. A full description of the Ext groups for simple modules is a natural byproduct of these arguments.

## 13.6. The Category $\mathcal{O}$

In the background of our study of modular representations is the better established theory of modules in the Bernstein–Gel'fand–Gel'fand category $\mathcal{O}$. This became influential through their papers of 1971–76 and the subsequent development of Kazhdan–Lusztig theory in 1979–80. The category $\mathcal{O}$ and its many relatives continue to be investigated extensively in the characteristic 0 literature. We recall here just the most basic situation involving integral weights, which leads to definitive results on Loewy series: see Irving [239, 240] and the sources he cites.

Some of the standard notation here conflicts with our conventions in characteristic $p$, so we adopt it only temporarily in the present section. Let $\mathfrak{g}$ be a semisimple Lie algebra over $\mathbb{C}$, with Cartan subalgebra $\mathfrak{h}$ and Borel subalgebra $\mathfrak{b}$ determining a set of positive roots. Denote by $W$ the Weyl group, which acts on the integral weight lattice $X$ associated with $\mathfrak{h}$. A module $M$ over the universal enveloping algebra $U(\mathfrak{g})$ lies in the **category** $\mathcal{O}$ if (a) $M$ is finitely generated, (b) each subspace $U(\mathfrak{b})x$ with $x \in M$ is finite dimensional, (c) $M$ is the direct sum of its weight spaces relative to $\mathfrak{h}$ (each of which is then necessarily finite dimensional).

The simple modules in $\mathcal{O}$ turn out to be highest weight modules $L(\lambda)$, parametrized here by all integral weights $\lambda \in X$, with $\dim L(\lambda) < \infty$ if and only if $\lambda \in X^+$. All modules in $\mathcal{O}$ have finite length, with well-determined composition factors of this type. The universal highest weight modules are also indexed by $X$ and denoted by $M(\lambda)$; these are called **Verma modules** (see 4.1). The module $M(\lambda)$ has $L(\lambda)$ as its unique simple quotient and also has a unique simple submodule. Moreover, $M(\lambda) = L(\lambda)$ if and only if $\lambda + \rho \in -X^+$; an example is $\lambda = -\rho$.

There are enough projective and injective objects in $\mathcal{O}$. Each $L(\lambda)$ has an indecomposable projective cover $P(\lambda)$ and an indecomposable injective hull $I(\lambda)$. These are typically distinct, coinciding if and only if $\lambda + \rho \in -X^+$. At the other extreme, for $\lambda + \rho \in X^+$ we have $P(\lambda) = M(\lambda)$. In particular, $P(-\rho) = M(-\rho) = L(-\rho)$ (a characteristic 0 analogue of the Steinberg module).

Each $P(\lambda)$ has a filtration with quotients of the type $M(\mu)$, satisfying **BGG reciprocity**: the number of quotients isomorphic to $M(\mu)$ equals the composition factor multiplicity $[M(\mu) : L(\lambda)]$.

Category $\mathcal{O}$ is a direct sum of subcategories $\mathcal{O}_\chi$ analogous to blocks of a group algebra, parametrized by characters of the center $Z(\mathfrak{g})$ of $U(\mathfrak{g})$. Each character $\chi$ corresponds naturally to a $W$-linkage class of weights in $X$ relative to the dot action $w \cdot \lambda = w(\lambda + \rho) - \rho$. Then $\mathcal{O}_\chi$ has only finitely many simple modules corresponding to these weights. Jantzen's translation functors (relative to $W$ in this case) relate the various blocks systematically, so for many purposes it is enough to consider just the "principal block" determined by the linkage class of $\lambda = 0$.

The formal characters of Verma modules are easily found, but not the characters $\operatorname{ch} L(\lambda)$. Determination of these for the principal block is equivalent to knowledge of all multiplicities $[M(w \cdot 0) : L(y \cdot 0)]$ for $w, y \in W$. In 1979 Kazhdan and Lusztig conjectured that this multiplicity is given by the value at 1 of a Kazhdan–Lusztig polynomial for $W$ determined by $w, y$. (The "inverse" polynomials are actually involved here, but they coincide with KL polynomials for related Weyl group elements.) The conjecture was soon proven using deep geometric methods, independently by Beilinson–Bernstein and Brylinski–Kashiwara, as was Jantzen's stronger earlier conjecture on filtrations of Verma modules.

Further consequences involving the coefficients of the polynomials were then explored by Gabber–Joseph, Irving, and others. In particular, Irving proved:

THEOREM. *Consider the category $\mathcal{O}$ for a semisimple Lie algebra $\mathfrak{g}$ over $\mathbb{C}$, involving modules with integral weights. Let $m$ be the number of positive roots.*

(a) *Each module $M(\lambda)$ is rigid, with Loewy length generically $m + 1$. The multiplicities of simple modules $L(\mu)$ in the Loewy layers can be computed from coefficients of (inverse) Kazhdan–Lusztig polynomials for $W$.*

(b) *When $\lambda + \rho \in -X^+$, $P(\lambda)$ coincides with $I(\lambda)$. This module is rigid, with Loewy length generically $2m + 1$.*

The connection between Kazhdan–Lusztig polynomials and Loewy series involves an independent approach to the Kazhdan–Lusztig conjecture due to Vogan. It turns out that the conjecture can be reformulated neatly as a statement about the Loewy length of certain modules.

## 13.7. Analogies in Characteristic $p$

The results just summarized for the category $\mathcal{O}$ have inspired similar ideas in characteristic $p$ (and to some extent *vice versa*), since the module categories for $\mathbf{G}, \mathbf{G}_r$, and $\mathbf{G}_r\mathbf{T}$ exhibit many parallels to $\mathcal{O}$. But most of the results explained below remain conjectural, depending as they do on the truth of Lusztig's Conjecture (3.11). Moreover, the transition from $r = 1$ to larger $r$ involves a systematic blowing up of the patterns of composition factors, as discussed in 3.10. So the current picture, though promising, is incomplete in some directions.

Before getting into the details, it may be helpful to give an informal overview. The best way to approach the finite groups of Lie type is to start with $\mathbf{G}_r\mathbf{T}$-modules, lift when possible to $\mathbf{G}$, and then restrict to $G$. This works well in the case of PIMs, which lift to $\mathbf{G}$ at least when $p \geq 2(h-1)$ (10.4). If one knows the Loewy layers of a (lifted) $\mathbf{G}$-module $Q_r(\lambda)$, one may expect in generic cases that these will remain semisimple on restriction to $G$ as twisted tensor products become ordinary tensor products: see 6.4. In this case the Loewy length for $KG$ might still decrease. To get lower bounds on Loewy length one needs further constraints coming from "intermediate" modules which mimic Verma modules in category $\mathcal{O}$.

The induced modules $\widehat{Z}_r(\lambda)$ are the obvious analogue of Verma modules in the category of $\mathbf{G}_r\mathbf{T}$-modules, but usually fail to have a compatible $\mathbf{G}$-module structure. On the other hand, Weyl modules $V(\lambda)$ behave like universal highest weight modules for $\mathbf{G}$ (3.3).

What are the appropriate intermediate modules in the category of $KG$-modules? We already encountered in 7.2 an elementary construction of *principal series modules*, induced from a 1-dimensional module for a Borel subgroup of $G$. These are a basic tool in the work of Pillen discussed below, but are not the only candidates for the intermediate role: reductions modulo $p$ of various irreducible representations over $\mathbb{C}$ need to be considered as well (see Chapter 17 below). For example, when $G = \mathrm{SL}(2,p)$, a typical PIM of dimension $2p$ cannot be filtered by principal series modules (which all have dimension $p+1$), so modules of dimension $p-1$ are needed.

In any case, the idea is to study the Loewy structure of the intermediate modules first, comparing the $\widehat{Z}_r(\lambda)$ systematically with Weyl modules of related highest weights and then with principal series modules. Once these comparisons are under control, the Loewy structure of the modules $\widehat{Q}_r(\lambda)$ can be understood in terms of "intermediate" modules which occur as submodules and as quotients (somewhat as in the category $\mathcal{O}$).

Ideally, one expects to find that all the modules studied are "generically" rigid, with a uniform Loewy length $rm + 1$ (in the intermediate case) or $2rm + 1$ (in the case of projectives). Moreover, the generic Loewy layers should be computable from coefficients of suitable inverse Kazhdan–Lusztig polynomials for $W_p$. This information should in turn restrict smoothly to the principal series and PIMs for $G$. Though the known results are still incomplete, they are all consistent with this ideal picture if "generic" is formulated correctly.

## 13.8. Frobenius Kernels and Algebraic Groups

Andersen and Kaneda [20] focus first on the category of $\mathbf{G}_1\mathbf{T}$-modules. Some of their arguments are closely analogous to those used by Irving [240] in the category $\mathcal{O}$; but numerous modifications are required. For a detailed account, see [**RAGS**,

II.D.4]. The key assumption needed here is Lusztig's Conjecture, as reformulated for the category of $\mathbf{G_1 T}$-modules: see [**RAGS**, II.C.11].

THEOREM. *Assume the truth of the* $\mathbf{G_1 T}$*-version of Lusztig's Conjecture.*

(a) *If* $\lambda \in X$ *is a p-regular weight, the* $\mathbf{G_1 T}$*-modules* $\widehat{Z}_1(\lambda)$ *and* $\widehat{Q}_1(\lambda)$ *are both rigid, with respective Loewy lengths* $m + 1$ *and* $2m + 1$.

(b) *If* $p \geq 3(h-1)$ *and* $\lambda \in X_1$ *is p-regular, the* $\mathbf{G}$*-module* $Q_1(\lambda)$ *is rigid, with Loewy length* $2m + 1$. *Its Loewy layers are derived from the* $\mathbf{G_1 T}$*-layers.*

(c) *Weyl modules* $V(\lambda)$ *are "generically" rigid, with Loewy length* $m + 1$.

Even without Lusztig's Conjecture, these values are shown (using translation functors and "wall crossings") to be *lower* bounds for Loewy length. Note that in earlier work of Andersen [**15**], the generic rigidity of Weyl modules had been deduced from the stronger Jantzen Conjecture and in turn used to show the rigidity of $\widehat{Z}_1(\lambda)$ for $p$ large enough. In [**21**] Andersen–Kaneda refine further their ideas on filtrations of $\mathbf{G_1 T}$-modules analogous to Jantzen filtrations.

Lin [**300**] extends these results by exploiting translation functors in a "truncated" category of modules. Working first with $p$-singular weights, he obtains by translation the socle series of injective modules in general. (Methods of his paper [**299**] are also applied here.)

### 13.9. Example: SL(3, $K$)

The rigid Loewy structure of a generic $\mathbf{G}$-module $Q_1(\lambda)$ can easily be illustrated for $\mathbf{G} = \mathrm{SL}(3, K)$. In Figure 2 we attach arbitrary labels $A, B, C, \ldots$ to some of the dominant $p$-alcoves for the affine Weyl group of type $A_2$. Then in Tables 1 and 2 we show the generic radical layers 1–7 of PIMs corresponding to weights $\lambda$ in the two lowest $p$-alcoves designated $A, B$. Here $Q_1(\lambda)$ has a filtration with Weyl modules as quotients; their composition factors are listed in the columns. The placement of the Weyl modules in layers reflects a kind of reciprocity, as in the category $\mathcal{O}$. For example, let $\lambda$ lie in the $B$ alcove and let $\mu$ be the $W_p$-linked weight in the $E$ alcove. Since the Weyl module $V(\mu)$ has $L(\lambda)$ in its third Loewy layer, a copy of $V(\mu)$ is placed in the third to fifth Loewy layers of $Q_1(\lambda)$.

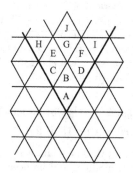

FIGURE 2. Labels for certain alcoves in type $A_2$

| 1 | B |     |     |     |     |     |       |
|---|---|-----|-----|-----|-----|-----|-------|
| 2 | A | C   | D   |     |     |     |       |
| 3 |   | B   | B   | E   | F   |     |       |
| 4 |   |     |     | ACD | ACD | G   |       |
| 5 |   |     |     | B   | B   | EF  |       |
| 6 |   |     |     |     |     | ACD |       |
| 7 |   |     |     |     |     | B   |       |

TABLE 1. Generic Loewy layers of $Q_1(\lambda)$ for SL$(3, K)$

| 1 | A |     |     |     |       |   |   |       |
|---|---|-----|-----|-----|-------|---|---|-------|
| 2 |   | B   | E   | F   |       |   |   |       |
| 3 |   | A   | ACD | ACD | G     | H | I |       |
| 4 |   |     | B   | B   | EF    | E | F | J     |
| 5 |   |     |     |     | ACD   | A | A | CDGHI |
| 6 |   |     |     |     | B     |   |   | EF    |
| 7 |   |     |     |     |       |   |   | A     |

TABLE 2. Generic Loewy layers of $Q_1(\lambda)$ for SL$(3, K)$

### 13.10. Principal Series Modules

In this and the following section we explain the main points in the work of Pillen [336] on Loewy series for finite Chevalley groups. He leaves aside the twisted groups and has to assume $r = 1$ for some of the stronger results, but similar results can be expected in general.

In 7.2 we constructed principal series modules in the context of groups with a $BN$-pair. Here we look more closely at that construction, seeking details on Loewy structure. For Pillen the group $G$ is a Chevalley group over a field of $p^r$ elements (usually with $r = 1$). Some of his arguments apply equally well to twisted groups. However, he has to use some earlier work of Jantzen [251, 253] for Chevalley groups which cannot readily be extended to twisted groups (see 17.13).

In the present context, principal series modules can be defined explicitly as spaces of functions in the spirit of the definition of the **G**-module $H^0(\lambda)$ in 3.3. There the opposite Borel subgroup $\mathbf{B}^-$ was preferred for the induction construction. For $G$ one can use either $B$ (coming from $\mathbf{B} = \mathbf{B}^+$) or $B^-$, but the resulting modules are the same up to labelling. Viewing $\lambda \in X_r$ as a character of $B$ or $B^-$, we define (with slightly different conventions than those used in 7.2):

$$M_r(\lambda) := \{f : G \to K \mid f(gb) = \lambda(b^{-1})f(g) \text{ for all } b \in B,\, g \in G\},$$
$$M'_r(\lambda) := \{f : G \to K \mid f(gb) = \lambda(b^{-1})f(g) \text{ for all } b \in B^-,\, g \in G\}.$$

It is easy to see that $M_r(\lambda) \cong M'_r(w_\circ\lambda)$. The main reason for the dual definition is to facilitate comparisons with Weyl modules and induced modules for **G**. Here some arguments used by Jantzen are invoked as well as results of Andersen–Kaneda. The notion of "sufficiently far" from walls in part (b) of the following theorem is made precise in [336].

THEOREM. *Let $G$ be a Chevalley group over a field of order $p^r > 2h - 1$.*

(a) *Set $\sigma_r := (p^r - 1)\rho$. If $\lambda \in X_r$, the $G$-submodule of $V(\sigma_r + \lambda)$ generated by a maximal vector is isomorphic to $M_r(\lambda)$ if and only if $\langle \lambda, \alpha^\vee \rangle > 0$ for all $\alpha \in \Delta$. Under the same conditions, $H^0(\sigma_r + \lambda)$ maps onto $M_r'(\lambda)$.*

(b) *For all $p$-regular $\lambda \in X_1$ sufficiently far from all walls of $p$-alcoves, $M_1(\lambda)$ is rigid and has Loewy length $m + 1$.*

## 13.11. Loewy Series for Chevalley Groups

The study of PIMs in Chapter 10 shows that $\dim U_r(\lambda)$ depends generically just on the $p$-alcove to which $\lambda$ belongs. For example, the typical dimensions for $SL(3, p)$ are $6p^3$ and $12p^3$. In spite of this variability of dimensions, the analogies with other module categories outlined above suggest a more uniform Loewy length. This is confirmed to some extent by Pillen, using Theorem 13.10 as a steppingstone.

THEOREM. *Assume the truth of Lusztig's Conjecture. Let $G$ be a Chevalley group over $\mathbb{F}_p$. For all $p$-regular weights $\lambda \in X_1$ sufficiently far from all walls of $p$-alcoves, the PIM $U_1(\lambda)$ is rigid and has Loewy length $2m + 1$.*

The hypothesis insures that $U_1(\lambda) = Q_1(\lambda)$ as $KG$-modules. Pillen builds on the argument of Andersen–Kaneda [20], which forces the Loewy length of the **G**-module $Q_1(\lambda)$ to be at least $2m + 1$ by embedding a Weyl module of length $m + 1$ into $Q_1(\lambda)$ and mapping $Q_1(\lambda)$ onto an induced module of the same length in such a way that the composite mapping is nonzero. In the context of $KG$-modules their sequence of maps is enlarged:

$$M_1(\sigma_1 + w_\circ \lambda) \hookrightarrow V(2\sigma_1 + w_\circ \lambda) \hookrightarrow Q_1(\lambda) \twoheadrightarrow H^0(2\sigma_1 + w_\circ \lambda) \twoheadrightarrow M_1'(\sigma_1 + w_\circ \lambda).$$

The key point is that the composite map takes the $B^-$-stable line on the left onto the similar line on the right. So the combined Loewy lengths of the two principal series modules force a generic lower bound for the Loewy length of $Q_1(\lambda)$. (A generic upper bound comes from the discussion in 13.7 above.)

As suggested earlier, a similar formulation should work in general:

CONJECTURE. *Let $G$ be a finite group of Lie type over a field of $p^r$ elements. Then for a $p$-regular weight $\lambda \in X_r$ sufficiently far from all walls of its $p$-alcove, the projective module $U_r(\lambda)$ is rigid and has Loewy length $2rm + 1$.*

Pillen's argument shows that the structure of each Loewy layer in generic cases can be adapted from **G** to $G$. This should ultimately depend on Kazhdan–Lusztig polynomials, still assuming Lusztig's Conjecture. For a Chevalley group having a twisted analogue, it is an intriguing problem to compare Loewy layers of PIMs for the two groups when restricted from a common (generic) **G**-module $Q_r(\lambda)$.

## 13.12. Example: SL(3, 2)

At the end of their paper, Andersen–Jørgensen–Landrock [19, 5.3] summarize the details of Loewy structure for $SL(3, 2)$. Here there are just three simple modules $L(0, 0), L(1, 0), L(0, 1)$ in the 1-block, along with the Steinberg module $L(1, 1)$. So it is not too difficult to work out the structure of the $Q_1(\lambda)$ for $SL(3, K)$ and then adapt to $SL(3, 2)$. For brevity we write just $Q(\lambda)$ and $U(\lambda)$ here.

Table 3 shows the radical layers of the **G**-modules and $G$-modules associated

| $Q(0,0)$ | $U(0,0)$ | $Q(1,0)$ | $U(1,0)$ |
|---|---|---|---|
| $(0,0)$ | $(0,0)$ | $(1,0)$ | $(1,0)$ |
| $(3,0) \oplus (0,3)$ | $(1,0) \oplus (0,1)$ | $(0,2)$ | $(0,1) \oplus (0,0)$ |
| $2(0,0) \oplus (2,2)$ | $(0,0)$ | $(1,0) \oplus (2,1)$ | $(1,0)$ |
| $(3,0) \oplus (0,3)$ | | $(0,2)$ | $(0,1)$ |
| $(0,0)$ | | $(1,0)$ | $(1,0)$ |

TABLE 3. Radical layers for SL(3, $K$) and SL(3, 2)

with the weights $(0,0)$ and $(1,0)$; the case of $(0,1)$ is symmetric. Here a notation such as $2(0,0) \oplus (2,2)$ indicates that the layer is of the form $L(0,0) \oplus L(0,0) \oplus L(2,2)$.

A comparison with socle series shows that the $Q(\lambda)$ are all rigid and (except for St) have Loewy length 5. On the other hand, $U(1,0)$ and $U(0,1)$ have Loewy length 5 but fail to be rigid, while $U(0,0)$ is rigid but has Loewy length 3.

### 13.13. Example: SL(3,3)

In 11.14 we exhibited some data for $G = SL(3,3)$, including the matrix of Cartan invariants. Koshitani [281] determines the radical series of PIMs for $G$ when $p = 3$. In this case it is not difficult to work out the basic dimension and multiplicity data, which we recall in Table 4. It is convenient here to assign dimension labels to the simple modules for use in Table 5.

| $\lambda$ | label | $\dim L(\lambda)$ | $d_\lambda$ | $\dim Q(\lambda)/3^3$ | $\dim U(\lambda)/3^3$ |
|---|---|---|---|---|---|
| $(0,0)$ | 1 | 1 | 6 | 12 | 7 |
| $(1,0)$ | 3 | 3 | 6 | 6 | 3 |
| $(0,1)$ | $3^*$ | 3 | 6 | 6 | 3 |
| $(2,0)$ | 6 | 6 | 3 | 3 | 2 |
| $(0,2)$ | $6^*$ | 6 | 3 | 3 | 2 |
| $(1,1)$ | 7 | 7 | 3 | 6 | 6 |
| $(1,2)$ | 15 | 15 | 3 | 3 | 3 |
| $(2,1)$ | $15^*$ | 15 | 3 | 3 | 3 |
| $(2,2)$ | 27 | 27 | 1 | 1 | 1 |

TABLE 4. Dimension data for SL(3, $K$), $p = 3$

The decomposition numbers $d_\lambda$ are 3 times as large as one would expect from studying primes $> 3$, a "ramification" phenomenon resulting from the fact that 3 is the index of connection: linkage classes are only a third as large as usual here.

Even for this rather small group, the details of the Loewy structure of PIMs require considerable work. Koshitani studies $G$ in tandem with its automorphism group $A := SL(3,3) \rtimes \mathbb{Z}_2$. (The outer automorphism is conjugate-transpose.) This is motivated by the fact that $A$ is a maximal subgroup (of index prime to 3) in the simple Tits group $^2F_4(2)'$. So the Loewy series of $A$ can be used to determine the Loewy series of the Tits group: see [282], where the results of [283] are announced. Along with some of the general techniques indicated above, Koshitani relies heavily on induction from a maximal subgroup of index 13: the parabolic subgroup $GL(2,3) \ltimes (\mathbb{Z}_3 \times \mathbb{Z}_3)$.

Table 5 exhibits the radical series of various PIMs; the dual cases behave sym-

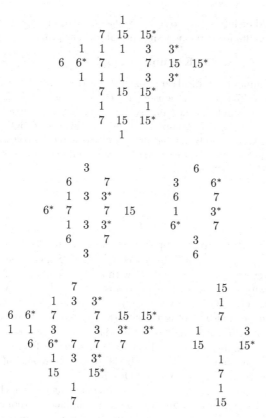

TABLE 5. Radical series of some PIMs for SL(3,3) [**281**]

metrically. But the situation is not sufficiently generic to produce rigidity, nor is the Loewy length usually equal to the expected generic value 7. (It is reassuring to note that independent calculations by Andersen in [**14**] confirm the dimensions of the various Ext groups exhibited here.)

## 13.14. Example: SL(4, 2)

In [**44**] Benson works out the radical series of the alternating group $A_8$ over a splitting field of characteristic 2, then goes on in a companion paper to apply the results to $A_9$. Since $A_8 \cong SL(4,2)$, this provides an example for a group of Lie type. Here there are eight simple modules: the Steinberg module and seven others lying in the principal block. Benson is able to exhibit the radical series explicitly in each case. It is noteworthy that in four cases (involving simple modules of dimensions 1, 14, 20, 20) the Loewy length is $13 = 2m + 1$, with $m = 6$ the number of positive roots. (The other Loewy lengths of nonprojective PIMs are 9, involving simple modules of dimensions $4, 4, 6$.) The methods used in the paper involve ordinary and modular character data, combined with a close study of modules induced from subgroups isomorphic to $A_6$ and $A_7$.

REMARK. It would be interesting to compare the above results for special linear groups with similar results for corresponding special unitary groups.

### 13.15. Example: SO(5,3)

Benson [46] determines the radical series of PIMs for the symmetry group of the famous configuration of 27 lines on a nonsingular cubic surface. This is probably the most complicated example worked out so far involving a group of Lie type in the defining characteristic. His group $G$ can be identified with SO(5,3). It is studied together with some closely related groups of the same Lie type. Here we use Benson's notation [46]; the Atlas notation for this family of groups is different: see the listing in [117] for $U_4(2) \cong S_4(3)$.

$$\begin{array}{ccc} \widehat{G}' & \hookrightarrow & \widehat{G} \\ \downarrow & & \downarrow \\ G' & \hookrightarrow & G \end{array}$$

Here $\widehat{G} := \mathrm{CSp}(4,3)$ is the group of symplectic similitudes. Its center has order 2 and lies in the derived group $\widehat{G}' = \mathrm{Sp}(4,3)$, which has index 2 in $\widehat{G}$. The respective quotients by the center are denoted $G$ and $G'$, with $G'$ simple of order $25,920$.

There are numerous realizations of these groups, described in the Atlas [117]. As a group of Lie type, $G = \mathrm{SO}(5,3)$ (which happens to be isomorphic to the Weyl group of type $E_6$). One can also identify $G'$ with the twisted group $\mathrm{SU}(4,2)$.

Note that $|\widehat{G}'| = 51,840 = |G|$, consistent with the fact that isogenous algebraic groups have the same number of rational points over a finite field (1.1). Here the isogeny combines an isomorphism of algebraic groups between $\mathrm{Sp}(4,K)$ and $\mathrm{Spin}(5,K)$ with the natural covering map to $\mathrm{SO}(5,K)$.

Benson's paper is something of a *tour de force*, interweaving methods from finite group theory and Lie theory. In the finite group setting, induction from a maximal subgroup of index 27 in $G$ plays a major role. In the algebraic group setting, he works out a considerable amount of detail about tensor products, Jantzen filtrations of Weyl modules, etc. This is done first for $\mathrm{Sp}(4,K)$ and then adapted to $G$.

As noted in 11.15, the 1-block of $\widehat{G}' = \mathrm{Sp}(4,3)$ has five simple modules of respective dimensions $1, 5, 10, 14, 25$ which induce simple modules for the quotient $G'$. Each of these in turn extends in two ways to a simple module in the 1-block of $G$. Benson labels them with superscripts $+$ and $-$. It turns out that the roles of these pairs of simple modules in the Loewy structure are symmetric. For example, with Benson's choice of labels, no simple module of one type ("plus" or "minus") can extend a simple module of the same type.

In Table 6 we exhibit the radical series of PIMs for $KG$; interchanging plus and minus signs gives the remaining PIMs in the 1-block. (Removing all signs yields the radical series of the PIMs when restricted to $G'$.) While Benson's methods are ad hoc, the results are reasonably close to what might have been predicted in the spirit of Conjecture 13.11. The group $G$ is far too small to afford generic configurations, but some PIMs do achieve the expected generic Loewy length $9 = 2m + 1$.

```
                    1⁺
              5⁻        14⁻
        1⁺  1⁺  5⁺  10⁺  14⁺
    1⁻  5⁻  5⁻      10⁻  14⁻  25⁻
        1⁺  1⁺  5⁺  10⁺  14⁺
              5⁻        14⁻
                    1⁺
```

```
                         5⁺
                   1⁻  5⁻  10⁻  14⁻  25⁻
            1⁺  5⁺  5⁺  5⁺  5⁺  10⁺  10⁺  14⁺  25⁺
    1⁻  1⁻  5⁻  5⁻  5⁻  10⁻      10⁻  10⁻  14⁻  14⁻  25⁻  25⁻
            1⁺  5⁺  5⁺  5⁺  5⁺  10⁺  10⁺  14⁺  25⁺
                   1⁻  5⁻  10⁻  14⁻  25⁻
                         5⁺
```

```
                         10⁺
                   5⁻  10⁻      14⁻  25⁻
            1⁺  5⁺  5⁺  10⁺  10⁺  10⁺  10⁺  14⁺  25⁺
    1⁻  5⁻  5⁻  5⁻  10⁻  10⁻      10⁻  10⁻  14⁻  14⁻  25⁻  25⁻
            1⁺  5⁺  5⁺  10⁺  10⁺  10⁺  10⁺  14⁺  25⁺  25⁺
                   5⁻  10⁻  10⁻  10⁻  14⁻  25⁻
                            10⁺  10⁺  25⁺
                            10⁻      25⁻
                                 10⁺
```

```
                         14⁺
                   1⁻  5⁻  10⁻
            1⁺  5⁺  10⁺      14⁺  14⁺  25⁺
            1⁻  5⁻  5⁻      10⁻  10⁻  14⁻
            1⁺  5⁺  10⁺      14⁺  14⁺  25⁺
                   1⁻  5⁻  10⁻
                         14⁺
```

```
                         25⁺
                   5⁻        10⁻
             5⁺  10⁺      14⁺  25⁺
       1⁻  5⁻  5⁻      10⁻  10⁻  25⁻
       5⁺  10⁺  10⁺      14⁺  25⁺  25⁺
                   5⁻  10⁻  25⁻
                        10⁺
                        10⁻
                        25⁺
```

TABLE 6. Radical series of some PIMs for SO(5, 3) [46]

CHAPTER 14

# Cohomology

The study of extensions between simple modules is only one aspect of the broader study of cohomology of finite groups, which encodes in a subtle way finer points about the category of $KG$-modules. This study leads in many directions, some originally motivated by algebraic topology or $K$-theory rather than representation theory (as in Adem–Milgram [1]). The subject tends to be open-ended, involving diverse methods and many special computations. In our setting cohomology might be thought of as "representation theory by other means". For groups of Lie type, we continue to emphasize the interaction with cohomology of algebraic groups and their Frobenius kernels, working always in the defining characteristic.

In 14.1 we recall some standard facts about the cohomology of an arbitrary finite group. Then we review briefly in 14.3–14.4 the parallel theory of rational cohomology for algebraic groups and Frobenius kernels, as presented in [**RAGS**]. Seminal work in the mid-1970s by Cline–Parshall–Scott, especially their joint paper with van der Kallen [**116**], relates rational cohomology indirectly to cohomology groups for finite groups of Lie type (14.5). This leads to a number of explicit computations, for example in the further work by Parshall and Friedlander (14.8). The case $G = \mathrm{SL}(2, q)$ already illustrates some of the complexities encountered here (14.7).

More recent work by Bendel, Lin, Nakano, Parshall, Pillen, and others approaches in different ways the interplay of cohomology for algebraic groups, Frobenius kernels, and finite groups of Lie type: some of this is discussed in 14.9.

Unless otherwise specified, modules will be *finite dimensional*, although some of the results for algebraic groups hold for all rational modules.

## 14.1. Cohomology of Finite Groups

Let $G$ be any finite group. Cohomology can be defined over an arbitrary field or over a base ring such as $\mathbb{Z}$ or $\mathbb{Z}/n\mathbb{Z}$, but we consider just the case when the base is our algebraically closed field $K$. Many techniques are employed in the literature: the long exact cohomology sequence associated with a short exact sequence of modules, various spectral sequences, etc. One soon observes that information about the cohomology of $G$ is intimately related to the cohomology of a Sylow $p$-subgroup (or its normalizer), which permits some explicit computations. A standard source for the basic theory is Evens [**167**]. The treatment by Benson in [**47**, Chap. 2] and [**48**, Chap. 2] introduces cohomology of groups first in the context of homological algebra and then in the context of algebraic topology.

Algebraically, cohomology of $G$ with coefficients in a module $M$ (often referred to simply as "cohomology of $M$") may be defined using right derived functors of the fixed point functor. This produces finite dimensional vector spaces $H^i(G, M)$, which have concrete interpretations when $i \leq 2$. A fundamental question is whether or not $H^i(G, M)$ is zero. If nonzero, what is its dimension?

When $M$ is the trivial module, abbreviated by $K$, there is additional structure: under the cup product, $H^*(G, K) := \bigoplus H^i(G, K)$ becomes a $K$-algebra. It is *graded commutative*, as explained in Benson [**47**, p. 52]. A key theorem, proved in different ways by Evens and Venkov, asserts that it is also *finitely generated*. This plays an essential role in the following chapter.

THEOREM. *If $G$ is an arbitrary finite group, its cohomology ring $H^*(G, K)$ is a finitely generated, graded commutative $K$-algebra.*

Later on we will need the following standard fact about the cohomology of "twisted" modules, relative to an automorphism $\sigma$ of $G$. An important special case involves the Frobenius morphism when applied to a finite group of Lie type. Each representation $G \to \mathrm{GL}(M)$ can be composed with $\sigma$ to yield a twisted representation. Denote the resulting $KG$-module by $M^\sigma$.

LEMMA. *Let $\sigma$ be an automorphism of a finite group $G$. Then $M \mapsto M^\sigma$ is an equivalence of the category of $KG$-modules with itself, permuting simple modules but preserving the regular representation and thus taking projectives to projectives. As a result, the vector spaces $H^n(G, M)$ and $H^n(G, M^\sigma)$ are isomorphic for all $n$.*

## 14.2. Finite Groups of Lie Type

Let $G = \mathbf{G}^F$ over a field of $p^r$ elements. Finite group methods combined with those of algebraic topology (but without use of algebraic groups) have yielded fragmentary results, involving for example the action of a classical group on its natural module or the trivial module. On the other hand, many of the results in the literature arise in characteristics different from $p$: see Adem–Milgram [**1**, Chap. VII], where $\mathbb{F}_2$ coefficients are the most interesting. Early computations in characteristic $p$ rely on specific features of particular groups and their root systems—notably the structure of Sylow $p$-subgroups.

- During the 1970s J. McLaughlin and his students at the University of Michigan computed $H^1$ and $H^2$ for a number of Chevalley groups and twisted groups. Usually the groups are of classical type, while cohomology relative to the natural module is studied. See Pollatsek [**341**, **342**], Landázuri [**294**], Bell [**35**, **36**], Avrunin [**25**, **26**]. For example, Avrunin studies $H^2$ of some unitary groups, using his vanishing criterion in [**25**], where it is shown that $H^2$ of $\mathrm{SU}(n, q)$ with coefficients in the standard $n$-dimensional module over $\mathbb{F}_{q^2}$ is generally 0. The possible exceptions are the cases $(n, q) = (2, 2^r)$ with $r > 1$, $(3, 4), (3, 3^r), (4, 4)$.
- Hiller [**206**] adapts methods of Friedlander and Quillen to obtain a vanishing range result for the cohomology of a finite Chevalley group $G$ over a field of $q = p^r$ elements. He defines integers $C(\Phi, q)$ and $\varepsilon(q)$ depending on $\Phi$ and/or $q$. The main result is that $H^i(G, \mathbb{F}_p) = 0$ for $i < (p-1)r/C(\Phi, q)\varepsilon(q)$. For example, when $\mathbf{G}$ is of type $A_\ell$, $\varepsilon(q) = 1$ and $C(\Phi, q) = (\ell + 1, q - 1)$.

From now on we focus just on comparisons with algebraic group cohomology. In particular, we consider only those $KG$-modules which come from rational $\mathbf{G}$-modules. As we saw in the case of Ext in Chapter 12, a naive restriction from $\mathbf{G}$ to $G$ can capture only part of the picture. More subtle methods are required.

What are the most natural problems along this line? Relative to the trivial module $L(0)$ (also denoted simply as $K$), we can ask for the full structure of the cohomology ring $H^*(G, K)$. Taking cohomology more generally of a simple module $L(\lambda)$ for $\lambda \in X_r$, we can ask for the dimension of each $H^i(G, L(\lambda))$. But in the absence of detailed knowledge of the $L(\lambda)$, it may be a more realistic first step to look at $H^*(G, H^0(\lambda))$.

## 14.3. Cohomology of Algebraic Groups

For algebraic groups there is a parallel cohomology theory, introduced by Hochschild. This involves the category of (not necessarily finite dimensional) rational **G**-modules and may be referred to as **rational cohomology**. It is defined as in the finite case via the derived functors of the fixed point functor, using the fact that there are enough injectives.

The theory generalizes well to group schemes such as Frobenius kernels, as explained in [**RAGS**, I.4]. But Jantzen places more emphasis on sheaf cohomology, for which there is a partial dictionary with rational cohomology. In a series of papers, Cline–Parshall–Scott and their collaborators develop the latter theory more fully and reinterpret Kempf's vanishing theorem (3.3) in their setting: see Cline–Parshall–Scott–van der Kallen [**116**, 1.2]. (Note that their **B** is Jantzen's **B**$^+$, so there is often a sign difference for weights.)

In our situation, we try to understand the vector spaces $H^i(\mathbf{G}, M)$ for $i \geq 0$, where $M$ is a finite dimensional rational **G**-module. In principle this could be done by starting with an injective resolution of $M$, taking fixed points to get a complex, and then working out the resulting cohomology in each degree. In practice this approach would be very difficult, in part because the injectives are infinite dimensional. It is shown in [**116**, 3.12] that $M$ actually has a finite resolution by finite dimensional *acyclic* **G**-modules, such as tensor products $H^0(\lambda) \otimes H^0(\mu)$. Thus $\dim H^i(\mathbf{G}, M) < \infty$. Even so, direct computation remains a daunting task.

First we assemble some basic information. Enumerate the simple roots in some way as $\alpha_1, \ldots, \alpha_\ell$. For an arbitrary weight $\lambda \in X$, write $\lambda = \sum_i r_i \alpha_i$ with $r_i \in \mathbb{Q}$. (When $\lambda \in X^+$ it is well-known that all $r_i > 0$ unless $\lambda = 0$.) Then define $h(\lambda) := \sum r_i$. For a rational **T**-module $M$, define $h(M)$ to be the maximum over all weights $\lambda$ of $M$ of the values $h(\lambda)$. Abbreviate by $\lambda$ the 1-dimensional **B**-module afforded by $\lambda \in X$, then write $H^i(\mathbf{B}, \lambda)$.

THEOREM. *Let $M$ be a finite dimensional rational **G**-module.*

(a) *All $H^i(\mathbf{G}, M) = 0$ unless some composition factor of $M$ has highest weight linked by $W_p$ to the 0 weight. In particular, $H^i(\mathbf{G}, L(\lambda)) = 0$ unless $\lambda$ lies in the root lattice.*

(b) *The restriction map from **G** to **B** induces an isomorphism $H^i(\mathbf{G}, M) \cong H^i(\mathbf{B}, M)$ for all $i \geq 0$.*

(c) *For any $\lambda \in X$, we have $H^i(\mathbf{B}, \lambda) = 0$ whenever $i > h(\lambda)$.*

(d) *For all $i > h(M)$, we have $H^i(\mathbf{G}, M) = 0$.*

Part (a) follows from the Linkage Principle: see [**RAGS**, II.7.3]. For the other parts, see [**RAGS**, II.4.7/10] or [**116**, §2], where the authors note that the case $\mathbf{G} = \mathrm{SL}(2, K)$ when $p = 2$ already shows that the bound in (d) is exact.

REMARK. Part (a) of the theorem shows that "most" simple modules $L(\lambda)$ have trivial cohomology in all degrees. This reflects the block structure of the category of rational **G**-modules, which differs sharply from the case of finite groups of Lie type (Chapter 8). In the latter situation, one can only be sure that the cohomology of $L(\lambda)$ vanishes when this module fails to lie in the 1-block of $KG$.

## 14.4. Twisting by Frobenius Maps

One of the main concerns in [**116**] is the relationship between the cohomology of a **G**-module $M$ and that of its $r$th Frobenius twist $M^{[r]}$ (obtained by composing the standard Frobenius map relative to $p^r$ with the given representation). There is a natural map $H^i(\mathbf{G}, M) \to H^i(\mathbf{G}, M^{[r]})$. Using nontrivial machinery, one can prove:

THEOREM. *If $M$ is a rational **G**-module, then the natural map $H^i(\mathbf{G}, M) \to H^i(\mathbf{G}, M^{[r]})$ is injective for all $i$.*

For a proof see [**RAGS**, II.10.14–16]. As a simple example, the nonvanishing of $H^1(\mathrm{SL}(2, K), L(2p - 2))$ (see 12.2) forces nonvanishing for the corresponding cohomology for all Frobenius twists of $L(2p-2)$. However, it is far from easy to see when the Frobenius map induces an *isomorphism* above. It is also not obvious that the sequence of injective maps $H^i(\mathbf{G}, M) \to H^i(\mathbf{G}, M^{[1]}) \to H^i(\mathbf{G}, M^{[2]}) \to \cdots$ must eventually stabilize. These are among the issues addressed in [**116**].

By way of contrast, it follows from the elementary Lemma 14.1 that on the level of finite groups of Lie type, the analogous map is always an isomorphism. Here the difference is that the Frobenius map is itself a group automorphism.

## 14.5. Rational and Generic Cohomology

Most of the groundwork needed to relate the cohomology of **G** or $\mathbf{G}_r$ to that of a finite group $G = \mathbf{G}^F$ over a field of $p^r$ elements has been laid by Cline, Friedlander, Parshall, Scott (in various combinations).

We begin by reviewing a fundamental paper by Cline, Parshall, Scott, and van der Kallen [**116**], which follows work on $H^1$ in [**112, 113**] as well as [**264**]. Here it is essential to let the power of $p$ vary, so we write $\mathbf{G}(p^r)$ in place of $G$. Initially only *Chevalley groups* are considered.

Roughly speaking, the idea is to show that, for a given **G**-module $M$ and a given range of degrees, when $M$ is twisted by a high enough power of the Frobenius map the cohomology stabilizes and becomes isomorphic to the $\mathbf{G}(p^r)$-cohomology of $M$ for all sufficiently large $r$. But the result obtained is not just qualitative; rather, it involves definite estimates which permit some specific calculations.

The main theorem of [**116**] is necessarily rather technical. In order to state it precisely, we have to introduce quite a bit of notation. Here we mostly follow their conventions, including the use of $m$ for a variable index in the theorem below. Enumerate the simple roots in $\Phi$ as $\alpha_1, \ldots, \alpha_\ell$. As usual, we use square brackets below to denote the greatest integer function.

$X/\mathbb{Z}\Phi$      weight lattice modulo root lattice

$t$      exponent of the finite abelian group $X/\mathbb{Z}\Phi$

$t(\lambda)$      order of image of $\lambda \in X$ in $X/\mathbb{Z}\Phi$

$t_p(\lambda)$      $p$-part of $t(\lambda)$

$c$      $\max n_i$ if the highest root $\tilde{\alpha} = \sum_i n_i \alpha_i$

$c(\zeta)$      $\max |n_i|$ if $\zeta = \sum_i n_i \alpha_i$ in $\mathbb{Z}\Phi$

$e : \mathbb{Z} \to \mathbb{Z}$    $e(n) := \left\lceil \frac{n-1}{p-1} \right\rceil$

$f : \mathbb{Z} \to \mathbb{Z}$    $f(n) := \left\lceil \log_p(|n| + 1) \right\rceil + 2$

Now the main theorem runs as follows:

THEOREM. *Assume that* $\mathbf{G}$ *is defined and split over* $\mathbb{F}_p$. *Let* $M$ *be a finite dimensional rational* $\mathbf{G}$-*module, and fix an integer* $m \geq 0$. *With* $c$ *and* $t$ *defined as above in terms of the root system* $\Phi$, *let* $e$ *and* $f$ *satisfy:*

- $e \geq e(ctm)$;
- $f \geq f(ct\lambda)$ *for all weights* $\lambda$ *of* $M$;
- *If* $p$ *is odd, then* $e \geq e(ct_p(\lambda)(m-1)) + 1$ *for all such* $\lambda$.

*Then for all* $n \leq m$,

$$res : H^n(\mathbf{G}, M^{[e]}) \to H^n(\mathbf{G}(p^{e+f}), M^{[e]})$$

*is an isomorphism. For* $n = m + 1$ *this map is at least injective.*

Taking into account Lemma 14.1, the finite group cohomology in the theorem is in turn isomorphic to $H^n(\mathbf{G}(p^{e+f}), M)$. Therefore, for the given $\mathbf{G}$-module $M$ and for $q$ larger than a fixed power of $p$ (relative to $M$ and $m$) we obtain a stable value for $H^n(\mathbf{G}(q), M)$. This is written

$$H^n_{\text{gen}}(\mathbf{G}, M)$$

and called the **generic cohomology** of $\mathbf{G}$ relative to $M$.

It is natural to ask what can be said about the twisted groups, including Ree and Suzuki groups. For these groups, Avrunin [27, Thm. 5] adapts the methods of [116] to obtain a modified but similarly precise statement giving bounds above which the restriction map induces an isomorphism of cohomology groups. Here the twisting introduces multiples of weights by various powers of $p$, which must be taken into account. (One could twist either by the standard Frobenius map or by a map involving a graph automorphism as well, without changing the end result.)

## 14.6. Discussion of the Proof

It is easy to outline the method of proof, but much harder to fill in the delicate arithmetic estimates needed. A central theme is the relationship between cohomology of $\mathbf{G}$ and cohomology of a Borel subgroup $\mathbf{B}$, which can in turn be reduced to the study of the $\mathbf{T}$-action on the cohomology of $\mathbf{U}$. To compute the latter, one exploits ultimately a detailed knowledge of the cohomology of the additive group.

After developing in §2 and §3 some of the foundational material summarized above in 14.3, the authors devote §4 to a computation of the cohomology ring of the additive group relative to an action by a torus split over $\mathbb{F}_p$. This is expressed in terms of certain exterior and symmetric algebras (with only the latter involved when $p = 2$). A key outcome is that restriction to the cohomology of the additive group over $\mathbb{F}_q$ is an isomorphism.

Turning to the Borel subgroup $\mathbf{B} = \mathbf{TU}$, the authors study the cohomology of $\mathbf{U}$ in terms of the cohomology of root subgroups isomorphic to the additive group. Spectral sequence arguments in §5 relate this cohomology (with its natural $\mathbf{T}$-action) to that of $\mathbf{B}$. The main task is to develop precise arithmetic conditions, depending on $p, i$, and the weights of a $\mathbf{B}$-module $M$, for the restriction map $H^i(\mathbf{B}, M) \to H^i(\mathbf{B}(q), M)$ to be an isomorphism (or at least to be injective). As in §4, the cases $p = 2$ and $p > 2$ have to be treated separately.

This culminates in (5.6) in the main comparison involving restriction maps:

$$
\begin{array}{ccc}
H^i(\mathbf{G}, M) & \to & H^i(\mathbf{G}(q), M) \\
\downarrow & & \downarrow \\
H^i(\mathbf{B}, M) & \to & H^i(\mathbf{B}(q), M)
\end{array}
$$

Assume that the bottom arrow is an isomorphism (resp. injective). The left vertical arrow is an isomorphism thanks to Theorem 14.3(b). On the other hand, the right vertical arrow is at least injective, since $p$ does not divide $[\mathbf{G}(q) : \mathbf{B}(q)]$. Thus the top arrow is an isomorphism (resp. injective).

Note that in the isomorphism case, it follows that the right vertical arrow must also be an isomorphism. This had been observed for sufficiently large $q$ in earlier work by Cline–Parshall-Scott, providing motivation for the approach here.

Finally, §6 develops the needed arithmetic conditions far enough to deduce the main theorem stated above.

## 14.7. Example: $\mathrm{SL}(2, q)$

Rank one groups already illustrate how challenging it can be to get definitive answers to questions about cohomology with various coefficient modules. First consider what can be said about $\mathbf{G} = \mathrm{SL}(2, K)$ and arbitrary simple modules $L(\lambda)$, $\lambda \in X^+$ (identified as usual with $\mathbb{Z}^+$). Thanks to Theorem 14.3, $H^i(\mathbf{G}, L(\lambda))$ can be nonzero only if $\lambda$ is of the form $2pn - 2$ or $2pn$ for some $n \geq 1$. The function $h$ defined there takes the value $h(\lambda) = \lambda/2$. Thus $H^i(\mathbf{G}, L(2pn - 2)) = 0$ for all $i > pn - 1$ and $H^i(\mathbf{G}, L(2pn)) = 0$ for all $i > pn$. Beyond these generalities there are only partial computations in the literature:

(A) In [**116**, 7.6a)] the authors illustrate their main theorem in the case of the standard module $L(1)$ for $\mathrm{SL}(2, K)$, when $p = 2$. Since the fundamental weight is not in the root lattice, all $H^i(\mathbf{G}, L(1)) = 0$. But twisting just once by the Frobenius map yields $H^1(\mathbf{G}, L(p)) \cong H^1_{\mathrm{gen}}(\mathbf{G}, L(1)) \cong K$. For higher degrees, further twisting is needed to reach generic cohomology.

(B) When $0 \leq n < p$ and $i < p - 2$, Friedlander–Parshall [**180**, 6.7] deduce from their general theorems that $H^i_{\mathrm{gen}}(\mathbf{G}, L(n)) = 0$, while

$$
H^i_{\mathrm{gen}}(\mathbf{G}, L(2p - n - 2)) = \begin{cases} 0 & \text{if } n \text{ is odd}, \\ 0 & \text{if } n \neq i - 1 \text{ is even}, \\ K & \text{if } n = i - 1 \text{ is even}. \end{cases}
$$

(C) As discussed below in 14.8, Friedlander [**178**] computes the cohomology of $L(\lambda)$ for certain special weights. For type $A_\ell$ the method is made explicit.

(D) One can also approach $G = \mathrm{SL}(2, q)$ directly, using a mixture of finite group techniques. In [83] Carlson describes explicit (though increasingly complicated) bases for the various cohomology groups of simple modules, along with a description of the cup product of basis elements in the cohomology ring. More generally, he describes all $\mathrm{Ext}_G^i(L, M)$ for pairs of simple $KG$-modules $L, M$. The fact that a Sylow $p$-subgroup of $G$ is a trivial-intersection group makes it possible to use standard induction techniques. He finds in particular that for $i > 0$ one has $\mathrm{Ext}_G^i(L, M) \cong \mathrm{Ext}_B^i(L, M)$. The bases are made explicit for $\mathrm{Ext}^1$ (thereby recovering results of Andersen–Jørgensen–Landrock [19]) and $\mathrm{Ext}^2$. It is already clear at this stage that much of the cohomology is non-generic, with numerous special cases when $r$ is small.

## 14.8. Explicit Computations

A number of special cases have been worked out using the methods of Cline–Parshall–Scott–van der Kallen [116], together with refinements of their formulation.

(A) In [116, §7] the authors make their stability estimates more explicit for $H^1$ and $H^2$. In particular, they prove for any (finite dimensional) $\mathbf{G}$-module $M$ that $H^1(\mathbf{G}, M^{[1]}) \cong H^1_{\mathrm{gen}}(\mathbf{G}, M^{[1]})$, except possibly in type $C_\ell$ when $p = 2$. In any case, $H^2(\mathbf{G}, M) \to H^2_{\mathrm{gen}}(\mathbf{G}, M)$ is injective.

In another direction, they derive more precise conditions (typically satisfied by modules with "small" weights) for $H^0(\mathbf{G}, M) \cong H^0(\mathbf{G}(q), M)$ and for restriction $H^1(\mathbf{G}, M) \to H^1(\mathbf{G}(q), M)$ to be injective. In such special situations, it follows that all $\mathbf{G}(q)$-submodules of $M$ are also $\mathbf{G}$-submodules. The proof involves a result by Wong on Weyl modules which was approached from another direction in 5.9 above.

(B) Friedlander–Parshall explore many of the related issues about cohomology of algebraic groups, Frobenius kernels, and finite groups of Lie type.

In [180] they begin with a detailed computation of $H^i(\mathrm{GL}(n, q), M)$ when $M$ is the adjoint module, $p$ is odd, $q = p^r$, and $i < \min\{2p - 1, r(2p - 3) - 2\}$. Their main theorem (6.1) builds on [116] and treats the general case when $\mathbf{G}$ is split over $\mathbb{F}_p$ and $\lambda$ lies in the closure of the lowest $p$-alcove. For $i < p - 2h + 2$ (resp. $i < 2(p - 10)/3$ in type $G_2$) and $r \geq 1$, they prove that $\dim H^i(\mathbf{G}, L(\lambda)^{[r]})$ is the number of "generalized exponents" of $\lambda$ equal to $i/2$. (The generalized exponents arise in classical work of Kostant and Hesselink and are computable from root data. For example, when $\lambda$ is the highest root, these are just the usual exponents of $W$.) The argument shows also that this cohomology is isomorphic to $H^i_{\mathrm{gen}}(\mathbf{G}, L(\lambda))$ whenever $r \geq 1$. For the proof of the theorem, they make essential use of spectral sequences relating cohomology of $\mathbf{G}$ and $\mathbf{G}_1$.

Further work using similar methods in [182] yields more explicit descriptions of generic cohomology for Chevalley groups of type $A_\ell$, for a range of degrees roughly twice that covered by [180]. In the general case they strengthen the main theorem of [116]. (A few corrections to the proof are noted by Friedlander [178].) This leads to more explicit computations of finite group cohomology with coefficients in induced modules $H^0(\lambda)$. One broad conclusion is that for $p$ sufficiently large relative to $\lambda, i$, and the type of $\mathbf{G}$, the dimension of $H^i(\mathbf{G}(p), L(\lambda))$ is independent of $p$.

(C) Initial computations of rational and generic cohomology of the sort just described require that the highest weights involved be "small" relative to $p$. Using similar techniques, Friedlander [**178**] works out instead some special cases in which the weights are "large". With $\mathbf{G}$ as before and a fixed $n \geq 0$, he considers a weight $\lambda = \lambda_0 + p\lambda_1 + \cdots + p^{r-1}\lambda_{r-1}$ satisfying some technical inequalities which depend on $n, p$ and the root data. Then the odd cohomology $H^{2m-1}(\mathbf{G}(p^r), L(\lambda)) = 0$ in degrees $< n$, while the dimensions of all $H^{2m}$ in degrees $< n$ are specified explicitly in terms of combinatorial data involving $\lambda$ and $W$.

This is illustrated explicitly when $\mathbf{G}$ has type $A_\ell$ and $r = 2$, so $G = \mathrm{SL}(\ell+1, p^2)$. When $n < p - 2\ell - 4$, the weight $\lambda := \alpha + p\alpha$ (where $\alpha$ is the highest root) satisfies the inequalities. It follows that

$$\dim H^{2m}(G, L(\lambda)) = \begin{cases} m - 1 \text{ if } m \leq \ell + 1, \\ 2\ell + 1 - m \text{ if } \ell + 1 \leq m \leq 2\ell, \\ 0 \text{ otherwise.} \end{cases}$$

For example, when $G = \mathrm{SL}(2, p^2)$, one gets $\dim H^4(G, L(2p + 2)) = 1$.

## 14.9. Recent Developments

In the series of papers [**37**]–[**43**] already encountered in 12.8–12.10, Bendel–Nakano–Pillen have refined and extended considerably the earlier comparisons with finite group cohomology.

As in the case of Ext, their analysis of cohomology in higher degrees involves induction of $KG$-modules to $\mathbf{G}$ followed by truncation. To deal with degrees up to a specified bound $s$, they require that $p$ be large enough relative to both $h$ and $s$. They also modify their functor $\mathcal{G}$ to a functor $\mathcal{G}_s$ and find conditions for $\mathcal{G}_s(K)$ to be semisimple. Here we just formulate one of their resulting comparison theorems [**37**, Thm. 7.5]. To obtain an arbitrary $\lambda \in X_r$, they apply the extended affine Weyl group $\widetilde{W}_p$ to a weight in the lowest $p$-alcove. When there is more than one such weight in a $\widetilde{W}_p$-orbit, they take care to start with an especially small weight.

THEOREM. *Let $G = \mathbf{G}^F$ be a Chevalley group over a field of $p^r$ elements. Fix an integer $s \geq 1$ and $p \geq (4s + 1)(h - 1)$. Let $\nu \in X_r$ satisfy $\langle \nu, \alpha_0^\vee \rangle < 2s(h - 1)$, so in particular $\nu$ lies in the lowest $p$-alcove. Choose $w \in \widetilde{W}_p$ for which $w \cdot \nu \in X_r$. Then for all degrees $0 \leq i \leq s$,*

$$H^i(G, L(w \cdot \nu)) \cong H^i(\mathbf{G}, L(w \cdot 0 + p^r\nu)).$$

*Moreover, all nonvanishing cohomology groups for $G$ in this range are obtained by this process from $\mathbf{G}$-cohomology.*

These ideas are generalized further in [**39**, §4], where the comparison of $G$-cohomology and $\mathbf{G}$-cohomology is related more explicitly to the effect of translation functors on modules. Recall from 12.4 that $H^*(\mathbf{G}, L(\lambda)) = 0$ unless $\lambda$ is $W_p$-linked to 0, in contrast to the situation for $G$. This shows up in the theorem just quoted and makes the use of translation essential.

# CHAPTER 15

# Complexity and Support Varieties

In this chapter we look further at the use of cohomological methods to study the module category of a finite group. Unless otherwise indicated, all modules are assumed to be *finite dimensional.*

First we define a numerical measure of the "complexity" of a minimal projective resolution, then relate this to a geometric construction of "support varieties" based on the cohomology ring (15.1–15.2). In the case of finite groups of Lie type there is a useful indirect comparison with the parallel theory for Frobenius kernels, where more systematic results are available. This parallel has been worked out most fully in the special case of Chevalley groups over $\mathbb{F}_p$ (15.3–15.5), but is expected to generalize well.

Precise results have been obtained, using more special methods, in the case of modules of complexity $\leq 1$: those which admit periodic projective resolutions (see 15.6–15.7).

## 15.1. Complexity of a Module

Alperin [4] formulates the following definition of (cohomological) complexity. For an arbitrary finite group $G$ and $KG$-module $M$, consider a minimal projective resolution (whose terms are direct sums of PIMs, uniquely determined up to isomorphism):

$$\cdots P_3 \to P_2 \to P_1 \to P_0 \to M \to 0$$

Then the **complexity** $c_G(M)$ of $M$ is defined to be the least nonnegative integer $c$ (or $\infty$ if no such such $c$ exists) for which there is a constant $C$ such that

$$\dim P_n \leq Cn^{c-1} \text{ for all } n > 0.$$

Expressed a bit differently, $c_G(M)$ is the least $c$ for which

$$\lim_{n \to \infty} \frac{\dim P_n}{n^c} = 0.$$

For example, a projective module clearly has complexity 0. Going the other way, a module of complexity 0 is projective due to the fact that $KG$ is a self-injective algebra. Complexity of a module may be thought of as measuring how far the module is from being projective.

Here are some basic properties of complexity, due to Alperin–Evens and others. (See Benson [**45**, 2.24] or [**48**, Chap. 5], as well as Carlson [**82**].) Recall that the $p$-**rank** of a finite group $G$, denoted $r_p(G)$, is the largest rank of any elementary abelian $p$-subgroup.

THEOREM. *For an arbitrary finite group $G$, let $K$ be the trivial $KG$-module and let $M$ be an arbitrary $KG$-module.*

(a) *Letting $E$ range over all elementary abelian p-subgroups of $G$, $c_G(M) = \max_E c_E(M_E)$, where $M_E$ denotes the restriction of $M$ to $E$.*
(b) *$c_G(K)$ is equal to $r_p(G)$.*
(c) *$c_G(M) \leq c_G(K)$.*

The case $c_G(M) \leq 1$ will be discussed in some detail below: see 15.6.

## 15.2. Support Varieties

It turns out that $c_G(M)$ can be interpreted as the dimension of an affine variety $V_G(M)$, called the **support variety** of $M$. This idea was developed extensively by Carlson in a series of papers. When $M$ is the trivial module $K$, we just write $V_G$ in place of $V_G(K)$. (The alternate notation $|G|_M$ is also found in the literature.)

To construct such varieties, one exploits the fact that the cohomology ring $H^*(G, K)$ is a finitely generated, graded commutative $K$-algebra (14.1). Thus it is actually commutative when $p = 2$. When $p > 2$, it is known that the square of each element of *odd* degree is 0; so only the even cohomology contributes to the geometric interpretation. Accordingly we define $H^\bullet(G, K)$ to be $H^*(G, K)$ when $p = 2$ and to be the sum of all cohomology groups in even degrees when $p > 2$. As a $K$-algebra this is both commutative and finitely generated. Thus its maximal ideal spectrum defines an affine variety $V_G$, which is also *homogeneous* (closed under scalar multiples when realized in an affine space). In turn, if $M$ is a finite dimensional $KG$-module, then $H^\bullet(G, K)$ acts in a natural way on $\operatorname{Ext}_G^*(M, M)$. The annihilator ideal corresponds to a homogeneous subvariety $V_G(M)$ of $V_G$, the support variety of $M$.

Support varieties are defined in a somewhat abstract way in terms of cohomology. But they can be realized more concretely in terms of how modules restrict to elementary abelian $p$-subgroups $E$ of $G$, in the spirit of Theorem 15.1(a). For this Carlson introduced the **rank variety** of an $E$-module $M$, which plays for $E$ essentially the same role as $V_E(M)$. Say $E$ is elementary abelian of order $p^r$, with generators $x_1, \ldots, x_r$. Thus $\dim V_E = r$. The rank variety naturally isomorphic to $V_E$ is just the linear span $V_E^\sharp \subset KE$ of the (independent) vectors $1 - x_i$. In turn, $V_E(M)$ is identified with the subset consisting of 0 together with those $v \in V_E^\sharp$ such that the restriction of $M$ to the "shifted" subgroup of the multiplicative group of $KE$ generated by $1 + v$ is not free.

Then $V_G(M)$ is in natural bijection (under an inseparable isogeny) with the direct limit of rank varieties as $E$ varies. While the definition of rank variety is a bit technical, it can often be explicitly computed (compare the special cases discussed in 5.11 above).

The following theorem summarizes some basic features of support varieties, obtained in the context of rank varieties by a number of people; some of the proofs are easy, but others quite subtle. See Benson [**45**, 2.25–2.27] or [**48**, Chap. 5] for a thorough account.

THEOREM. *Let $G$ be any finite group. If $M$ and $N$ are $KG$-modules, then:*

(a) *$\dim V_G(M) = c_G(M)$. In particular, $\dim V_G = r_p(G)$.*
(b) *$V_G(M)$ is a homogeneous subvariety of $V_G$, hence defines an associated projective variety $\overline{V_G(M)}$.*
(c) *If $p$ does not divide $\dim M$, then $V_G(M) = V_G$.*
(d) *$V_G(M \oplus N) = V_G(M) \cup V_G(N)$.*

(e) $V_G(M \otimes N) = V_G(M) \cap V_G(N)$.

(f) *Every homogeneous subvariety of $V_G$ has the form $V_G(M)$ for some $M$. If the variety is irreducible, then $M$ can be chosen to be indecomposable.*

(g) *If $M$ is an indecomposable $KG$-module, then $\overline{V_G(M)}$ is connected.*

As an exercise, Benson [**45**, p. 139] states a more precise version of part (c): If $s := r_p(G) - c_G(M)$, then $p^s \mid \dim M$.

## 15.3. Support Varieties for Restricted Lie Algebras

A closely analogous theory of support varieties has been developed for arbitrary restricted Lie algebras and their restricted representations, mainly by Friedlander–Parshall [**181**]–[**183**] and Jantzen [**255**]–[**256**]. For a useful survey, see Nakano [**325**]. Here we outline some of the high points as they apply to $\mathfrak{g}$ (or $\mathbf{G}_1$). This leads to subtle indirect comparisons with the theory for Chevalley groups over $\mathbb{F}_p$ through work of Lin–Nakano [**301**], Nakano–Parshall–Vella [**326**], and Carlson–Lin–Nakano [**84**]. Current work is extending many of the ideas to higher Frobenius kernels $\mathbf{G}_r$, where comparisons can be made with Chevalley groups (and potentially twisted groups) over arbitrary finite fields. But we limit the discussion to the prime field, where the strongest results have so far been obtained.

As in the finite group case, the starting point is a general theorem due to Friedlander–Parshall on the finite generation of the cohomology algebra of a restricted Lie algebra $\mathfrak{g}$. Call its restricted enveloping algebra $u(\mathfrak{g})$. Then the maximal ideal spectrum of the commutative algebra obtained from the even degree part of $H^*(u(\mathfrak{g}), K)$ is a homogeneous affine variety, sometimes called the **cohomology variety** $X_{\mathfrak{g}}$. In our situation $\mathfrak{g} = \mathrm{Lie}(\mathbf{G})$, which insures that $\mathbf{G}$ acts naturally on $X_{\mathfrak{g}}$. Given a finite dimensional $\mathfrak{g}$-module $M$, one gets a subvariety $X_{\mathfrak{g}}(M)$ much as in the finite group case. When $M$ is a rational $\mathbf{G}$-module, such as $L(\lambda)$ or $H^0(\lambda)$, there is also a natural $\mathbf{G}$-action on this subvariety.

The situation can be made more concrete by mapping $X_{\mathfrak{g}}$ into $\mathfrak{g}$. Denote by $\mathcal{N}$ the **nilpotent variety** (also called **nullcone**) in $\mathfrak{g}$, consisting of all nilpotent elements. This is a much-studied affine variety of dimension $2m = |\Phi|$, comprising a finite number of $\mathbf{G}$-orbits. What motivates the introduction of $\mathcal{N}$ here is the fact (valid only for $p > h$) that odd degree cohomology vanishes and $H^*(\mathfrak{g}, K)$ is isomorphic to the Frobenius twist of the algebra of regular functions on $\mathcal{N}$.

For arbitrary $p$ a further refinement is needed. Set $\mathcal{N}_1 := \{x \in \mathcal{N} \mid x^{[p]} = 0\}$, sometimes called the **restricted nullcone**: it coincides with $\mathcal{N}$ just when $p \geq h$. There is a *finite* morphism from $X_{\mathfrak{g}}$ onto $\mathcal{N}_1$, so we denote $\mathcal{N}_1$ here by $V_{\mathfrak{g}}$ and call it the **support variety** of $\mathfrak{g}$. Similarly, we denote by $V_{\mathfrak{g}}(M)$ the image of $X_{\mathfrak{g}}(M)$ and call it the support variety of $M$. It is closed, homogeneous, and (if $M$ comes from a $\mathbf{G}$-module) also $\mathbf{G}$-stable. For most purposes it is enough to study these subvarieties of $\mathcal{N}$, which satisfy the same properties listed in Theorem 15.2.

Considerable progress has been made in the study of support varieties for $\mathfrak{g}$. When the characteristic $p$ is good, Nakano–Parshall–Vella [**326**] show that $\mathcal{N}_1$ is actually the closure of a single nilpotent orbit (and thus is an irreducible variety). They also determine the support varieties $V_{\mathfrak{g}}(H^0(\lambda))$, confirming an earlier conjecture by Jantzen. The orbit having $\mathcal{N}_1$ as closure is computed in each case by Carlson, Lin, Nakano, and Parshall. More recently, the University of Georgia VIGRE Algebra Group has extended these results to bad primes.

As in the finite group case, the variety $V_{\mathfrak{g}}(M)$ has a concrete description as a "rank variety": it consists of 0 together with all nonzero elements $x \in \mathcal{N}_1$ for which the restriction of $M$ to the restricted subalgebra $Kx$ is not projective (or equivalently, not free).

These ideas can potentially be generalized to higher Frobenius kernels, but here we just look at $\mathbf{G}_1$ (whose cohomology and representation theory are equivalent in the restricted situation to those of $\mathfrak{g}$). Initially Lin–Nakano [**301**, Lemma 3.4] are able to work out using homological methods some relationships involving complexity as follows:

PROPOSITION. *If $M$ is a finite dimensional $\mathbf{G}$-module, then* $c_{\mathbf{U}_1}(M) = c_{\mathbf{B}_1}(M) = \frac{1}{2}c_{\mathbf{G}_1}(M)$.

## 15.4. Support Varieties for Groups of Lie Type

Here we focus on a Chevalley group $G$ over $\mathbb{F}_p$ and on those $KG$-modules which are restrictions of $\mathbf{G}$-modules. It is not clear at first what connection might exist between support varieties for $G$ and for $\mathfrak{g}$. But it is easy to compare the dimensions, since $\dim V_G = r_p(G)$ and $\dim V_{\mathfrak{g}} = 2m$: see Table 1. Here the $p$-ranks of Chevalley

| Type | $m$ | $p$-rank of $\mathbf{G}(\mathbb{F}_{p^r})$ |
|------|------|------|
| $A_\ell$ | $\binom{\ell+1}{2}$ | $\left[\left(\frac{\ell+1}{2}\right)^2\right]r$ |
| $B_\ell$ | $\ell^2$ | $\begin{cases} \binom{\ell+1}{2}r & \text{if } p = 2 \\ (2\ell - 1)r & \text{if } p \text{ is odd}, \ell \leq 3 \\ \left[1 + \binom{\ell}{2}\right]r & \text{if } p \text{ is odd}, \ell \geq 4 \end{cases}$ |
| $C_\ell$ | $\ell^2$ | $\binom{\ell+1}{2}r$ |
| $D_\ell$ | $\ell^2 - \ell$ | $\binom{\ell}{2}r$ |
| $E_6$ | 36 | $16r$ |
| $E_7$ | 63 | $27r$ |
| $E_8$ | 120 | $36r$ |
| $F_4$ | 24 | $\begin{cases} 11r & \text{if } p = 2 \\ 9r & \text{if } p \text{ is odd} \end{cases}$ |
| $G_2$ | 6 | $\begin{cases} 4r & \text{if } p = 3 \\ 3r & \text{if } p \neq 3 \end{cases}$ |

TABLE 1. Number of positive roots and $p$-rank of a Chevalley group

groups over finite fields are taken from Gorenstein–Lyons–Solomon [**190**, Table

3.3.1], where the defining prime is denoted by $r$ rather than $p$. They also list the $p$-ranks of twisted groups (including the Suzuki and Ree groups). For example, the $p$-rank of a group of type $^2E_6$ over a field of $p^r$ elements is $13r$ when $p = 2$ and $12r$ when $p$ is odd.

The following complexity comparisons are worked out by Lin–Nakano [301, Thm. 3.4]:

THEOREM. *Let $G = \mathbf{G}(\mathbb{F}_p)$ be a Chevalley group over $\mathbb{F}_p$, with $B = \mathbf{B}(\mathbb{F}_p)$.*

(a) *If $M$ is a $\mathbf{B}$-module, then $c_B(M) \leq c_{\mathbf{B}_1}(M)$.*

(b) *If $M$ is a $\mathbf{G}$-module, then $c_G(M) \leq \frac{1}{2}c_{\mathbf{G}_1}(M)$.*

Part (b) follows immediately from part (a), thanks to Proposition 15.3. In order to compare the complexities of $KB$-modules and $\mathbf{B}_1$-modules, they first compare $KU$-modules with modules for the Lie algebra $\mathfrak{n}^+$ associated with $\mathbf{U}_1$. Methods of Quillen and Jennings combine with work on support varieties by Friedlander–Parshall and a spectral sequence due to May to relate the cohomology of $\mathbf{U}_1$ with that of $U$.

As a special case of the theorem, consider the trivial module $M = L(0)$. The complexities in part (b) are just the respective dimensions of the support varieties, so the inequality becomes: $r_p(G) \leq 2m/2 = m$. This is visible in Table 1.

As remarked in 10.4 above, the theorem answers a question of Parshall:

COROLLARY. *Let $G$ be as in the theorem. If $M$ is a finite dimensional rational $\mathbf{G}$-module which restricts to a projective $\mathbf{G}_1$-module, then $M$ is a projective $KG$-module.*

Easy examples show that the converse statement can fail. But a partial converse is proved in [301, Thm. 4.4], in the spirit of 12.8: Assume all weights $\mu$ of a $\mathbf{G}$-module $M$ satisfy the condition $\langle \mu, \alpha^\vee \rangle < p(p-1)$ for the highest root $\alpha \in \Phi$. Then $M$ is a projective $KG$-module if and only if $M$ is projective as a $\mathbf{G}_1$-module. For the proof, one has to compare closely the action on $M$ of a root subgroup in $\mathbf{G}$ and its Lie algebra.

## 15.5. Further Refinements

More recent work by Carlson–Lin–Nakano [84] compares the actual support varieties for the first Frobenius kernel $\mathbf{G}_1$ with support varieties for the corresponding Chevalley group over the prime field, assuming $p \geq 2h - 1$. The comparison here is rather subtle, but makes possible more explicit treatment of support varieties for the finite groups. This builds on the previous work of Nakano–Parshall–Vella [326] and Lin–Nakano [301].

As an example of what can be computed by combining the available techniques, Carlson–Lin–Nakano work out the dimensions of all support varieties of simple modules $L(\lambda)$ for Chevalley groups of rank 2 over $\mathbb{F}_p$ and weights $\lambda \in X_1$. Except for the Steinberg module, whose support is 0, each $L(\lambda)$ has full support: the dimension is the $p$-rank, equal to either 2 or 3. This is so even though $p$ divides the dimensions of some of these simple modules.

The paper [84] also extends the work of Alperin–Mason described in 5.11 above to arbitrary root systems and computes support varieties of various modules $H^0(\lambda)$ for finite general linear groups in low ranks.

## 15.6. Resolutions and Periodicity

Since the early work of Swan and others, periodicity phenomena in the cohomology of groups have attracted special interest, helping to inspire the notion of complexity of a module. For a finite group $G$ over $K$, Alperin [4] calls a (finite-dimensional) $KG$-module $M$ **periodic** if there exists an exact sequence of $KG$-modules

$$0 \to M \to P_{n-1} \to \ldots \to P_0 \to M \to 0,$$

with all $P_i$ projective. Equivalently, the minimal projective resolution of $M$ is periodic. Clearly projective modules themselves are periodic.

We assume below that $K$ is an algebraic closure of $\mathbb{F}_p$, though this condition can sometimes be weakened.

Work of Alperin, Carlson, Evens, and others leads to a number of precise results about periodic modules within the more general theory of complexity (see Carlson [82], Benson [48, 5.10]):

THEOREM. *Let $G$ be any finite group.*

  (a) *A $KG$-module $M$ is periodic if and only if its complexity $c_G(M)$ is $\leq 1$.*

  (b) *A $KG$-module $M$ is periodic if and only if it is **bounded**: for any $KG$-module $N$, there exists $n$ such that $\dim \operatorname{Ext}^i_{KG}(M, N) \leq n$ for all $i$.*

  (c) *If $r$ is the p-rank of $G$, then $p^{r-1}$ divides the dimension of any periodic module.*

  (d) *A $KG$-module is periodic if and only if its restriction to a Sylow p-subgroup is periodic.*

  (e) *If $G$ has cyclic Sylow p-subgroups, then all of the (finitely many) indecomposable $KG$-modules are periodic.*

  (f) *If the Sylow p-subgroups of $G$ are not cyclic, then there exist infinitely many isomorphism classes of indecomposable periodic $KG$-modules.*

## 15.7. Periodic Modules for Groups of Lie Type

When $G = \operatorname{SL}(2, p)$, Theorem 15.6(e) shows that all simple (or other indecomposable modules) are periodic. This was illustrated already in 13.4 for the simple modules. But beyond this case, it requires more work to sort out which simple modules are periodic in the case of groups of Lie type. Unless $G$ has a split $BN$-pair of rank 1, it turns out that periodic simple modules must be projective. The work of a number of people culminates in the following definitive result, which adapts readily to groups not of universal type:

THEOREM. *Let $G = \mathbf{G}^F$ over a field of $q = p^r$ elements. If $G = \operatorname{SL}(2, q)$, then the simple periodic $KG$-modules are precisely those of dimension divisible by $p^{r-1}$. In case $\mathbf{G}$ has rank $\ell > 1$, then:*

  (a) *If $G$ is a Chevalley group, there are no simple periodic $KG$-modules other than the Steinberg module.*

  (b) *If $G$ is twisted of type $A, D,$ or $E_6$, non-projective simple periodic modules exist only when $G = \operatorname{SU}(3, q)$.*

  (c) *Among the Ree and Suzuki groups, only the Suzuki groups ${}^2B_2(q)$ in characteristic 2 have non-projective simple periodic modules.*

When $G = \operatorname{SL}(2, q)$, Jeyakumar [262] determines the simple periodic modules, generalizing Alperin's earlier treatment of $\operatorname{SL}(2, 2^r)$. These are just the modules of dimension divisible by $p^{r-1}$ (a necessary condition by Theorem 15.6(c) since $r$ is the $p$-rank here). By the Tensor Product Theorem, which goes back to Brauer–Nesbitt in this case, the modules of this type involve at most one factor different from $L((p-1)\rho)$. Jeyakumar's proof involves examining concretely how such a module restricts to a Sylow $p$-subgroup of $G$.

The remainder of the theorem follows from the combined work of Fleischmann, Janiszczak, and Jantzen: Janiszczak [**241**] deals with special linear groups of rank at least 2 (by a method which also covers many other Chevalley groups). Then Janiszczak–Jantzen [**242**] finish the treatment of Chevalley groups, which mainly requires a close study of types $B_2$ and $G_2$. (Further results in this direction have been obtained by Carlson–Lin–Nakano in [**84**, 4.2].) Meanwhile, Fleischmann [**172, 173**] deals with the case $G = \mathrm{SU}(3,q)$, after which Fleischmann–Jantzen [**174**] complete the story for all twisted types.

The strategy in these papers is to reduce the problem to the case of rank 2. This is done first for Chevalley groups by invoking the theorem of Smith [**377**] (see 5.10 above) on restriction of simple modules to Levi subgroups. If $L(\lambda) \neq L((q-1)\rho)$ is a periodic non-projective module for $G$, it will also be periodic for any standard rank 2 Levi subgroup (or its derived group): otherwise one gets complexity $> 1$ on restriction to some elementary abelian $p$-subgroup of the latter. By Smith's theorem, the restricted module has as direct summand a simple module whose weight involves some coordinates of $\lambda$. Since the support of a direct sum of modules is the union of the supports, the simple module for the Levi subgroup must also be periodic. But for at least one of these rank 2 subgroups, the restricted module cannot be projective because of the assumption on $\lambda$.

Now in rank 2 one has very detailed information on simple modules, including the powers of $p$ dividing dimensions. Here it is possible to describe the relevant rank varieties concretely. In the case of Chevalley groups, the proof of (a) in the theorem is completed by ruling out non-projective periodic modules for each rank 2 group.

For twisted groups, a similar program is carried out. Here the problem is first reduced to $\mathrm{SU}(4,q)$, where concrete calculations can be done to exclude non-projective periodic modules. Then Fleischmann's determination of all rank varieties for simple modules in the case of $\mathrm{SU}(3,q)$ is combined with a close study of the Ree and Suzuki groups to work out the exceptional periodic cases in (b) and (c).

# CHAPTER 16

# Ordinary and Modular Representations

Until now we have limited our discussion to representations in the defining characteristic $p$. But in this and the following chapter we look at connections with representations in characteristic 0. For convenience we use the term **ordinary representation** (or character) to mean a $\mathbb{C}G$-module (or its character) and the term **modular representation** (or character) to mean a $KG$-module (or its Brauer character).

Brauer's study of modular representations was largely intended to provide new tools for the study of ordinary representations. There turn out to be intimate connections between the two theories, encapsulated in the decomposition matrix. We review the general theory briefly in 16.1–16.4 before turning to groups of Lie type. Blocks and their defect groups play an essential organizing role throughout. The decomposition behavior for blocks with a cyclic (possibly trivial) defect group is well understood for arbitrary finite groups, but has limited applicability to groups of Lie type (16.6–16.7).

To put the comparison of ordinary and modular characters in perspective, we take a detailed look at the family $\mathrm{SL}(2, q)$ in 16.8–16.10. This is a useful though oversimplified prototype for other groups of Lie type.

One obstacle to studying higher rank groups is the far greater complexity of their ordinary character theory. We recall in 16.11 some special cases in which characters have been worked out, mainly prior to the Deligne–Lusztig theory of the mid-1970s (to which the next chapter is devoted).

To round out our preliminary discussion of the decomposition problem, we examine the least generic possibility: ordinary characters of degree greater than 1 which remain irreducible modulo $p$. These are quite rare and occur mainly in type $C_\ell$ (16.13). Recent work by Tiep and Zalesskii almost completely determines them (16.14).

## 16.1. The Decomposition Matrix

In some cases all irreducible representations of a finite group $G$ can be realized over $\mathbb{Z}$: for example, when $G$ is a symmetric group. Then there is an obvious procedure for reducing matrix entries modulo a prime $p$ and extending the base field to $K$ to obtain modular representations. But in general it is necessary to work over suitable rings of algebraic (or $p$-adic) integers. For details we refer to Curtis–Reiner [**125**, Chap. XII] and [**126**, §16].

In the latter book, the standard set-up involves a *p-modular system*, which we denote $(L, R, k)$ in order to avoid notational conflict with our other conventions. Here $L$ is a field of characteristic 0, $R$ a discrete valuation ring (with unique maximal ideal $\mathfrak{p}$) having $L$ as field of fractions, and $k$ the residue field $R/\mathfrak{p}$ of characteristic $p > 0$. Typically $L$ is either a number field or its $\mathfrak{p}$-adic completion. For most arguments one requires that $L$ and $k$ have enough roots of unity to serve as splitting fields for a given finite group $G$. For example, it suffices for $L$ to contain the $n$th roots of 1 if $n$ is the least common multiple of orders of $p$-regular elements of $G$.

Given an $LG$-module $M$, one first observes that there exists an $R$-lattice $M_0$ in $M$ which is $G$-stable. Thus $k \otimes_R M_0$ becomes a $kG$-module, coming by reduction modulo $p$ from $M$; its dimension over $k$ is equal to $\dim_L(M)$. Even though the lattice $M_0$ is not uniquely determined, it turns out that $M$ determines the composition factor multiplicities of the resulting $kG$-module:

PROPOSITION. *Let $G$ be any finite group and $M$ any $LG$-module. If $M_0$ is any $G$-stable $R$-lattice in $M$, then the multiplicities of the composition factors of the $kG$-module $k \otimes_R M_0$ (and therefore its Brauer character) depend only on $M$.*

This allows us to associate unambiguously with a simple $LG$-module $S$ and a simple $kG$-module $T$ the multiplicity $d_{ST}$ with which $T$ occurs as a composition factor for any reduction modulo $p$ of $S$. The language of ordinary and modular characters is more commonly used here. The result is a rectangular matrix $D$, with rows traditionally indexed by the ordinary characters and columns indexed by the modular characters (in some chosen order). The trivial character of each type is usually placed first. Apart from the ordering of row and column indices, $D$ is uniquely determined by $G$. It is called the **decomposition matrix** of $G$ relative to $p$.

In more conceptual language, the proposition allows us to define unambiguously a **decomposition homomorphism** $d : R_L(G) \to R_k(G)$ between Grothendieck groups. Relative to the standard bases coming from simple modules, the matrix of $d$ is then the *transpose* of $D$. (In consulting the literature, the reader should be aware of the conflicting conventions about the row and column indexing of the decomposition matrix.)

## 16.2. Brauer Characters

The process of reduction modulo $p$ yields a comparison of ordinary characters with Brauer characters: see Curtis–Reiner [**125**, §82] or Serre [**366**, 18.1]. This uses the fact that $L$ contains sufficiently many roots of unity. Those of order prime to $p$ lie in $R$ and map bijectively under the residue map to the corresponding roots of unity in $k$. So the choice of a $p$-modular system $(L, R, k)$ gives a natural lifting of roots of unity for the definition of Brauer characters.

In this setting, let the simple $LG$-module $S$ afford the ordinary character $\zeta$. After reduction modulo $p$, the Brauer character values on the set $G_{\text{reg}}$ of $p$-regular elements are just the values of $\zeta$ on $G_{\text{reg}}$. It is convenient to write $\tilde{\zeta}$ for the Brauer character associated with $\zeta$. (This is unrelated to our earlier notation $\tilde{\nu} = \pi(\nu)$ for $\nu \in X$.)

REMARK. In the passage from $\zeta$ to the corresponding Brauer character, it is sometimes useful to define $\tilde{\zeta}$ to be 0 on elements of $g$ not in $G_{\text{reg}}$. While $\zeta$ itself usually has some nonzero values outside $G_{\text{reg}}$, it turns out that in one important special case $\zeta = \tilde{\zeta}$ everywhere: the case when the simple module $S$ has dimension divisible by the full power of $p$ dividing $|G|$ (see Theorem 16.6 below).

## 16.3. Blocks

Continuing in the framework of reduction modulo $p$, we observe next that the process is compatible with the block structure of $kG$:

PROPOSITION. *Let $G$ be any finite group and fix a p-modular system $(L, R, k)$ as above. If $S$ is a simple LG-module, then all composition factors of a kG-module obtained from $S$ by reduction modulo p lie in the same block of kG.*

This shows that, with suitable ordering of row and column indices, the decomposition matrix $D$ has a natural block diagonal form corresponding to the blocks of $KG$. So comparisons between ordinary and modular characters can be made block by block.

In Chapter 8 we looked at the blocks just in the context of modular theory and defined the notion of *defect*. This can be characterized in a new way in terms of ordinary characters. If $p^e$ is the power of $p$ dividing $|G|$, consider the largest power $p^f$ which divides the degree of every irreducible character in $\mathcal{B}$; here $f \le e$, since these degrees divide $|G|$. Then $d = e - f$ [**125**, 86.3]. Blocks of defect 0 are characterized in 16.6 below.

## 16.4. The Cartan–Brauer Triangle

From early work of Brauer–Nesbitt emerges a striking connection between the Cartan matrix $C$ and the decomposition matrix $D$, already quoted in Chapter 11:

THEOREM. *For an arbitrary finite group $G$, the matrices $C$ and $D$ are related by the equation $C = D^t D$.*

The equation $C = D^t D$ expresses what is now called **Brauer reciprocity**. It provides an indirect way to compute $C$. The original proof was carried out in the language of character theory: see Curtis–Reiner [**125**, 83.9]. But Serre [**366**, Part III] reformulated the ideas in a more conceptual way, which is incorporated in the later account by Curtis–Reiner [**126**, Chap. 2].

As before, let $(L, R, k)$ be a $p$-modular system for $G$. Denote by $P_k(G)$ the Grothendieck group of (finitely generated) projective $kG$-modules Then we have a map $c : P_k(G) \to R_k(G)$ recording the multiplicities of composition factors of PIMs. It is called the **Cartan homomorphism** and has as matrix relative to standard bases the Cartan matrix $C$. (Strictly speaking, the matrix should be the *transpose* of the earlier $C$, but since $C$ is symmetric this makes no difference.)

The proof of the above theorem involves a third map $e : P_k(G) \to R_L(G)$, defined by "lifting" a PIM to an $RG$-module, then tensoring this module with $L$ and passing to its class in $R_L(G)$. Combining these three maps, one gets the **Cartan–Brauer triangle** of $G$:

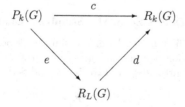

The triangle is commutative and has some other striking properties. The matrix of the map $e$ relative to standard bases of the Grothendieck groups turns out to be the transpose of the matrix for $d$ (the maps $e$ and $d$ being adjoints of each other in a suitable sense). Moreover, $d$ is *surjective* and $e$ is a split injection; its image consists of the classes of (virtual) representations whose character vanishes outside $G_{reg}$. The theorem follows, though in this modern format we are actually dealing with transpose matrices and should write $C = D D^t$ (as Serre does). But we prefer to follow Brauer's traditional way of writing $C$ and $D$.

Concretely, the triangle suggests an indirect approach to computing $D$: first determine the Brauer characters of PIMs in some way, then lift these to ordinary characters and express them in terms of irreducible characters. This actually leads to an effective algorithm for groups of Lie type, as we will explain in the next chapter.

The general theory just outlined is valid whenever the fields involved contain sufficiently many roots of unity, but for our limited purposes it is enough to consider only representations over the algebraically closed fields $\mathbb{C}$ and $K$. So we make no further use of the $p$-modular system $(L, R, k)$. We also interchange freely the language of modules and characters.

## 16.5. Groups of Lie Type

Return now to the study of finite groups of Lie type. Explicit character data for some of the smaller groups may be found in the two atlases [117] and [243]. In isolated cases the matrix $D$ has been computed explicitly for individual groups, often in conjunction with the computation of the Cartan matrix $C$: see the summary in 11.12 as well as 16.12 below.

But beyond devising computational algorithms, we are interested in finding patterns in the decomposition behavior of characters for an entire family of groups such as $SL(n, q)$, with $q$ as a parameter. While individual character tables and decomposition matrices become unwieldy in size as $q$ grows, the underlying theory exhibits a considerable amount of uniformity. This is already visible in the early work of Frobenius and Schur on the characters of $SL(2, q)$, which we review below in 16.8 as a potential model for the general case. For all $G$ the conjugacy classes have been well-studied: see Carter [89], Humphreys [237, Chap. 8], Springer–Steinberg [386]. For large enough $q$, most semisimple (= $p$-regular) classes lie in families sharing a common size and centralizer type. But the characters are much more difficult to find.

Eventually the work of Deligne–Lusztig (described in the following chapter) leads to almost definitive information about the characters of arbitrary groups of Lie type. Here we just examine some initial steps based on Brauer theory.

## 16.6. Blocks of Defect Zero: the Steinberg Character

It is instructive to look first at blocks of the simplest type. Brauer-Nesbitt [61] establish a natural bijection between the blocks of defect 0 in $KG$ and certain irreducible characters. (See Curtis–Reiner [126, 18.28] and Serre [366, 16.4].)

THEOREM. *Let $G$ be an arbitrary finite group, and let $p^e$ be the largest power of $p$ dividing $|G|$.*

(a) *If $M$ is a simple $\mathbb{C}G$-module whose dimension is divisible by $p^e$, then it remains simple after reduction modulo $p$ and becomes a projective $KG$-module. Thus it belongs to a block of defect 0. Moreover, the character afforded by $M$ vanishes outside the set $G_{reg}$ of $p$-regular elements of $G$.*

(b) *Conversely, every block of defect 0 arises in this way.*

What does this imply about a finite group of Lie type? As always we assume that $G = \mathbf{G}^F$ over a field of $p^r$ elements, with $\mathbf{G}$ simply connected. We already saw in 3.7 that any finite group of Lie type has a unique simple module over $K$ of dimension exactly the power of $p$ dividing $|G|$: the *Steinberg module* $\mathrm{St}_r$. It belongs to the unique block of defect 0 in $KG$. This all comes via the representation theory of algebraic groups. But earlier Steinberg [**392**] had shown that $\mathbb{C}G$ also possesses a simple module of the same dimension: it affords the **Steinberg character**. From the theorem it follows at once that its reduction modulo $p$ must be $\mathrm{St}_r$.

The Steinberg character has become ubiquitous in the ordinary character theory of $G$. The underlying representation can actually be constructed over $\mathbb{Z}$, and the character value at a $p$-regular element $s \in G$ is (up to a specified sign) the power of $p$ dividing $|C_G(s)|$. By abuse of notation, we write $\mathrm{St}_r$ (or $\mathrm{St}$ when $r = 1$) for this character. It agrees with the Brauer character of the Steinberg module on $G_{\mathrm{reg}}$ and vanishes elsewhere. See Carter [**89**, Chap. 6] for a comprehensive treatment and Humphreys [**231**] for a survey and references.

## 16.7. Cyclic Blocks and Brauer Trees

Apart from blocks of defect 0, the best understood blocks are those which have (nontrivial) cyclic defect groups. Such a block is often called a **cyclic block**. After Brauer initially studied the case when a defect group has order $p$, Dade and others developed a rich theory applying to all blocks with a cyclic defect group. Modern accounts may be found in Curtis–Reiner [**127**, §62] and Feit [**169**, Chap. VII]. (See also Rouquier [**352**], where $\mathrm{PSL}(2, p)$ serves as an example.) While this theory rarely applies to groups of Lie type, it provides a model for the ongoing study of more complicated blocks such as those having an arbitrary abelian defect group.

Suppose a group algebra $KG$ has a cyclic block $\mathcal{B}$. Then the numbers of irreducible characters (ordinary and modular) can be specified in terms of group data, while the decomposition behavior can be visualized in terms of a **Brauer tree**. We already looked at this tree in the modular setting for the blocks of highest defect of $\mathrm{SL}(2, p)$ in 9.9 and will complete the picture below after reviewing the ordinary characters. In general, the vertices of the tree are labelled with ordinary characters and the edges with modular characters (or equally well with the Brauer characters of PIMs). An "exceptional" vertex, labelling a family of characters of equal degree, is shaded. Then the decomposition of a (lifted) PIM into ordinary characters involves just the characters attached to adjacent vertices. By reciprocity, the reduction modulo $p$ of a character attached to a vertex involves just the modular characters attached to neighboring edges.

One very special feature of a cyclic block, already encountered in 8.9, is that it has finite representation type. The finitely many indecomposable modules can in fact be described precisely in terms of the Brauer tree, as shown by Green [**195**]. For $\mathrm{SL}(2, p)$, see the discussion in 9.10.

## 16.8. Characters of $\mathrm{SL}(2,q)$

Now we look in more detail at the family $\mathrm{SL}(2,q)$. First we recall in Table 1 the well-known character table computed by Schur. References for this include Digne–Michel [**134**, 15.9], Dornhoff [**145**, §38], Humphreys [**219**], Srinivasan [**387**].

|  | $1$ | $z$ | $a^l$ | $b^m$ | $c$ | $d$ |
|---|---|---|---|---|---|---|
| $1_G$ | $1$ | $1$ | $1$ | $1$ | $1$ | $1$ |
| $\psi$ | $q$ | $q$ | $1$ | $-1$ | $0$ | $0$ |
| $\zeta_i$ | $q+1$ | $(-1)^i(q+1)$ | $\tau^{il}+\tau^{-il}$ | $0$ | $1$ | $1$ |
| $\xi_1$ | $\frac{1}{2}(q+1)$ | $\frac{1}{2}\varepsilon(q+1)$ | $(-1)^l$ | $0$ | $\frac{1}{2}(1+\sqrt{\varepsilon q})$ | $\frac{1}{2}(1-\sqrt{\varepsilon q})$ |
| $\xi_2$ | $\frac{1}{2}(q+1)$ | $\frac{1}{2}\varepsilon(q+1)$ | $(-1)^l$ | $0$ | $\frac{1}{2}(1-\sqrt{\varepsilon q})$ | $\frac{1}{2}(1+\sqrt{\varepsilon q})$ |
| $\theta_j$ | $q-1$ | $(-1)^j(q-1)$ | $0$ | $-(\sigma^{jm}+\sigma^{-jm})$ | $-1$ | $-1$ |
| $\eta_1$ | $\frac{1}{2}(q-1)$ | $-\frac{1}{2}\varepsilon(q-1)$ | $0$ | $(-1)^{m+1}$ | $\frac{1}{2}(-1+\sqrt{\varepsilon q})$ | $\frac{1}{2}(-1-\sqrt{\varepsilon q})$ |
| $\eta_2$ | $\frac{1}{2}(q-1)$ | $-\frac{1}{2}\varepsilon(q-1)$ | $0$ | $(-1)^{m+1}$ | $\frac{1}{2}(-1-\sqrt{\varepsilon q})$ | $\frac{1}{2}(-1+\sqrt{\varepsilon q})$ |

TABLE 1. Characters of $\mathrm{SL}(2,q)$, $q$ odd

For notational convenience we assume that $p$ is *odd*. In the easier case $p=2$, there are $q+1$ classes and characters; apart from the trivial character and the Steinberg character, the characters are of degree either $q+1$ or $q-1$ (see [**145**]).

The table involves a number of parameters. Set $\varepsilon := (-1)^{(q-1)/2}$. Fix a primitive $(q-1)$th root $\tau$ of 1, along with a primitive $(q+1)$th root $\sigma$ of 1 (both in $\mathbb{C}$). Let $\nu$ be a fixed generator of the multiplicative group of $\mathbb{F}_p$.

Class representatives include the identity matrix and its negative $z$, along with two unipotent matrices

$$c := \begin{pmatrix} 1 & 1 \\ 0 & 1 \end{pmatrix}, \, d := \begin{pmatrix} 1 & \nu \\ 0 & 1 \end{pmatrix}$$

Two families of $p$-regular elements are represented by powers of the semisimple matrix $a := \mathrm{diag}(\tau, \tau^{-1})$ and a semisimple matrix $b$ of order $q+1$ (which requires a quadratic extension of $\mathbb{F}_q$ to be diagonalized).

The indices for powers of $a$ and $b$ run as follows: $1 \leq l \leq (q-3)/2$ and $1 \leq m \leq (q-1)/2$. Similarly, the indices for the families of characters $\zeta_i$ and $\theta_j$ take values: $1 \leq i \leq (q-3)/2$ and $1 \leq j \leq (q-1)/2$.

In the table we omit the classes of $zc$ and $zd$, for which the character values are easily computed, as follows. For any irreducible character $\chi$, the matrix representing

$z$ is just the scalar $\chi(z)/\chi(1)$. So for arbitrary $g \in G$,

$$\chi(zg) = \frac{\chi(z)}{\chi(1)}\chi(g).$$

## 16.9. The Brauer Tree of SL(2, $p$)

While the character theory of SL(2, $q$) behaves uniformly, it is only when $q = p$ that $G$ has a cyclic Sylow $p$-subgroup. Here we look more closely at how the general theory sketched in 16.7 applies to $G = $ SL(2, $p$). Earlier we considered only the modular aspect of the theory (9.9), as presented for example in Alperin [6, §17]. The full story is given in Curtis–Reiner as well as Dornhoff [146, §71].

The situation when $p = 2$ is easy to describe directly, since here $G$ is isomorphic to the symmetric group of order 6. In this case $KG$ has two blocks, one containing just the Steinberg module and the other containing just the trivial module.

If $p$ is odd, there are two blocks having the Sylow subgroup (cyclic of order $p$) as defect group, along with the block of defect 0 containing St. The respective Brauer trees are given in Figures 1 and 2, for the two cases $p \equiv 1 \pmod 4$ and $p \equiv 3 \pmod 4$. One can work out in an ad hoc way the missing subscripts of the $\zeta$ and $\theta$ characters.

$$
\overset{0}{\circ}\!\!\underset{1_G}{\overset{}{\rule{1.5em}{0.4pt}}}\!\!\overset{p-3}{\underset{\theta}{\circ}}\!\!\underset{}{\overset{2}{\rule{1.5em}{0.4pt}}}\!\!\overset{}{\underset{\zeta}{\circ}}\!\!\overset{p-5}{\underset{\theta}{\rule{1.5em}{0.4pt}}}\!\!\underset{}{\overset{4}{\circ}}\!\!\overset{}{\underset{\zeta}{\rule{1.5em}{0.4pt}}}\!\!\overset{p-7}{\underset{\theta}{\circ}}\!\! \cdots \circ \overset{(p-1)/2}{\rule{1.5em}{0.4pt}}\!\!\underset{\xi}{\bullet}
$$

$$
\overset{}{\circ}\!\!\overset{p-2}{\underset{\theta}{\rule{1.5em}{0.4pt}}}\!\!\overset{1}{\underset{\zeta}{\circ}}\!\!\overset{p-4}{\underset{\theta}{\rule{1.5em}{0.4pt}}}\!\!\overset{3}{\underset{\zeta}{\circ}}\!\!\overset{p-6}{\underset{\theta}{\rule{1.5em}{0.4pt}}}\!\!\overset{5}{\underset{\zeta}{\circ}}\!\! \cdots \circ \overset{(p-3)/2}{\rule{1.5em}{0.4pt}}\!\!\underset{\eta}{\bullet}
$$

FIGURE 1. Brauer trees for SL(2, $p$), $p \equiv 1 \pmod 4$

$$
\overset{0}{\circ}\!\!\underset{1_G}{\overset{}{\rule{1.5em}{0.4pt}}}\!\!\overset{p-3}{\underset{\theta}{\circ}}\!\!\overset{2}{\underset{\zeta}{\rule{1.5em}{0.4pt}}}\!\!\overset{}{\circ}\!\!\overset{p-5}{\underset{\theta}{\rule{1.5em}{0.4pt}}}\!\!\overset{4}{\underset{\zeta}{\circ}}\!\!\overset{p-7}{\underset{\theta}{\rule{1.5em}{0.4pt}}}\!\! \cdots \circ \overset{(p-3)/2}{\rule{1.5em}{0.4pt}}\!\!\underset{\eta}{\bullet}
$$

$$
\overset{}{\circ}\!\!\overset{p-2}{\underset{\theta}{\rule{1.5em}{0.4pt}}}\!\!\overset{1}{\underset{\zeta}{\circ}}\!\!\overset{p-4}{\underset{\theta}{\rule{1.5em}{0.4pt}}}\!\!\overset{3}{\underset{\zeta}{\circ}}\!\!\overset{p-6}{\underset{\theta}{\rule{1.5em}{0.4pt}}}\!\!\overset{5}{\underset{\zeta}{\circ}}\!\! \cdots \circ \overset{(p-1)/2}{\rule{1.5em}{0.4pt}}\!\!\underset{\xi}{\bullet}
$$

FIGURE 2. Brauer trees for SL(2, $p$), $p \equiv 3 \pmod 4$

## 16.10. Decomposition Numbers of SL(2, $q$)

Brauer–Nesbitt [61] work out the matrices $D$ and $C$ for SL(2, $p$) by comparing the ordinary and modular characters directly. They also express the irreducible modular characters of SL(2, $q$) as twisted tensor products of those for SL(2, $p$) (the prototype of Steinberg's tensor product theorem). But they do not describe the matrix $D$ in detail. Indeed, over a field of order $q = p^r$ with $r > 1$, it requires considerably more effort to work out the decomposition numbers. When $G = $ SL(2, $q$), Srinivasan [387] obtains $D$ inductively by an explicit comparison of ordinary and modular characters. In the process she computes the (lifted) principal indecomposable characters.

THEOREM. *Let $G = \mathrm{SL}(2, q)$, with $p$ odd and $q = p^r$. Then all decomposition numbers of $KG$ are equal to 0 or 1, and can be computed explicitly. A "typical" row of $D$ has $2^r$ entries equal to 1.*

The meaning of "typical" becomes clear when one looks at the details of the algorithm. In particular, typical cases predominate when $p$ gets large (with $r$ fixed). Here is an outline, using some of our own notation.

First one has to label both ordinary and modular characters in a compatible way using $p$-adic expansions of nonnegative integers. We symbolize a $p^r$-restricted dominant weight $a_0 + a_1 p + \cdots + a_{r-1} p^{r-1}$ by the $r$-tuple $(a_0, a_1, \ldots, a_{r-1})$. As seen in the character table, ordinary characters are also naturally associated with the restrictions of weights to the finite torus—but with some ambiguities.

For each of the two large series of characters (having respective degrees $q+1$ and $q-1$), the procedure is similar. Take for example the case $q + 1$, parametrized by a weight $(a_0, a_1, \ldots, a_{r-1})$. Srinivasan organizes the list of modular constituents into two bunches called $\Phi(a_0, \ldots, a_{r-1})$ and $\Phi(p-1-a_0, \ldots, p-1-a_{r-1})$. Each $\Phi$ expression is itself a sum of $2^{r-1}$ modular characters, but those having a coordinate $-1$ are interpreted as 0. The recipe is inductive in nature.

Here is a concrete example when $p = 5$ and $r = 3$. Start with the restricted weight $37 = (2, 2, 1)$. This leads to two lists $\Phi(2, 2, 1)$ and $\Phi(2, 2, 3)$ (corresponding to the restricted weight 87). A given weight $(a_0, a_1, a_2)$ determines one modular character $E_0(a_0, a_1, a_2)$. At the next stage one gets a sum

$$E_1(a_0, a_1, a_2) = E_0(a_0, a_1, a_2) + E_0(p - 2 - a_0, a_1 - 1, a_2),$$

followed by

$$E_2(a_0, a_1, a_2) = E_1(a_0, a_1, a_2) + E_1(p - 2 - a_0, p - 1 - a_1, a_2 - 1).$$

For $r = 3$ the process stops here, with $\Phi = E_2$. In our example, we end up with 8 distinct weights in two bunches: $(2, 2, 1)$, $(1, 1, 1)$, $(1, 2, 0)$, $(2, 1, 0)$ and $(2, 2, 3)$, $(1, 1, 3)$, $(1, 2, 2)$, $(2, 1, 2)$.

All of this seems at first glance somewhat opaque, but there is a suggestive pattern. In our example, the four weights in each bunch correspond to the composition factors of the Weyl module with highest weight 37 or 87 (or equivalently, the induced $\mathbf{G}_3\mathbf{T}$-module with that highest weight). Notice how the Weyl dimensions add up: $38 + 88 = 126 = q + 1$. This interpretation is implicit in [**387**, Lemma 1].

For the ordinary characters of degree $q-1$, the procedure is essentially the same, but with $p - 3$ replacing $p - 1$ in the recipe for the second $\Phi$ expression above. For the remaining somewhat degenerate characters, there are modified versions.

Of course, both $\zeta_1 = 1_G$ and $\zeta_2 = \mathrm{St}$ remain irreducible modulo $p$, which translates here into $\zeta_1 = \Phi(0, 0, \ldots, 0)$ and $\zeta_2 = \Phi(p - 1, p - 1, \ldots, p - 1)$. Each of the characters $\zeta_3, \zeta_4, \zeta_5, \zeta_6$ is equal to a single $\Phi$ expression. For example, when $q \equiv 1 \pmod 4$, we have:

$$\zeta_3 = \zeta_4 = \Phi\left(\frac{p-1}{2}, \frac{p-1}{2}, \ldots, \frac{p-1}{2}\right),$$

whereas

$$\zeta_5 = \zeta_6 = \Phi\left(\frac{p-3}{2}, \frac{p-1}{2}, \ldots, \frac{p-1}{2}\right).$$

When $q \equiv -1 \pmod 4$, the recipes are reversed.

REMARK. Although it is usually impossible to recover $D$ from the Cartan matrix, this can be done for $\mathrm{SL}(2,q)$ if one knows in advance that all entries of $D$ are $0, 1$. This is explored by Johnson [263].

## 16.11. Character Computations for Lie Families

We have seen in earlier chapters that irreducible Brauer characters in the defining characteristic for finite groups of Lie type are far from being completely determined. The computation of ordinary characters for families beyond $\mathrm{SL}(2,q)$ has also been extremely challenging. But the breakthrough in the mid-1970s by Deligne and Lusztig [129], followed by extensive work of Lusztig and others, has brought this computation within reach—at least in an algorithmic sense. To provide some historical context, we survey the specific calculations done for Lie-type families, mainly prior to the work of Deligne–Lusztig (to be discussed in Chapter 17). For the Suzuki and Ree groups, see Chapter 20.

- Steinberg [390] develops an approach to the construction of representations of $\mathrm{GL}(n,q)$. In [391] he works out the characters of $\mathrm{GL}(3,q)$ (with steps toward determining the characters of $\mathrm{GL}(4,q)$).
- Green [193] gives a complete combinatorial algorithm for computing the character values of $\mathrm{GL}(n,q)$. (See Springer [383] for an exposition.) Green's approach inspired the general conjectures of Macdonald for other reductive groups over finite fields.
- Ennola [163] adapts Green's ideas to the unitary groups, suggesting in the process that systematic sign changes (in effect replacing $q$ by $-q$) can relate the character tables of general linear and unitary groups. He finds in particular the characters of $\mathrm{U}(3,q^2)$.
- Simpson–Frame [367] determine explicitly the characters of $\mathrm{SL}(3,q)$ and $\mathrm{SU}(3,q)$ (but with incorrect parameters for one family of characters in the latter case, as noted in [233]). They confirm Ennola's expectations for these groups.
- Srinivasan [388] works out the character table for $\mathrm{Sp}(4,q)$ in case $p$ is odd. (There are a few small errors involving signs or parameters: see Przygocki [346]).) This is compared by Jantzen [257] with the known modular data.
- Enomoto [164] completes the determination of the characters of $\mathrm{Sp}(4,q)$ in case $p = 2$.
- Chang–Ree [94] compute the characters of $\mathrm{G}_2(q)$ when $p \neq 2, 3$. The cases $p = 2, 3$ are worked out separately by Enomoto [165, 166].
- Nozawa [327, 328] works out the characters of $\mathrm{U}(4,q^2)$ and $\mathrm{U}(5,q^2)$.
- Looker [305] determines the characters of $\mathrm{Sp}(6,q)$ for $q$ even, following the Deligne–Lusztig method, while Downes [157] partially determines the characters of $\mathrm{Sp}(6,q)$ for $q$ odd. In an unpublished paper, F. Lübeck finds the characters of $\mathrm{CSp}_6(q)$ for $q$ odd and $\mathrm{Sp}_6(q)$ for $q$ even: IWR/SFB-Preprint 93-61, Univ. Heidelberg, 1993.
- Deriziotis–Michler [131] compute the characters of ${}^3D_4(q)$, also following Deligne–Lusztig.

All of these examples exhibit a pattern like that found for $\mathrm{SL}(2,q)$: for large enough $p$, most of the $p$-regular classes fall into families sharing the same size and centralizer order, while most of the characters belong to families (of sizes similar to the class sizes) sharing a common degree roughly equal to $q^m$ ($m = |\Phi^+|$). This pattern occurs quite generally, as we shall see in Chapter 17.

## 16.12. Some Explicit Decomposition Matrices

Apart from Srinivasan's systematic determination of $D$ for $SL(2,q)$ (16.10), decomposition matrices have been found explicitly only for some groups of low rank over very small fields. Many of these are mentioned (with references) in 11.12 and 11.16 in connection with the computation of the Cartan matrix $C$. Except for some of the cases computed by Zaslawsky, these decomposition matrices have been obtained using algebraic group data. But direct comparison of ordinary and modular character tables also yields most of them along with further examples such as the group of type $G_2$ over $\mathbb{F}_5$ and the group $SL(6,2)$. These matrices are currently available at the Modular Atlas homepage [243].

There also exist some fragmentary results about specific decomposition numbers, usually in the least generic cases: see Zalesskii [454, 455] and Zalesskii–Suprunenko [459]. See Chapter 20 for Suzuki and Ree groups.

## 16.13. Special Characters of $Sp(2n, q)$

Apart from the individual cases over small fields just enumerated, some special families of ordinary representations for groups of type C in odd characteristic exhibit interesting decomposition behavior. It is convenient to regard $SL(2,q)$ as a group of type $C_1$.

One family of examples in fact occurs only for $G = SL(2,p)$. Here we identify weights as usual with integers. Notice that one ordinary character of degree $p - 1$ placed at an end of the Brauer tree for the non-principal block has as reduction mod $p$ the simple $KG$-module of highest weight $p - 2$. At the other end of each Brauer tree is a pair of exceptional characters of equal degree $(p \pm 1)/2$ which have as reduction mod $p$ the simple module of highest weight $(p-1)/2$ or $(p-3)/2$.

Similar pairs occur in higher rank for $Sp(2n, q)$ (with $p$ odd), usually called **Weil representations**. These have been constructed and studied over local and finite fields by Weil and later authors. In [457] Zalesskii and Suprunenko prove:

THEOREM. *Let* $\mathbf{G} = Sp(2n, K)$ *with $p$ odd. (In case $n = 1$, $\mathbf{G} = SL(2, K)$.) Use the standard ordering of vertices in the Dynkin diagram, so the last vertex corresponds to the long simple root. Then:*

(a) *The simple $\mathbf{G}$-modules $L(\lambda)$ of respective highest weights*

$$\lambda = \varpi_{n-1} + \frac{p-3}{2}\varpi_n \text{ and } \frac{p-1}{2}\varpi_n$$

*have respective dimensions $(p^n - 1)/2$ and $(p^n + 1)/2$, with all weight multiplicities being 1. (In case $n = 1$, omit $\varpi_{n-1}$ here.)*

(b) *The restrictions of these $\mathbf{G}$-modules to $Sp(2n, p)$ are obtained by reduction modulo $p$ from the pairs of ordinary irreducible characters of the indicated degrees.*

The proof goes by induction on $n$, using explicit descriptions of the way the representations restrict to certain natural subgroups such as products of smaller symplectic groups.

The characters in (b) exist for all $q = p^r$, but have more complicated decompositions modulo $p$ when $r > 1$. In Srinivasan's character table [388] for $Sp(4, q)$, these characters are labelled: $\theta_3$ and $\theta_4$ of degree $(q^2+1)/2$ (in the 1-block); $\theta_7$ and $\theta_8$ of degree $(q^2-1)/2$ (in the other block of highest defect).

## 16.14. Irreducibility Modulo $p$

In the next chapter we will investigate conditions for "generic" decomposition behavior of ordinary characters. At the opposite extreme, the examples just discussed for type C suggest a natural question:

> Which ordinary irreducible representations of a finite group of
> Lie type remain irreducible modulo the defining characteristic?

Obvious examples include the trivial representation and the Steinberg representation. (If one works with a finite group such as $\mathrm{PGL}(n,q)$ unequal to its commutator subgroup, there will be some additional variants of these representations having the same degrees.) But the generic picture developed in the following chapter provides heuristic evidence that representations having this property should be rather rare.

While apparently straightforward, the above question has not (at this writing) been completely answered. However, recent work of Tiep and Zalesskii [**412, 413, 414**], related to the search for "globally irreducible" lattices (see Gross [**197**]), comes very close. Apart from the examples in type C (for odd $p$) discussed in 16.13, and the obvious cases mentioned at the outset, their work strongly suggests that there will only be isolated further examples involving ranks $\leq 2$ and primes $p \leq 5$. For example, $\mathrm{Sp}(4,p)$ has two characters of degree $(p-1)(p^2+1)/2$ which remain irreducible modulo $p$ just when $p = 3,5$. Other such cases include $\mathrm{SU}(3,3)$, $\mathrm{G}_2(2)'$, $^2\mathrm{G}_2(3)'$, $^3\mathrm{D}_4(2)$, and $^2\mathrm{F}_4(2)'$.

The methods used in these papers are somewhat ad hoc, since there is no obvious way to exploit the algorithmic approach of Jantzen described in the next chapter. Many techniques come into play: integral representations, structure theory of groups of Lie type, arithmetic properties of characters, Deligne–Lusztig theory, and representations of algebraic groups. For example, irreducibility modulo $p$ can most readily be ruled out for characters of degree $> 1$ (not in a block of defect 0) whose values lie in a number field unramified over $\mathbb{Q}$ at $p$ [**412**], while the contrary cases require more elaborate study [**414**].

Tiep and Zalesskii are sometimes able to exclude further examples of representations which remain irreducible mod $p$ by showing that dimensions of ordinary and modular representations fail to overlap except in the obvious cases. For an asymptotic result of this type (involving large enough powers of $p$), see [**413**, Thm. 2.2].

Complementing these methods, a close but subtle connection is found between the irreducibility problem and the self-extension problem considered in 12.10 above. In [**412**, Prop. 1.4] they arrive at the following precise formulation, which reinforces the special role of groups of type C in both problems:

PROPOSITION. *Let $p > 3$, with $G$ coming from a simple algebraic group, but exclude the simple types $\mathrm{A}_1, \mathrm{G}_2, \mathrm{F}_4$. If an ordinary character of $G$ of degree $> 1$ remains irreducible modulo the defining characteristic $p$ but does not lie in a block of defect 0, then the modular representation in question must have a nontrivial self-extension.*

# CHAPTER 17

# Deligne–Lusztig Characters

Following Green's successful combinatorial attack on the ordinary characters of $GL(n,q)$, the further character computations cited in 16.11 dealt mainly with groups of very small rank. Even though Macdonald was able to predict the main features of the character theory for arbitrary groups of Lie type, finite group techniques alone seemed inadequate to prove his conjectures. Then a landmark paper by Deligne–Lusztig [**129**] showed how to construct a large number of (virtual) characters from actions of a finite group of Lie type on the étale cohomology of certain subvarieties of the ambient algebraic group. All irreducible characters occur here as constituents. Further work, especially by Lusztig, has determined the irreducible constituents of these virtual characters. Most of the character values have by now been recursively determined; this too requires sophisticated homological and geometric techniques.

We shall not attempt to expose this theory in detail. Instead we rely on the accounts given by Carter [**89**] and Digne–Michel [**134**]; some of the highlights are sketched below. Our focus here is on the way in which Deligne–Lusztig characters (which we call for short **DL characters**) reduce modulo $p$. Combined with Lusztig's methods for relating ordinary characters to DL characters, this yields considerable insight into the decomposition patterns.

Broadly speaking, the ordinary character theory imitates the Harish-Chandra theory of infinite dimensional representations of Lie groups, with its organization of characters into "series" depending on different types of maximal tori. On the other hand, the modular theory imitates the Cartan–Weyl highest weight theory for compact Lie groups. So the decomposition problem mixes these two very different flavors of Lie theory.

One surprising outcome is the fact that, generically, all ordinary characters have the same pattern of composition factors modulo $p$, in spite of the disparate origins of the various series of characters. This pattern agrees with that of a generic $\mathbf{G}_r\mathbf{T}$-module $\widehat{Z}_r(\lambda)$ (or a corresponding generic Weyl module): for example, when $G = SL(3, p^r)$ one typically gets $9^r$ composition factors with multiplicity 1.

## 17.1. Reductive Groups and Frobenius Maps

In this chapter we have to reconsider our basic framework for working with algebraic groups. When dealing only with the modular theory, we have streamlined the formulations by assuming that $\mathbf{G}$ is simple and simply connected. As a rule the results extend routinely to the class of connected reductive groups of all isogeny types. Indeed, the proofs in [**RAGS**] often require this greater generality.

On the other hand, the ordinary character theory can vary substantially for finite groups emanating from a single root system. For example, the character tables of $GL(n, q)$ and $PGL(n, q)$ look much less complicated than that of $SL(n, q)$.

Moreover, the Deligne–Lusztig theory is inherently inductive in nature, requiring equal status for various reductive subgroups of $\mathbf{G}$ including maximal tori and Levi subgroups of parabolics. (More subtle refinements involve nonconnected reductive groups: for example, centralizers of semisimple elements need not be connected when $\mathbf{G}$ is semisimple but fails to be simply connected or to have a connected center.)

Thus in this chapter we allow $\mathbf{G}$ to denote an arbitrary *connected reductive* group defined over $\mathbb{F}_q$; it may for example be a torus. Its rational structure over $\mathbb{F}_q$ corresponds to a Frobenius map $F : \mathbf{G} \to \mathbf{G}$. While $\mathbf{G}$ must be quasisplit over $\mathbb{F}_q$ (by Lang's Theorem 1.4), it need not be split. (When we discuss work of Jantzen and others on reduction modulo $p$, we revert to our standard assumption that $\mathbf{G}$ is simple, simply connected, and split over $\mathbb{F}_p$.)

## 17.2. $F$-Stable Maximal Tori

There is a further complication in Deligne–Lusztig theory not present in the characteristic $p$ setting. Since simple modules over $K$ for a finite group of Lie type are restrictions of rational modules for the ambient algebraic group, they can be studied in terms of weights for a fixed maximal torus $\mathbf{T}$ which is defined and split over $\mathbb{F}_p$. But the character theory of $G$ over $\mathbb{C}$ involves all types of $F$-stable maximal tori. This shows up already in the description of $p$-regular classes in $G$ and later becomes essential for the organization of the characters themselves into series.

By Lang's Theorem, some $F$-stable maximal torus $\mathbf{T}_0$ lies in an $F$-stable Borel subgroup of $\mathbf{G}$; all such tori are in fact $\mathbf{G}^F$-conjugate. Such a torus is called **maximally split**: it contains a subtorus defined and split over $\mathbb{F}_q$ which has largest possible dimension, called the $\mathbb{F}_q$-**rank** of $\mathbf{G}$. Since $F$ also stabilizes $N_{\mathbf{G}}(\mathbf{T}_0)$, there is an induced action of $F$ on $W = N_{\mathbf{G}}(\mathbf{T}_0)/\mathbf{T}_0$. (This action is trivial if $F$ is a standard Frobenius map relative to $q$.)

In general the $\mathbf{G}^F$-conjugacy classes of $F$-stable maximal tori are classified in terms of the Weyl group. Define an equivalence relation on $W$ by calling $w$ and $w'$ $F$-**conjugate** whenever $w' = x^{-1}wF(x)$ for some $x \in W$.

Now given an arbitrary $F$-stable maximal torus $\mathbf{T}$, there exists $g \in \mathbf{G}$ such that $g\mathbf{T}_0 g^{-1} = \mathbf{T}$. Because $\mathbf{T}$ is $F$-stable, $g^{-1}F(g)$ must lie in $N_{\mathbf{G}}(\mathbf{T}_0)$; it therefore projects onto some $w \in W$. We say that $\mathbf{T}$ is obtained from $\mathbf{T}_0$ by "twisting by $w$". Lang's Theorem can be used to show that $\mathbf{T} \mapsto w$ induces the bijection in part (a) of the following theorem. (All of this is explained thoroughly in Carter [**89**, Chap. 3], Springer–Steinberg [**386**].)

THEOREM. *Let $\mathbf{G}$ be a connected reductive group defined over $\mathbb{F}_q$, with corresponding Frobenius map $F$ and finite fixed point subgroup $G = \mathbf{G}^F$. Denote by $Y_0 := \mathrm{Hom}(K^\times, \mathbf{T}_0)$ the cocharacter group of a fixed maximally split torus $\mathbf{T}_0$.*

(a) *There is a natural bijection between the $G$-conjugacy classes of $F$-stable maximal tori in $\mathbf{G}$ and the collection of $F$-conjugacy classes in $W$, under which the class of $\mathbf{T}_0$ corresponds to the class of the trivial element in $W$.*

(b) *Let $\mathbf{T}_w$ be obtained from $\mathbf{T}_0$ by twisting by $w \in W$. Set $T_w := \mathbf{T}_w^F$. If $f(x)$ is the characteristic polynomial of $\pi^{-1} \circ w$ on $Y_0 \otimes \mathbb{R}$, then $|T_w|$ is the absolute value of $f(q)$, a polynomial in $q$ of degree $\ell$.*

EXAMPLES. In the setting of (b), it is not too difficult to compute the orders of finite tori in special cases. For example, when $G = \mathrm{SL}(2, q)$ a finite torus coming from $\mathbf{T}_0$ has order $q-1$ while a torus which splits only over a quadratic extension of $\mathbb{F}_q$ yields a finite torus of order $q+1$. When $G = \mathrm{SL}(3,q)$ there are three classes in $W$ involving elements of orders $1, 3, 2$, yielding corresponding finite tori of orders $(q-1)^2$, $q^2+q+1$, $q^2-1$.

## 17.3. DL Characters

The problem solved initially by Deligne–Lusztig [129] is to construct "cuspidal characters", which by definition are not constituents of representations induced from proper parabolic subgroups. In a small case like $\mathrm{SL}(2,q)$, these missing characters of degree $q-1$ or $(q-1)/2$ can be supplied by ad hoc means involving orthogonality relations. Although Green [193] was able in a somewhat similar spirit to find a complete combinatorial description of the characters of $\mathrm{GL}(n,q)$, one soon reaches the limits of finite group techniques.

In their 1976 paper, Deligne and Lusztig introduced a more sophisticated construction using étale cohomology (with compact support) of certain subvarieties $\widetilde{X}$ of $\mathbf{G}$. Coefficients are taken in a field $\overline{\mathbb{Q}}_l$ of $l$-adic numbers (or its algebraic closure), where $l$ is any prime number different from $p$. Ultimately the choice of $l$ is immaterial, since the resulting character values for $G$ are algebraic integers in $\overline{\mathbb{Q}}_l$ which may be identified with complex numbers.

In the Deligne–Lusztig construction, the finite group $G$ acts on cohomology in each degree, but one gets no direct information about the individual cohomology groups. Instead a character formula is found for the alternating sum, based on a Leftschetz fixed point formula. This is a **virtual** (or **generalized**) character: a $\mathbb{Z}$-linear combination of irreducible characters. Such a character is called **proper** if all coefficients are nonnegative.

Fix a connected reductive group $\mathbf{G}$ defined over $\mathbb{F}_q$ and its Frobenius map $F$. The DL (virtual) characters of $G = \mathbf{G}^F$ are parametrized by pairs $(\mathbf{T}, \theta)$, where $\mathbf{T}$ is an $F$-stable maximal torus of $\mathbf{G}$ and $\theta \in \widehat{T} = \mathrm{Hom}(T, \mathbb{C}^\times)$ is a $\mathbb{C}$-valued character of $T = \mathbf{T}^F$. There is a natural equivalence relation on such pairs, called **geometric conjugacy**, which plays an essential role in organizing irreducible characters into families.

For each pair $(\mathbf{T}, \theta)$, where $\mathbf{T}$ corresponds to an $F$-conjugacy class in $W$, one gets a DL character. Its value at $x \in G$ is given by an Euler character:

$$R_{\mathbf{T}}^{\mathbf{G}}(\theta)(x) := \sum_{i \geq 0} (-1)^i \, \mathrm{Tr}[x, H_c^i(\widetilde{X}, \overline{\mathbb{Q}}_l)_\theta].$$

Here the subscript $\theta$ indicates the subspace on which $T$ acts via $\theta$.

Notation for DL characters varies in the literature, other choices being for example $R_{\mathbf{T},\theta}^{\mathbf{G}}$ or $R_{\mathbf{T}}^{\mathbf{G}}\theta$. Besides [129], accounts of the theory are found in Carter [89], Digne–Michel [134], Cabanes–Enguehard [78]. For shorter surveys, see Carter [87, 91], Curtis [124], Srinivasan [389]. Important sources for later developments include Lusztig [309, 312] together with a large number of research papers by Lusztig and others.

REMARK. Most of the Deligne–Lusztig theory adapts well to the groups of Suzuki and Ree, as shown in [129, §11]. We return to this in Chapter 20.

## 17.4. Basic Properties of DL Characters

For any connected reductive group $\mathbf{G}$ defined over $\mathbb{F}_q$, let $d$ be its $\mathbb{F}_q$-rank (17.1) and set $\varepsilon_{\mathbf{G}} := (-1)^d$. Here are some of the key features of the theory for $G = \mathbf{G}^F$ developed in [**129**]:

- The value of $R_{\mathbf{T}}^{\mathbf{G}}(\theta)$ at 1 is $\varepsilon_{\mathbf{G}}\varepsilon_{\mathbf{T}}|G|_{p'}/|T|$, where $|G|_{p'}$ is the part of $|G|$ relatively prime to $p$. This is a polynomial in $q$ of degree $m = |\Phi^+|$. Although it is sometimes negative, we refer to it as the "degree" of the DL character. For example, when $G = \mathrm{SL}(2,q)$ one has degrees $q+1$ and $1-q$. When $G = \mathrm{SL}(3,p)$, the degrees are $q^3+2q^2+2q+1$, $1-q^3$, $q^3-q^2-q+1$.

- At a unipotent element $u \in G$, the value of $R_{\mathbf{T}}^{\mathbf{G}}(\theta)$ is independent of $\theta$ and is therefore given by the value of $R_{\mathbf{T}}^{\mathbf{G}}(1)$ at $u$. The resulting function $Q_{\mathbf{T}}^{\mathbf{G}}$ on the unipotent subset of $G$ is called a **Green function**.

- More generally, the value of $R_{\mathbf{T}}^{\mathbf{G}}(\theta)$ at an element $x = su$ of $G$ (written in terms of its Jordan decomposition) is given recursively as follows. Denote by $\mathbf{H}$ the identity component of $C_{\mathbf{G}}(s)$; this is a connected reductive group, equal to $C_{\mathbf{G}}(s)$ if the derived group of $\mathbf{G}$ is simply connected. Then the value of $R_{\mathbf{T}}^{\mathbf{G}}(\theta)$ depends just on the data: $|\mathbf{H}^F|$, the values of $\theta$ on $T$, and the values of various Green functions for $\mathbf{H}$.

- When restricted to the set $G_{\mathrm{reg}}$ of $p$-regular elements, the DL characters span the space of all class functions on $G_{\mathrm{reg}}$.

- When $\theta$ is in "general position", $R_{\mathbf{T}}^{\mathbf{G}}(\theta)$ is up to sign an irreducible character. For arbitrary $\theta$, there are $\leq |W|$ irreducible constituents. (This follows from inspection of inner products of DL characters.)

- Every ordinary character occurs as a constituent of some DL character. (However, the DL characters do not usually form a $\mathbb{Z}$-basis of $R_{\mathbb{C}}(G)$.)

- If the collection of DL characters is partitioned according to the geometric conjugacy classes of pairs $(\mathbf{T}, \theta)$, then no ordinary character occurs as a constituent of DL characters lying in distinct classes. Thus the ordinary characters may also be partitioned into **geometric conjugacy classes**.

- If the $F$-stable maximal torus $\mathbf{T}$ lies in an $F$-stable Levi subgroup $\mathbf{L}$ of an $F$-stable parabolic subgroup $\mathbf{P}$ of $\mathbf{G}$, then $R_{\mathbf{T}}^{\mathbf{G}}(\theta)$ is obtained by inducing from $\mathbf{P}$ to $\mathbf{G}$ the lift from $\mathbf{L}$ to $\mathbf{P}$ of the virtual character $R_{\mathbf{T}}^{\mathbf{L}}(\theta)$.

The theory exhibits a strong analogy with the Harish-Chandra theory for a semisimple Lie group. Representations occur in series, obtained in many cases by induction from proper parabolic subgroups. For example, inducing a 1-dimensional representation of a Borel subgroup results in a **principal series** representation. But some representations cannot be extracted from the conventional induction process and belong instead to **discrete series**; their characters are called **cuspidal**. For finite groups of Lie type this distinction shows up already in the case of $\mathrm{SL}(2,q)$: the characters of degree $q + 1$ are constructed by induction from a Borel subgroup, whereas the characters of degree $q - 1$ have more mysterious origins.

REMARK. There is a formal duality on the collection of ordinary characters of a finite group of Lie type. This **Curtis–Alvis duality** has been studied from a number of viewpoints by Alvis, Curtis, Kawanaka, and Deligne–Lusztig: see Carter [**89**, 8.2], Curtis–Reiner [**127**, §71], Digne–Michel [**134**, Chap. 8]. It interchanges an irreducible character with $\pm$ an irreducible character, sending for example the trivial character to the Steinberg character. Moreover, the $p'$-part of the degree of a character is the same as the $p'$-part of the degree of its dual. This duality turns out to fit naturally into Deligne–Lusztig theory.

## 17.5. The Decomposition Problem

In what follows we fix for a given finite group $G$ an isomorphism between a suitable group of roots of unity in $K^\times$ and a corresponding subgroup in $\mathbb{C}^\times$, compatible with reduction modulo $p$ of characters and formation of Brauer characters (see 5.5 and 16.2). For an individual finite group, the decomposition matrix can in principle be computed once the tables of ordinary and Brauer characters are known. But for a Lie family it is extremely difficult to study the decomposition behavior in this spirit. Here, as in the related study of Cartan invariants, we are looking for patterns rather than just computational algorithms. The decomposition matrix itself will get unmanageably large (and at the same time increasingly sparse) as the power of $p$ grows.

In our situation there is no obvious way to look directly at how ordinary characters decompose modulo $p$ in terms of Brauer characters of simple $KG$-modules. An indirect route turns out to be more promising, based on comparison of DL characters and Weyl characters. To streamline the parametrization of modular characters by weights, we revert here and later to the assumption that $\mathbf{G}$ is *simple, simply connected*, and $\mathbb{F}_p$-*split*.

Given a DL character $R^{\mathbf{G}}_{\mathbf{T}}(\theta)$, the Brauer character obtained after reduction modulo $p$ agrees with the original character on $G_{\mathrm{reg}}$ and is denoted by $\widetilde{R}^{\mathbf{G}}_{\mathbf{T}}(\theta)$. On the other hand, the Brauer character $\mathrm{br}\,\chi$ of a formal character $\chi \in \mathcal{X} = \mathbb{Z}[X]^W$ may be written briefly as $\widetilde{\chi}$. Since the Weyl characters $\chi(\lambda)$ for $\lambda \in X^+$ form a $\mathbb{Z}$-basis of $\mathcal{X}$, we can ask how to express $\widetilde{R}^{\mathbf{G}}_{\mathbf{T}}(\theta)$ as a $\mathbb{Z}$-linear combination of the various $\widetilde{\chi}(\lambda)$. If this is known, two further steps are needed to solve the decomposition problem:

(1) For each irreducible character $\zeta$, write $\widetilde{\zeta}$ as a $\mathbb{C}$-linear combination of DL characters restricted to $G_{\mathrm{reg}}$. This can (at least in principle) be done using Lusztig's work. Moreover, the coefficients lie in $\mathbb{Q}$: this follows from the discussion of "uniform functions" in Digne–Michel [**134**, Chap. 12], thanks to the explicit projection formula 12.12.

(2) Write each relevant $\widetilde{\chi}(\lambda)$ as a $\mathbb{Z}$-linear combination of Brauer characters of simple $KG$-modules. Doing this depends of course on knowing the standard character data for $\mathbf{G}$. It also becomes harder to work out in practice as the power of $p$ grows. This underscores the contrast between the Deligne–Lusztig theory, where everything is fairly uniform relative to the parameter $q$, and the modular theory, where there is a hierarchical structure depending on $r$.

Below we explain Jantzen's approach, which involves a further transition. He invokes Brauer reciprocity in order to reformulate the decomposition problem in terms of PIMs and ordinary characters. In turn, we have seen in Chapter 10 how to use standard character data for $\mathbf{G}$ to compare PIMs $U_r(\lambda)$ for $KG$ and $\widehat{Q}_r(\lambda)$ for $\mathbf{G}_r\mathbf{T}$. So we can start instead with the modules $\widehat{Q}_r(\lambda)$ and try to work out the relationship of their Brauer characters with DL characters. One ingredient still missing is a bridge between these data.

## 17.6. PIMs and DL Characters

Consider $G = \mathbf{G}^F$ as above, over a field of $p^r$ elements. We can associate to $\mu \in X$ and $w \in W$ a character $\theta_\mu : T_w \to \mathbb{C}^\times$, as follows. If $\mathbf{T}$ is conjugate to $\mathbf{T}_0$ by $g \in \mathbf{G}$, compose the resulting isomorphism $\mathbf{T} \to \mathbf{T}_0$ with $\mu : \mathbf{T}_0 \to K^\times$, then restrict to $T_w$. Follow this with the chosen isomorphism between roots of unity in $K^\times$ and roots of unity in $\mathbb{C}^\times$.

For brevity, define for each $\mu \in X$

$$R_w(\mu) := \varepsilon_{\mathbf{G}} \varepsilon_{\mathbf{T}_w} R_{\mathbf{T}_w}^{\mathbf{G}}(\theta_\mu).$$

This depends on the choice of the power $r$ of $p$ involved in $F$, which we keep fixed. The sign adjustment insures that $R_w(\mu)$ has positive degree.

Recall from 9.6 that for $\mu \in X$, the formal character $s(\mu)$ yields a formal Brauer character denoted $s_\mu$. The following theorem is an easy consequence of the character calculations in [129] (see Humphreys [225]).

THEOREM. *With notation as above, let $\mu \in X$ and denote by $W_\mu$ the stabilizer of $\mu$ in $W$. Then*

$$\sum_{w \in W} R_w(\mu) = |W_\mu|\, s_\mu \operatorname{St}$$

*on $G_{reg}$ and vanishes elsewhere. In particular, this is a $\mathbb{Z}$-linear combination of Brauer characters of PIMs of $G$.*

EXAMPLE. In the case $G = \operatorname{SL}(2,p)$, most PIMs for $KG$ have dimension $2p$, with Brauer characters of the form $s_\mu \operatorname{St}$; they typically involve two DL characters. This has been seen directly in 9.8.

## 17.7. DL Characters and Weyl Characters

Now we can outline the approach taken by Jantzen [251] (which he explains informally in [257]). We continue to assume that $\mathbf{G}$ is simple and simply connected.

Recall that we add a tilde to denote the Brauer character associated with a given formal character in $\mathcal{X} = \mathbb{Z}[X]^W$. For example, $\tilde{\chi}(\lambda)$ abbreviates $\operatorname{br}\chi(\lambda)$ when $\lambda \in X$. As in the previous section, we write $R_w(\mu) := \pm R_{\mathbf{T}_w}^{\mathbf{G}}(\theta_\mu)$ for $w \in W$ and $\mu \in X$, with the sign chosen to make the degree of $R_w(\mu)$ positive. Here $r$ is understood to be fixed.

The Weyl group plays a leading role in what follows. For each $w \in W$, define $\rho_w := \sum \varpi_\alpha$, with the sum taken over simple roots $\alpha$ for which $w^{-1}(\alpha) < 0$. Then set $\varepsilon_w := w^{-1}\rho_w$. Following a conjecture of Verma [426], Hulsurkar worked out the following formalism for the Weyl group: Consider the $W \times W$ matrix having as $(w, w')$-entry the formal character $\det(w')\chi(-\varepsilon_{w_0 w} + \varepsilon_{w'} - \rho)$. For a suitable ordering, this matrix is upper triangular unipotent. Thus it has an inverse matrix with entries $\gamma_{w,w'}$ in $\mathcal{X} = \mathbb{Z}[X]^W$, satisfying $\gamma_{w,w} = 1$ for all $w \in W$.

With this notation we can formulate Jantzen's general comparison of Brauer characters associated with DL characters and Weyl characters [251, 3.4]:

THEOREM. *For all $w \in W$ and $\mu \in X$, there is an equality of Brauer characters:*

$$\tilde{R}_w(\mu) = \sum_{w_1, w_2 \in W} \tilde{\gamma}_{w_1, w_2}^{[r]} \tilde{\chi}\big(w_1(\mu - w\pi(\varepsilon_{w_0} w_2)) + p^r \rho_{w_1} - \rho\big).$$

The proof involves a considerable amount of formal manipulation, but the strategy can be explained as follows. By working out the ideas in 10.3 more completely, one gets a character formula in the setting of $\mathbf{G}_r\mathbf{T}$-modules:

$$\operatorname{ch}\widehat{Q}_r(\lambda) = \sum_{\mu \in X^+} [\widehat{Z}_r(\mu - \rho) : \widehat{L}_r(\lambda - p^r\rho)]\, s(\mu) \operatorname{ch}\operatorname{St}_r.$$

This uses the fact that the formal characters of the $\widehat{Z}_r(\nu)$ are just translates of $\operatorname{ch}\operatorname{St}_r$. Passing to Brauer characters, one can now use Theorem 17.6 to bring various $\widetilde{R}_w(\mu)$ into the formula. Eventually Brauer reciprocity is invoked, leading to an inverted formula as in the statement of the theorem. Here is where Hulsurkar's inversion formula is needed.

EXAMPLES. It is by no means easy to implement the formula in the theorem. But in rank 2 cases the Hulsurkar matrix is not so difficult to invert. For type $A_2$ and type $C_2$ respectively, see Jantzen [**253**, §5] and [**257**, App. 3]. In the latter case, the irreducible characters (restricted to $G_{\mathrm{reg}}$) are expressed as $\mathbb{Q}$-linear combinations of various DL characters, then as $\mathbb{Z}$-linear combinations of various $\widetilde{\chi}(\mu)$. For example, the cuspidal unipotent character called $\theta_{10}$ in Srinivasan's character table leads to a decomposition

$$\widetilde{\theta}_{10} = \frac{1}{4}[\widetilde{R}_{s_1 s_2}(0) - \widetilde{R}_{w_\circ}(0)] = \widetilde{\chi}(q - 3, 0) + \widetilde{\chi}(0, q - 2)$$

when $q$ is odd. A similar treatment for type $G_2$ is given by Mertens [**319**, Kapitel 4].

REMARK. In several papers [**101, 102, 103**], Chastkofsky has explored what happens when a weight $\lambda$ is fixed but the power $p^r$ is allowed to grow. How do the decomposition numbers involving $L(\lambda)$ behave as $r$ increases? For example, in [**101**] he applies some of his and Jantzen's previous formalism for PIMs of Chevalley groups when weights $\lambda, \mu$ are fixed. For large enough $p$, the multiplicity of a fixed DL character $\widetilde{R}_w(\mu)$ of $G$ in $\operatorname{br}\widehat{Q}_r(\lambda)$ becomes stable as $r$ increases. (Generically, the PIM $\widehat{Q}_r(\lambda)$ for $\mathbf{G}_r\mathbf{T}$ agrees with the corresponding PIM $U_r(\lambda)$.)

## 17.8. Generic Decomposition Patterns

In the framework of Deligne–Lusztig theory, Jantzen [**251**, §4] refines earlier ideas of Verma [**426**] and Humphreys [**218**] on generic decomposition behavior of ordinary characters. His theorem is formulated as a description of the Brauer character of $\widetilde{R}_w(\mu + \rho)$ when $\mu$ lies in the lowest $p^r$-alcove and has a trivial stabilizer in the extended affine Weyl group $\widetilde{W}_p = W \ltimes pX$ (see 10.1). To state the theorem, set $D_r := \{x \in \widetilde{W}_p \mid x \cdot \mu \in X_r\}$.

THEOREM. *Let $\mu \in X$ lie in the lowest $p^r$-alcove. If $\alpha > 0$, write $\langle \mu + \rho, \alpha^\vee \rangle = m_\alpha p + d_\alpha$ with $0 \le d_\alpha < p$. Suppose the isotropy group of $\mu$ in the extended affine Weyl group $\widetilde{W}_p$ is trivial, while $\min(d_\alpha, p - d_\alpha) \ge 2(h - 1)$ for all $\alpha > 0$. Then*

$$\widetilde{R}_w(\mu + \rho) = \sum_{x \in D_r} \sum_{\nu \in X} [\widehat{Z}_r(\mu + p^r\rho) : \widehat{L}_r(x \cdot \mu + p^r\nu)]\, \widetilde{\chi}_p(x \cdot (\mu + w\pi(\nu)).$$

This amounts to saying that the decomposition pattern for DL characters is generically uniform and imitates that for a suitable induced $\mathbf{G}_r\mathbf{T}$-module. The conditions here require something like $p > 4(h-1)$ in order for such generic behavior to occur. While Jantzen derives the theorem from the more general framework outlined in 17.7, Pillen [**334**, §8] offers a more direct proof.

## 17.9. Twisted groups

Suppose the root system of **G** is of type $A_\ell$ (with $\ell > 1$), $D_\ell$, or $E_6$, and let $G$ be the associated Chevalley group over a field of order $p^r$. We have already encountered close relationships between modular representations of $G$ and of a twisted analogue (call it for convenience $G'$):

- The restrictions to $G$ or $G'$ of the simple **G**-modules $L(\lambda)$ with $\lambda \in X_r$ are simple and nonisomorphic. They represent all isomorphism classes of simple modules for the finite groups. (See 2.11.)
- For large enough $p$, "most" PIMs for $G$ or $G'$ are obtained by restricting lifts to **G** of the PIMS $Q_r(\lambda)$ for $\mathbf{G}_r$ (where $\lambda \in X_r$). In general, Brauer characters of PIMs for the finite groups are obtained in similar ways by decomposing Brauer characters attached to the $Q_r(\lambda)$. (See 10.11, as well as 10.14, where PIMs for $\mathrm{SL}(3, p)$ and $\mathrm{SU}(3, p)$ are compared.)

On the other hand, strong resemblances have been shown between the character tables of $G$ and $G'$, starting with the observations of Ennola [163] on general linear and unitary groups. (Roughly speaking, the transition from $G$ to $G'$ involves replacement of $q$ by $-q$.)

The results of Jantzen just discussed point to similar uniformities in the decomposition behavior of characters for the two groups. If the Frobenius map involves a nontrivial graph automorphism $\pi$, the effect in Theorem 17.8 is fairly minor: $\pi$ twists some very small weights $\nu$ there. So one concludes that the generic decomposition patterns for the groups are the "same", in terms of the distribution and multiplicities of composition factors in the reduction modulo $p$ of an ordinary character.

This suggests that one might be able to predict the decomposition matrix of $G'$ from that of $G$. But this is not entirely straightforward, since typically the number of $p$-blocks of highest defect will differ for the two groups. There are some intriguing indications of hidden similarities in the study of $\mathrm{SL}(3, p)$ and $\mathrm{SU}(3, p)$. For example Humphreys [233] observes that the projective covers of the trivial module differ in dimension but nonetheless both involve the same number of ordinary characters.

## 17.10. Unipotent Characters

In Steinberg's work the conjugacy classes of **G** (or $G$) are organized around the concepts of semisimple, unipotent, and regular. (See [237] for an exposition.) Analogous notions arise in the character theory of $G$. This begins with the definition of **unipotent characters** by Deligne–Lusztig [129, §7]: these are the irreducible constituents of the various DL characters $R_{\mathbf{T}}^{\mathbf{G}}(1)$, where 1 denotes as usual the trivial character of $T$. Their values at unipotent elements are a key ingredient in the formula quoted above for values of DL characters. (See Carter [89, Chap. 12], Digne–Michel [134, Chap. 13].)

For example, all constituents of the induced character $\mathrm{Ind}_B^G(1)$ (with 1 the trivial character of $B$) are unipotent; these include the trivial character and the Steinberg character. When $G = \mathrm{SL}(2, q)$ or $\mathrm{SL}(3, q)$, all unipotent characters result from parabolic induction. But starting with $\mathrm{Sp}(4, q)$ one begins to encounter examples of cuspidal unipotent characters not arising in this way.

Lusztig [309] works out the degrees of unipotent characters for all simple types. As in the case of unipotent classes, the classification of these characters is independent of the isogeny class of $\mathbf{G}$. Moreover, each unipotent character can be expressed as a $\mathbb{Q}$-linear combination of the $R_{\mathbf{T}}^{\mathbf{G}}(1)$. For example, when $\mathbf{G} = \mathrm{SL}(2, K)$, with split torus $\mathbf{T}_0$ and nonsplit torus $\mathbf{T}$, we can write

$$1 = \frac{1}{2}[R_{\mathbf{T}_0}^{\mathbf{G}}(1) + R_{\mathbf{T}}^{\mathbf{G}}(1)] \text{ and } \mathrm{St} = \frac{1}{2}[R_{\mathbf{T}_0}^{\mathbf{G}}(1) - R_{\mathbf{T}}^{\mathbf{G}}(1)].$$

REMARK. While unipotent characters appear quite naturally in the context of Deligne–Lusztig theory, there does not seem to be a simple way to recognize them intrinsically among all irreducible characters.

When $\mathbf{G}$ is simple and simply connected, Jantzen [252] investigates thoroughly the natural question: *Which modular characters are involved in the decomposition of unipotent characters?* His basic observation is that the weights occurring must be linked under the extended affine Weyl group $\widetilde{W}_p$ to "very small" weights.

When a Brauer character $\varphi$ is written in terms of the reductions modulo $p$ of irreducible characters, denote by $\varphi_{\mathrm{un}}$ the summand involving just unipotent characters.

THEOREM. *With notation as above, assume $p > h$. For a subset $I \subset \Delta$, set $\rho_I := \sum_{\alpha \in I} \varpi_\alpha$ and let $W_I$ be the subgroup of $W$ generated by the reflections $s_\alpha$ for $\alpha \in I$. Assume $\pi(I) = I$ if $F$ involves a nontrivial graph automorphism $\pi$. Then*

$$\mathrm{br\,ch}\,\widehat{Q}_r((p^r - 1)\rho_I)_{\mathrm{un}} = \frac{1}{|W_I|} \sum_{w \in W_I} \widetilde{R}_w(0).$$

This follows from a more general analysis of the "unipotent part" of a typical $\mathrm{br\,ch}\,\widehat{Q}_r(\lambda)$. In [252, §3] there is a further discussion of how to isolate the contributions of various PIMs $U_r(\mu)$ to the above formula. Here the unipotent part is seen to come from various $U_r(p^r-1)\rho_J)_{\mathrm{un}}$ for $\pi$-stable subsets $J \subset I$. The combinatorics here can be made explicit in case $G$ is a classical group.

In [265], Kaneda extends this work of Jantzen by treating not necessarily unipotent characters whose reduction modulo $p$ involves weights close to $(p^r - 1)\rho$, again with special attention to the classical groups.

## 17.11. Semisimple and Regular Characters

In Deligne–Lusztig theory, unipotent characters play a role analogous to that of unipotent classes. It is natural to ask whether there are also analogies with semisimple classes and regular classes. This is taken up in [129, §10] under the restrictive assumption that the center of $\mathbf{G}$ is *connected*, which is true for example when $\mathbf{G} = \mathrm{GL}(n, K)$ or $\mathrm{PGL}(n, K)$, but not when $\mathbf{G} = \mathrm{SL}(n, K)$. Detailed accounts are given by Carter [89, Chap. 8] and Digne–Michel [134, Chap. 13–14].

An irreducible character of $G$ is called **semisimple** if its average value on the set of regular unipotent elements is nonzero (in which case this value is $\pm 1$). In principle one can decide which characters meet this criterion by inspecting the character table of $G$. But there is a much easier criterion when $p$ is **good** for $G$. Recall that $p = 2$ is a bad prime for all simple types except $A_\ell$, while $p = 3$ is bad for all exceptional types and $p = 5$ is bad for type $E_8$; otherwise $p$ is good. In general, $p$ is good for $\mathbf{G}$ just when it is good for all simple factors of the derived group of $\mathbf{G}$.

PROPOSITION. *Suppose $Z(\mathbf{G})$ is connected, and let $p$ be a good prime for $\mathbf{G}$. Then an irreducible character $\zeta$ of $G$ is semisimple if and only if either of these equivalent conditions holds:*

(a) *The value of $\zeta$ at every regular unipotent element of $G$ is nonzero.*

(b) *The degree $\zeta(1)$ is not divisible by $p$.*

*When $p$ is arbitrary, statement $(b)$ still holds for all semisimple characters.*

Pursuing the analogy with conjugacy classes, one can also define a notion of "regular character". This requires the construction of **Gel'fand–Graev characters**. Start with an $F$-stable Borel subgroup $\mathbf{B} = \mathbf{TU}$, where $\mathbf{T}$ is maximally split over $\mathbb{F}_q$. There is a natural notion of "regular" (or "nondegenerate") linear character $\xi$ of $U = \mathbf{U}^F$. Then the Gel'fand–Graev character $\Gamma := \operatorname{Ind}_U^G(\xi)$ turns out to be independent of the choice of $\xi$ (under our assumption that $Z(\mathbf{G})$ is *connected*). Its irreducible constituents all occur with multiplicity 1 and are called **regular**.

These ideas fit well with the classification of irreducible characters into geometric conjugacy classes. In a given class there is a unique semisimple character and a unique regular character. Moreover, these are interchanged by Curtis–Alvis duality (Remark 17.4).

What can be said when $Z(\mathbf{G})$ is disconnected? Much but not all of the above discussion remains valid, as shown for example in Lusztig [312, Chap. 14] and Digne–Lehrer–Michel [133]. One can construct multiplicity-free Gel'fand–Graev characters as before, then define a character to be "regular" precisely when it occurs as a constituent of one of these. (But now there is more than one possible Gel'fand–Graev character.) Motivated by the preceding remarks, one can then define a character to be "semisimple" if its dual is $\pm$ a regular character. But easy examples show that a geometric conjugacy class need no longer be so well-organized as before: take for example the class for $\operatorname{SL}(2, q)$ consisting of two characters both of degree $(q + 1)/2$. On the other hand, for $\operatorname{SL}(3, q)$ one has a family of geometric conjugacy classes, each consisting of a regular character of degree $q^3 + q^2 + q$ and a semisimple character of degree $q^2 + q + 1$ (whose sum is a degenerate principal series character).

REMARK. Lusztig's classification of characters [312] leads even further, to a kind of "Jordan decomposition" of characters (see Carter [89, 12.9], Digne–Michel [134, 13.23]). It is again convenient to assume that the center of $\mathbf{G}$ is *connected*, though the treatment in [134] explains how to relax this assumption.

The Langlands dual group $\mathbf{G}^*$ plays an essential role here. This duality interchanges simply connected with adjoint groups having dual root systems (so types $\mathrm{B}_\ell$ and $\mathrm{C}_\ell$ get switched). For example, $\operatorname{SL}(n, K)$ is dual to $\operatorname{PGL}(n, K)$, while $\operatorname{GL}(n, K)$ is dual to itself. To a Frobenius map for $\mathbf{G}$ then corresponds a Frobenius map for $\mathbf{G}^*$ and resulting finite subgroup. (See Carter [89, 4.5], Digne–Michel [134, 13.10], Humphreys [237, 2.9].)

The basic idea is to use semisimple classes in the dual group to parametrize semisimple characters of $G$. Say $\mathbf{G}$ is of adjoint type. With a semisimple character one can then associate the unipotent characters of the fixed point subgroup of $\mathbf{G}^*$ coming from the centralizer of a semisimple element. The degree of a typical irreducible character of $G$ is then the product of degrees of a semisimple character and such an associated unipotent character.

## 17.12.  Geometric Conjugacy Classes of Cardinality 2

Beyond the best behaved situation in which a geometric conjugacy class consists of a single irreducible character whose decomposition behavior is "generic", there appear to be many further regularities. In [**334**] Pillen investigates some "generic $p$-singular" cases, under the assumption that $p \geq 2(h-1)$. His genericity assumptions require that certain weights lie inside an alcove wall, not too close to any other wall. This occurs for many geometric conjugacy classes which involve precisely two DL characters and two irreducible characters. A simple reflection $s$ then plays a key role in relating them. The DL character of larger degree is a proper character denoted $R_w(\mu + \rho)$, while the other is denoted $R_{ws}(\mu + \rho)$. In this situation, $R_w(\mu + \rho) = \zeta_r + \zeta_{ss}$ with $\zeta_r$ regular and $\zeta_{ss}$ semisimple. On the other hand, $R_{ws}(\mu + \rho) = \zeta_r - \zeta_{ss}$.

Examples are already plentiful in the known character tables (16.11). As noted above, $SL(3, q)$ has a family of geometric conjugacy classes, each of which involves a pair of DL characters having respective (normalized) degrees $q^3 + 2q^2 + 2q + 1$ and $q^3 - 1$. The DL characters share the same constituents: a regular character of degree $q^3 + q^2 + q$ and a semisimple character of degree $q^2 + q + 1$. (Similar examples for a group of type $G_2$ are indicated in 18.6.)

THEOREM. *Let* **G** *be simple and simply connected, with* $G = \mathbf{G}^F$ *a Chevalley group. Under Pillen's genericity assumptions, all composition factors of the semisimple character* $\zeta_{ss}$ *have $p$-regular highest weights.*

The arguments in [**334**] should carry over to twisted groups as well.

## 17.13.  Jantzen Filtration of a Principal Series Module

The reduction modulo $p$ of a "generic" DL character has the same pattern of composition factors (counting multiplicities) as a generic $\mathbf{G}_r\mathbf{T}$-module $\widehat{Z}_r(\lambda)$ or a generic Weyl module for **G** with highest weight in the lowest $p^{r+1}$-alcove. This makes it reasonable to look for further structural analogies with the latter two situations. The starting point should be a naturally defined $KG$-module resulting from the reduction modulo $p$ of a $\mathbb{C}G$-module affording a proper DL character. The principal series modules discussed in 7.2 and 13.10, or analogues involving induction from larger parabolic subgroups, provide a natural setting for study of corresponding Deligne–Lusztig series. But no such models for the discrete series are available in characteristic $p$.

In case $G$ is a *Chevalley group*, Jantzen [**253**] constructs a filtration in an arbitrary principal series module having a sum formula analogous to that for a Weyl module (3.9). Following Andersen's strategy for the Weyl module case (and higher sheaf cohomology modules as well), he constructs everything over a discrete valuation ring and studies the various intertwining maps among principal series modules (as in Carter–Lusztig [**93**]). The filtrations are constructed here in terms of the powers of $p$ dividing images or kernels of the maps. After reducing modulo $p$, one gets a filtration with a sum formula describing the Brauer character of the sum of all proper submodules in the filtration.

A special feature of this sum formula is that it involves all series of DL characters. This adds to the previous combinatorial evidence that these series should have equal status in the theory.

There remain serious obstacles to extending Jantzen's work to twisted groups, starting with the more complicated possibilities for "rank 1" subgroups such as $SU(3, q)$.

As indicated in 13.10, Pillen [**336**] has explored the filtrations further in connection with Loewy series of principal series modules.

## 17.14. Extremal Composition Factors

We have been emphasizing parallels in the generic composition factor patterns of "standard" modules in several settings: Weyl modules for **G**, induced modules $\widehat{Z}_r(\lambda)$ for $\mathbf{G}_r\mathbf{T}$, reduction modulo $p$ of modules affording DL characters for finite groups of Lie type. These patterns (counting multiplicities) become more and more complicated as the rank or the power of $p$ grows, making the combinatorics look quite intricate. But a family of composition factors indexed by the Weyl group may play a leading role here. These are called **extremal composition factors** by Humphreys [**235**].

The easiest (and most general) definition goes as follows: Start with the Steinberg module $\mathrm{St}_r = \widehat{Z}_r((p^r - 1)\rho)$ for $\mathbf{G}_r\mathbf{T}$. Its extremal weights $(p^r - 1)w\rho$ for $w \in W$ occur with multiplicity 1. The weight diagram of any $\widehat{Z}_r(\lambda)$ is obtained from that of $\mathrm{St}_r$ simply by translation (10.3). The translates of the extremal weights then determine the desired composition factors of multiplicity 1 in $\widehat{Z}_r(\lambda)$.

In the case of DL characters, one can identify a corresponding family of extremal composition factors in Jantzen's formalism: set $w_1 = w_2$ in [**251**, 3.3(2)].

In Pillen's study of reduction modulo $p$ for DL characters involving just two irreducible constituents, one regular and one semisimple (17.12), he has observed a tendency for equal numbers of extremal composition factors to occur in each. One might ask in general: For a fixed DL character, does the reduction modulo $p$ of each of its irreducible constituents involve at least one extremal composition factor? This should be compatible with the role played by $W$ (or its Hecke algebra) in organizing the constituents of parabolically induced characters.

Study of principal series characters and their associated modules in characteristic $p$ suggests a precise conjecture along this line. Here there are Jantzen filtrations, arising from images and kernels of intertwining maps (17.13). Given an intertwining map relative to a simple root, examples support the conjecture that the image and the kernel of the map will each involve half of the extremal composition factors in the respective principal series module.

Apart from experimental evidence in low ranks, there has been little progress in studying these questions systematically.

## 17.15. Another Look at the Brauer Tree of $SL(2, p)$

In the case of a cyclic block, the decomposition behavior of ordinary characters can easily be visualized in terms of a graph (16.7). This is illustrated for $SL(2, p)$ in 16.9. Nothing quite so simple can be done for $SL(2, p^r)$ when $r > 1$, though the decomposition pattern for $r = 1$ propagates in an orderly way.

What can be done over $\mathbb{F}_p$ when the 1-block no longer has a cyclic defect group? Examples in rank 2 show immediately that the decomposition behavior is too complicated to be conveyed by a graph, so we look for a higher-dimensional substitute. To get started, we reformulate the Brauer tree as in 13.4: reverse the role of vertices and edges, using weights to label vertices and ordinary characters to label edges. Figure 1 shows how the resulting graph looks for the 1-block. (The

$$\circ \overset{1_G}{\underset{0}{\rule{1.2em}{0.4pt}}} \circ \overset{\theta}{\underset{p-3}{\rule{1.2em}{0.4pt}}} \circ \overset{\zeta}{\underset{2}{\rule{1.2em}{0.4pt}}} \circ \overset{\theta}{\underset{p-5}{\rule{1.2em}{0.4pt}}} \circ \overset{\zeta}{\underset{4}{\rule{1.2em}{0.4pt}}} \circ \cdots \circ \overset{\zeta}{\underset{4}{\rule{1.2em}{0.4pt}}} \circ \overset{\theta}{\underset{p-5}{\rule{1.2em}{0.4pt}}} \circ \overset{\zeta}{\underset{2}{\rule{1.2em}{0.4pt}}} \circ \overset{\theta}{\underset{p-3}{\rule{1.2em}{0.4pt}}} \circ \overset{1_G}{\underset{0}{\rule{1.2em}{0.4pt}}} \circ$$

FIGURE 1. Graph for 1-block of $\mathrm{SL}(2,p)$, $p$ odd

reader can easily imitate the labelling for the other block of highest defect.) Now the vertices attached to an edge indicate the (usually two) composition factors in the reduction modulo $p$ of the ordinary character assigned to that edge. Dually, the lifted PIM attached to a vertex involves the ordinary characters on adjacent edges.

By repeating most of the data we achieve greater symmetry here. The underlying reason for this repetition is the fact that $\mathrm{SL}(2,K)$ has a disconnected center of order 2 when $p$ is odd. For the moment just note that the total number of edges in our picture is $p$. Thus we can interpret the graph as a single large $p$-alcove divided into $p$ small ones.

In place of the special vertex at the end of the Brauer tree, labelled by a pair of characters $\xi_i$ or $\eta_i$, we now have a center edge connecting two vertices with the same weight label. This edge is labelled by a pair $\xi_i$ or $\eta_i$, depending on the residue class of $p \bmod 4$. When $p \equiv 1 \pmod 4$, the weight $(p-1)/2$ labels the vertices while the characters $\eta_1$ and $\eta_2$ label the middle edge. When $p \equiv 3 \pmod 4$, the weight $(p+1)/2$ labels the vertices while the characters $\xi_1$ and $\xi_2$ label the middle edge. In these cases the ordinary character remains irreducible modulo $p$.

## 17.16. The Brauer Complex

Working initially with Chevalley groups over $\mathbb{F}_p$, Humphreys [221] proposed an $\ell$-dimensional analogue of the reformulated Brauer tree for $\mathrm{SL}(2,p)$. Fix a $p$-block of $KG$ having highest defect, such as the 1-block. The idea is to start with a single large alcove for the associated affine Weyl group $W_p$, oriented for example in the same way as the top $p$-alcove in the $p$-restricted region. Then use translated root hyperplanes of all types to subdivide this alcove into $p^\ell$ smaller copies.

The vertices are labelled with those weights $\lambda$ in the top $p$-alcove for which $L(\lambda)$ lies in the chosen block, while the $p^\ell$ small alcoves are labelled with geometric conjugacy classes of characters in this block. This should be done in such a way that characters occurring in the lift of $U(\lambda)$ are placed in alcoves with $\lambda$ as a vertex. For example, if an interior vertex has the full number $|W|$ of surrounding alcoves, then $U(\lambda)$ should have dimension $|W|p^m$ (10.7). In this situation, the $|W|$ (irreducible) DL characters occurring in $U(\lambda)$ should be assigned to these alcoves.

This ideal picture cannot always be realized when $\mathbf{G}$ is simply connected but has a nontrivial center. Then there may be multiple blocks of highest defect (Theorem 8.3), leading to repetitions of this data as in the reformulated Brauer tree of $\mathrm{SL}(2,p)$. This extra complication is avoided in type $G_2$: see the example for $p=5$ in 18.8.

The involvement of weights lying in other restricted alcoves is only indirectly seen in the complex. If $\mu$ is linked to a weight $\lambda$ in the top alcove, its PIM $U(\mu)$ typically gets much larger than $U(\lambda)$ as $\mu$ gets smaller and may involve DL characters with multiplicity $> 1$. This is pictured as a larger (generically symmetric) array of alcoves surrounding the vertex $\lambda$. Each alcove in the array (taken with some multiplicity) corresponds to a DL character involved in $U(\mu)$. These symmetric arrays of alcoves also arise naturally in connection with Lusztig's Conjecture; they are pictured in rank 2 in [**311**]. Some experimental work with the Brauer complex in rank 2 is found in papers by Humphreys [**221, 223, 226, 230, 233**].

The main theoretical problem here is to insure the consistency of the labelling when alcove patterns for many PIMs overlap. This is verified by Chastkofsky [**99**]. For this he has to make explicit the assignment of DL character data to alcoves, which involves a careful correlation of weight and Weyl group parameters.

## 17.17. Dual Formulation

In general the representation-theoretic interpretation of the Brauer complex works best when $\mathbf{G}$ is replaced by its dual group $\mathbf{G}^*$, which is of adjoint type. For a simply connected group $\mathbf{G}$, Deriziotis [**130**] introduces a rigorously defined dual version of the Brauer complex to organize geometrically the study of semisimple classes and their centralizers. This recovers Steinberg's theorem that the universal Chevalley group $G$ over $\mathbb{F}_q$ (or a twisted analogue) has $q^\ell$ semisimple classes. The corresponding finite subgroup of $\mathbf{G}^*$ has $q^\ell$ semisimple characters (one for each geometric conjugacy class). Small cases in rank 2 when $q = 5$ are pictured in [**130**, p. 268].

The Deriziotis construction of a "Brauer complex" is explained by Carter [**89**, 3.8] and illustrated in some small cases. It applies to all finite groups of Lie type, including the Suzuki and Ree groups. The idea is to start as above with a large alcove divided into $q^\ell$ small alcoves. The closure of each small alcove has **facets** ranging from vertices of dimension 0 to faces of dimension $\ell$. Some of these (generically just the face) will lie on corresponding facets of the closure of the large alcove. There is a unique such facet of smallest dimension, which is then labelled in a natural way with a $p$-regular class of $G$. This assignment of classes to alcoves turns out to be bijective. Classes assigned to faces are precisely those containing regular semisimple elements. For an element in any class, the structure of its centralizer is encoded efficiently in the geometry of the complex.

In either guise, the Brauer complex has been useful mainly as an aid in visualizing complicated combinatorial relationships. For example, it adds some intuition to Pillen's work described in 17.12; there the DL characters having just a semisimple and a regular constituent are placed systematically along outer walls of the complex.

One open question is whether (as in the work of Deriziotis) one can get insight from the complex into what happens over finite fields larger than the prime field. Here the combinatorial data get more and more complicated, while the reduction modulo $p$ as studied by Jantzen shows considerable uniformity.

# CHAPTER 18

# The Groups $G_2(q)$

The groups of type $G_2$ illustrate well many of the complexities encountered in higher ranks. Since the standard character data for **G** are known in this case, it is possible to work out many details about representations of the corresponding finite groups. At the same time, the larger Weyl group and unusual asymmetry of the root system make the results much richer than those found in types $A_2$ and $B_2$.

In this chapter we draw together the information developed in earlier chapters, along with a number of concrete examples found in the unpublished Diplomarbeit of Mertens [**319**]. While the characteristic $p$ results in this case are essentially independent of Lusztig's Conjecture, the groups of type $G_2$ provide excellent confirmation of the conjecture and its consequences when $p \geq h\,(= 6)$. The comparison of ordinary and modular representations offers a good model for higher ranks, but with some technical issues avoided because $Z(\mathbf{G}) = 1$.

## 18.1. The Groups

Throughout the chapter, **G** denotes the algebraic group of type $G_2$ over $K$; it is simultaneously of simply connected and adjoint type. In particular, $Z(\mathbf{G}) = 1$. Moreover, **G** is its own Langlands dual group. The Weyl group $W$ is dihedral of order 12. Our convention is that the first simple root is *short*. Some authors, for example Carter [**89**] and Gilkey–Seitz [**186**], use the opposite convention.

There is a unique Chevalley group $G = \mathbf{G}(\mathbb{F}_q)$, often denoted $G_2(q)$; its order is $q^6(q^6 - 1)(q^2 - 1)$. Recall from 1.5 that $G$ is simple unless $q = 2$, in which case the derived group has index 2 and is isomorphic to the simple group $^2A_2(3)$. Except in this case, the fact that $Z(G) = 1$ implies that $KG$ has just two blocks: the 1-block and the block containing the Steinberg module (see 8.3 and 8.5). In fact the nonsimple group $G_2(2)$ also has just these two blocks, as seen below in the computations by Mertens.

In line with Lusztig's Conjecture, the representation theory of $G$ shows considerable uniformity when $p \geq 6$. The primes $2, 3$ (both "bad" for the root system), and 5 require case-by-case treatment in [**319**]. Note that when $p = 3$, the representations of $G$ exhibit extra symmetry due to the special isogeny $\mathbf{G} \to \mathbf{G}$ (5.3). They are closely related to those of the Ree group of type $^2G_2$ (Chapter 20).

## 18.2. The Affine Weyl Group

Figure 1 shows a portion of the alcove geometry, including the 12 alcoves which contain $p$-restricted weights. The numbering 1–16 of alcoves comes from [**319**] (where some other alcoves figuring in tensor product calculations and the like are also numbered). It is compatible with the natural partial ordering of alcoves. Here as in higher ranks there are some nonrestricted weights lying below restricted ones in the partial ordering; these occur in alcoves $9, 10, 12, 14$.

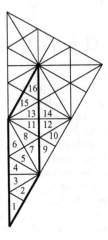

FIGURE 1. Some alcove geometry for type $G_2$

This alcove numbering is used in various tables below. Translation functors typically allow one to focus just on alcoves (or their facets), where weights behave uniformly. For weights lying in a wall between two alcoves, notation such as $2/3$ will be used. To compare the behavior of weights inside an alcove with those in the closure, Jantzen's methods require a distinction between weights lying in the "upper closure" or "lower closure" of an alcove. For example, $\widehat{2}$ indicates the upper closure of alcove 2.

## 18.3. Weyl Modules and Simple Modules

First we recall the sources of information for the representations of $\mathbf{G}$. Thanks to the Tensor Product Theorem, we only need to know the composition factors of Weyl modules whose highest weights lie in the closures of alcoves 1–16. Then the formal characters $\chi_p(\lambda)$ for $\lambda \in X_1$ can be expressed in terms of Weyl characters.

- Springer [382] makes the character data explicit for $G_2$ when $p = 3$, using the special isogeny in this case (5.3).
- Jantzen uses his Sum Formula (see 3.9) and related arguments to compute for $p \neq 3, 5$ the composition factors of Weyl modules when $\mathbf{G}$ has type $G_2$. See [246, §7] as well as the related discussion of generic decomposition behavior in [247].
- Hagelskjaer [201] computes the character data for $G_2$ when $p = 5$, thereby showing consistency with Lusztig's approach even though $p$ is smaller than the Coxeter number 6.
- Some of the dimensions of simple modules with small highest weights are computed by Gilkey–Seitz [186].

| $\lambda$ | $\chi_p(\lambda) = \sum b_\mu \chi(\mu)$ |
|---|---|
| $(1,2)$ | $(1,2) - (2,1)$ |
| $(0,3)$ | $(0,3) - (3,0)$ |
| $(3,1)$ | $(3,1) - (0,2)$ |
| $(2,2)$ | $(2,2) - (4,0) - (1,1)$ |
| $(4,1)$ | $(4,1) - (5,0) - (0,2)$ |
| $(1,3)$ | $(1,3) - (3,0) - (2,0)$ |
| $(3,2)$ | $(3,2) - (2,1) - (0,1)$ |
| $(2,3)$ | $(2,3) - (1,1) - (1,0)$ |
| $(4,2)$ | $(4,2) - (4,1) + (5,0)$ |
| $(0,4)$ | $(0,4) - (2,0)$ |
| $(1,4)$ | $(1,4) - (0,1)$ |
| $(3,3)$ | $(3,3) - (4,2) + (4,1) - (5,0) - (0,0)$ |
| $(2,4)$ | $(2,4) - (1,4) + (0,1)$ |
| $(3,4)$ | $(3,4) - (0,4) + (2,0)$ |
| $(4,3)$ | $(4,3) - (6,1) + (2,2) - (4,0) - (1,0)$ |

TABLE 1. Character of $L(\lambda)$ in terms of Weyl characters, $p = 5$

Here we review the details of formal character computations for $p \geq 5$ (leaving to the following sections the special cases $p = 2, 3$). Consider the case $p = 5$. This is discussed by Mertens [**319**, 3.7], using the explicit calculations of composition factors of Weyl modules by Hagelskjær and Jantzen. In Table 1 we list all of the 5-restricted weights $\lambda$ for which $L(\lambda) \neq V(\lambda)$, showing how the character $\chi_5(\lambda)$ of $L(\lambda)$ is expressed as a $\mathbb{Z}$-linear combination of Weyl characters $\chi(\mu)$ (where $\mu$ is sometimes nonrestricted). Here all coefficients happen to be $\pm 1$. For example, $\chi_5(1,2) = \chi(1,2) - \chi(2,1)$.

When $p \geq 7$ it is enough to know the standard character data for **G** for $p$-regular weights in alcoves 1–16. (Then translation to upper closures provides the remaining data.) We summarize in Tables 2 and 3 the composition factor multiplicities of Weyl modules and the resulting inverse data.

## 18.4. Projective Modules for $p = 2, 3, 5$

A major part of [**319**] is devoted to working out the Chastkofsky–Jantzen formula 10.11 for $G_2(p)$. This requires computation of composition factor multiplicities for numerous tensor products $L(\lambda) \otimes L(\mu)$. Here we look individually at the results for small primes $2, 3, 5$. We write $Q(\lambda)$ and $U(\lambda)$ here, dropping the subscript 1. In each case the reader can check that

$$\sum_{\lambda \in X_1} \dim L(\lambda) \dim U(\lambda) = |G| = p^6(p^6 - 1)(p^2 - 1).$$

(A) The case $G = G_2(2)$ is worked out in [**319**, 3.5]. Here the group order is $2^6 3^3 7$. When $p = 2$, the first anomaly one encounters for **G** is the drop in dimension of the first fundamental representation from 7 to 6. After computing the composition factor multiplicities of relevant tensor products, Mertens determines the dimensions of the PIM's as in Table 4. Here and in similar tables below, $d_\lambda$ is the "decomposition number" introduced in 10.1.

| $i\backslash j$ | 1 | 2 | 3 | 4 | 5 | 6 | 7 | 8 | 9 | 10 | 11 | 12 | 13 | 14 | 15 | 16 |
|---|---|---|---|---|---|---|---|---|---|---|---|---|---|---|---|---|
| 1 | 1 | | | | | | | | | | | | | | | |
| 2 | 1 | 1 | | | | | | | | | | | | | | |
| 3 | 0 | 1 | 1 | | | | | | | | | | | | | |
| 4 | 0 | 0 | 1 | 1 | | | | | | | | | | | | |
| 5 | 0 | 0 | 0 | 1 | 1 | | | | | | | | | | | |
| 6 | 0 | 0 | 0 | 0 | 1 | 1 | | | | | | | | | | |
| 7 | 0 | 0 | 0 | 0 | 1 | 0 | 1 | | | | | | | | | |
| 8 | 0 | 0 | 0 | 1 | 1 | 1 | 1 | 1 | | | | | | | | |
| 9 | 0 | 0 | 0 | 0 | 0 | 0 | 0 | 1 | 0 | 1 | | | | | | |
| 10 | 0 | 0 | 0 | 0 | 0 | 0 | 1 | 1 | 1 | 1 | | | | | | |
| 11 | 0 | 0 | 1 | 1 | 1 | 0 | 1 | 1 | 1 | 0 | 1 | | | | | |
| 12 | 0 | 0 | 0 | 0 | 1 | 1 | 1 | 1 | 1 | 1 | 1 | 1 | | | | |
| 13 | 0 | 1 | 1 | 1 | 1 | 0 | 0 | 0 | 0 | 0 | 0 | 1 | 0 | 1 | | |
| 14 | 0 | 0 | 0 | 1 | 1 | 1 | 0 | 1 | 0 | 0 | 1 | 1 | 1 | 1 | | |
| 15 | 1 | 1 | 1 | 1 | 0 | 0 | 0 | 0 | 0 | 0 | 0 | 0 | 0 | 1 | 0 | 1 |
| 16 | 1 | 1 | 0 | 0 | 0 | 0 | 0 | 0 | 0 | 0 | 0 | 0 | 1 | 1 | 1 | 1 |

TABLE 2. $[V(\lambda_i) : L(\lambda_j)]$ for $G_2$, $p \geq 7$ [**246**]

| $i\backslash j$ | 1 | 2 | 3 | 4 | 5 | 6 | 7 | 8 | 9 | 10 | 11 | 12 | 13 | 14 | 15 | 16 |
|---|---|---|---|---|---|---|---|---|---|---|---|---|---|---|---|---|
| 1 | 1 | | | | | | | | | | | | | | | |
| 2 | −1 | 1 | | | | | | | | | | | | | | |
| 3 | 1 | −1 | 1 | | | | | | | | | | | | | |
| 4 | −1 | 1 | −1 | 1 | | | | | | | | | | | | |
| 5 | 1 | −1 | 1 | −1 | 1 | | | | | | | | | | | |
| 6 | −1 | 1 | −1 | 1 | −1 | 1 | | | | | | | | | | |
| 7 | −1 | 1 | −1 | 1 | −1 | 0 | 1 | | | | | | | | | |
| 8 | 2 | −2 | 2 | −2 | 1 | −1 | −1 | 1 | | | | | | | | |
| 9 | 1 | −1 | 1 | −1 | 1 | 0 | −1 | 0 | 1 | | | | | | | |
| 10 | −2 | 2 | −2 | 2 | −1 | 1 | 1 | −1 | −1 | 1 | | | | | | |
| 11 | −3 | 3 | −3 | 2 | −2 | 1 | 1 | −1 | −1 | 0 | 1 | | | | | |
| 12 | 3 | −3 | 3 | −2 | 2 | −2 | −1 | 1 | 1 | −1 | −1 | 1 | | | | |
| 13 | 3 | −3 | 2 | −2 | 1 | −1 | −1 | 1 | 1 | 0 | −1 | 0 | 1 | | | |
| 14 | −4 | 4 | −3 | 3 | −2 | 2 | 2 | −2 | −1 | 1 | 1 | −1 | −1 | 1 | | |
| 15 | −3 | 2 | −2 | 1 | −1 | 1 | 1 | −1 | −1 | 0 | 1 | 0 | −1 | 0 | 1 | |
| 16 | 4 | −4 | 3 | −2 | 2 | −2 | −2 | 2 | 1 | −1 | −1 | 1 | 1 | −1 | −1 | 1 |

TABLE 3. $[L(\lambda_i) : V(\lambda_j)]$ for $G_2$, $p \geq 7$

Further calculations for $\mathfrak{g}$ yield more detailed information about the composition factor multiplicities of the $Q(\lambda)$, hence (indirectly) about the Cartan invariants for $G_2(2)$. For the principal block of $G_2(2)$, the resulting Cartan invariants are given in Table 5. Here the determinant is $2^{10}$ and the elementary divisors are $2^6, 2^3, 2$.

(B) Next let $p = 3$. The finite group $G_2(3)$ has order $2^6 \cdot 3^6 \cdot 7 \cdot 13$. Using Springer's results, Mertens computes the dimensions of PIMs $U(\lambda)$ as shown in Table 6. Note here the symmetry in the dimensions when coordinates are interchanged.

| $\lambda$ | alcove | $\dim L(\lambda)$ | $d_\lambda$ | $\dim Q(\lambda)/2^6$ | $\dim U(\lambda)/2^6$ |
|-----------|--------|-------------------|-------------|------------------------|------------------------|
| $(0,0)$ | $\widehat{5}$ | 1 | 12 | 36 | 7 |
| $(1,0)$ | $\widehat{11}$ | 6 | 4 | 12 | 8 |
| $(0,1)$ | $\widehat{15}$ | 14 | 2 | 6 | 5 |
| $(1,1)$ | $\widehat{16}$ | 64 | 1 | 1 | 1 |

TABLE 4. Dimension data for $G_2(2)$

| $\lambda\backslash\mu$ | $(0,0)$ | $(1,0)$ | $(0,1)$ |
|------------------------|---------|---------|---------|
| $(0,0)$ | 32 | 32 | 16 |
| $(1,0)$ | 32 | 38 | 18 |
| $(0,1)$ | 16 | 18 | 14 |

TABLE 5. Cartan invariants of 1-block of $G_2(2)$, $p = 2$ [**319**, p. 127]

| $\lambda$ | alcove | $\dim L(\lambda)$ | $d_\lambda$ | $\dim Q(\lambda)/3^6$ | $\dim U(\lambda)/3^6$ |
|-----------|--------|-------------------|-------------|------------------------|------------------------|
| $(0,0)$ | $\widehat{3}$ | 1 | 24 | 144 | 21 |
| $(1,0)$ | $\widehat{5}$ | 7 | 12 | 72 | 46 |
| $(0,1)$ | $\widehat{6}$ | 7 | 12 | 72 | 46 |
| $(1,1)$ | $\widehat{11}$ | 49 | 6 | 36 | 32 |
| $(2,0)$ | $\widehat{7}$ | 27 | 6 | 12 | 11 |
| $(0,2)$ | $\widehat{15}$ | 27 | 2 | 12 | 11 |
| $(2,1)$ | $\widehat{13}$ | 189 | 3 | 6 | 6 |
| $(1,2)$ | $\widehat{16}$ | 189 | 1 | 6 | 6 |
| $(2,2)$ | $\widehat{16}$ | 729 | 1 | 1 | 1 |

TABLE 6. Dimension data for $G_2(3)$

| $\lambda\backslash\mu$ | $(0,0)$ | $(1,0)$ | $(0,1)$ | $(2,0)$ | $(1,1)$ | $(0,2)$ | $(2,1)$ | $(1,2)$ |
|------------------------|---------|---------|---------|---------|---------|---------|---------|---------|
| $(0,0)$ | 78 | 138 | 138 | 39 | 105 | 39 | 16 | 16 |
| $(1,0)$ | 138 | 316 | 314 | 71 | 213 | 70 | 40 | 38 |
| $(0,1)$ | 138 | 314 | 316 | 70 | 213 | 71 | 38 | 40 |
| $(2,0)$ | 39 | 71 | 70 | 22 | 54 | 20 | 9 | 8 |
| $(1,1)$ | 105 | 213 | 213 | 54 | 153 | 54 | 26 | 26 |
| $(0,2)$ | 39 | 70 | 71 | 20 | 54 | 22 | 8 | 9 |
| $(2,1)$ | 16 | 40 | 38 | 9 | 26 | 8 | 7 | 4 |
| $(1,2)$ | 16 | 38 | 40 | 8 | 26 | 9 | 4 | 7 |

TABLE 7. Cartan invariants of 1-block of $G_2(3)$, $p = 3$ [**319**, p. 142]

This results from the special isogeny $\varphi : \mathbf{G} \to \mathbf{G}$, which becomes a group automorphism on restriction to $G$. Similarly, we have $[U(a,b) : L(c,d)] = [U(b,a) : L(d,c)]$ for all $a, b, c, d$. (See the discussion of the related Ree groups in Chapter 20 below.)

| | |
|---|---|
| $(0,0)$ | $(0,0) + (4,0) + 2(3,1) + (3,2) + (0,4) + (3,3) + (2,4) + (4,4)$ |
| $(1,0)$ | $(1,0 + (1,4) + 2(4,3) + (3,4) + 5(4,4)$ |
| $(0,1)$ | $(0,1) + (4,1) + (3,2) + (3,4)$ |
| $(2,0)$ | $(2,0) + (2,4) + 4(3,4) + 10(4,4)$ |
| $(1,1)$ | $(1,1) + (4,2) + (2,4) + (3,4) + 3(4,4)$ |
| $(0,2)$ | $(0,2) + (4,2)$ |
| $(3,0)$ | $(3,0) + 2(3,4) + 8(4,4)$ |
| $(1,2)$ | $(1,2) + (4,3)$ |
| $(0,3)$ | $(0,3) + (4,3) + 2(3,4) + 5(4,4)$ |
| $(4,0)$ | $(4,0) + (4,4)$ |
| $(1,3)$ | $(1,3) + 3(4,4)$ |
| $(0,4)$ | $(0,4) + (4,4)$ |

TABLE 8. Decomposition of $Q(\lambda)$ for $p = 5$

| $\lambda$ | alcove | $\dim V(\lambda)$ | $\dim L(\lambda)$ | $d_\lambda$ | $\dim Q(\lambda)/p^6$ | $\dim U(\lambda)/p^6$ |
|---|---|---|---|---|---|---|
| $(0,0)$ | $\widehat{1}$ | 1 | 1 | 29 | 174 | 13 |
| $(1,0)$ | $\widehat{2}$ | 7 | 7 | 16 | 96 | 61 |
| $(0,1)$ | $\widehat{3}$ | 14 | 14 | 17 | 102 | 48 |
| $(2,0)$ | $\widehat{3}$ | 27 | 27 | 17 | 102 | 62 |
| $(1,1)$ | $\widehat{4}$ | 64 | 64 | 18 | 108 | 75 |
| $(0,2)$ | $\widehat{5}$ | 77 | 77 | 12 | 72 | 54 |
| $(3,0)$ | $\widehat{5}$ | 77 | 77 | 12 | 72 | 52 |
| $(2,1)$ | $\widehat{5}$ | 189 | 189 | 12 | 72 | 72 |
| $(1,2)$ | $\widehat{6}$ | 286 | 97 | 6 | 36 | 30 |
| $(0,3)$ | $\widehat{6}$ | 273 | 196 | 6 | 36 | 13 |
| $(4,0)$ | $\widehat{7}$ | 182 | 182 | 6 | 36 | 35 |
| $(3,1)$ | $\widehat{7}$ | 448 | 371 | 6 | 36 | 36 |
| $(2,2)$ | $\widehat{8}$ | 729 | 483 | 5 | 30 | 30 |
| $(4,1)$ | $\widehat{11}$ | 924 | 469 | 4 | 24 | 24 |
| $(1,3)$ | $\widehat{11}$ | 896 | 792 | 4 | 24 | 21 |
| $(3,2)$ | $\widehat{11}$ | 1547 | 1344 | 4 | 24 | 24 |
| $(2,3)$ | $\widehat{13}$ | 2079 | 2008 | 3 | 18 | 18 |
| $(4,2)$ | $\widehat{13}$ | 2926 | 2380 | 3 | 18 | 18 |
| $(0,4)$ | $\widehat{15}$ | 748 | 721 | 2 | 12 | 11 |
| $(1,4)$ | $\widehat{15}$ | 2261 | 2247 | 2 | 12 | 12 |
| $(3,3)$ | $\widehat{15}$ | 4096 | 1715 | 2 | 12 | 12 |
| $(2,4)$ | $\widehat{16}$ | 4914 | 2667 | 1 | 6 | 6 |
| $(3,4)$ | $\widehat{16}$ | 9177 | 8456 | 1 | 6 | 6 |
| $(4,3)$ | $\widehat{16}$ | 7293 | 4830 | 1 | 6 | 6 |
| $(4,4)$ | $\widehat{16}$ | 15625 | 15625 | 1 | 1 | 1 |

TABLE 9. Dimension data for $G_2(5)$

For $p = 3$ the Cartan invariants in Table 7 are computed by Mertens [319] (as well as by Cheng [106], with one misprint). Here $\det = 3^{10}$, with elementary divisors $3^6, 3^2, 3, 3$.

(C) Now consider the group $G = G_2(5)$. Mertens again applies the Chastkofsky–Jantzen formula to obtain all multiplicities $[Q(\lambda) : U(\mu)]$. Table 8 lists all cases in which the dimensions of $Q(\lambda)$ and $U(\lambda)$ differ. The notation indicates, for example, that $Q(1,3)$ has Brauer character equal to that of the direct sum of $U(1,3)$ and three copies of $U(4,4)$.

Table 9 summarizes the resulting dimension data. Although Mertens does not exhibit the Cartan matrix of $G_2(5)$, this is given in a paper by Hu and Ye [211]. They compute $\det C = 5^{16}$. (Here Brauer theory predicts elementary divisors $5^6, 5^3, 5^2, 5, 5, 5, 5, 5$.) More generally, for $p = 5$ and $r \geq 1$, Hu and Ye work out a closed formula: $\dim U_r(0) = (174^r - a^r - b^r + 1)5^{6r}$, where $a$ and $b$ are the roots of $x^2 - 162x + 792$.

REMARK. The 5-modular characters of $G_2(5)$ enter into the study of the Lyons group Ly by Meyer–Neutsch [320]. Here $G_2(5)$ occurs naturally as the subgroup stabilizing a point in the associated geometry.

## 18.5. Projective Modules for $p \geq 7$

When $p \geq 7$, the behavior of PIMs becomes more uniform. But it is convenient to look separately at the small case $p = 7$. Table 10 summarizes the numerical data for $p = 7$ (omitting St). The dimensions of Weyl modules are readily computable by hand, but can also be found in the tables of McKay–Patera–Rand [316].

When $p > 7$ Mertens works out the Chastkofsky–Jantzen formula in a uniform fashion. In this situation, the tensor products which need to be studied mostly involve a factor having a fundamental weight as highest weight. His tables were never formally published. Independently but later, Ye [447] obtained the same results in the course of investigating extensions of simple modules for the algebraic and finite groups of type $G_2$. Our comparison of the original version of his Table 2 with the table of Mertens revealed a few minor errors in each, which were easily corrected. It is very likely that the version here is accurate, but to have complete confidence one would have to reproduce all of the calculations. When $p > 7$ the only cases in which the Brauer characters of $Q(\lambda)$ and $U(\lambda)$ fail to agree are those listed in Table 11.

## 18.6. Characters of $G_2(q)$

The conjugacy classes and irreducible characters of $G$ have been worked out by Chang–Ree [94] for $p > 3$ and by Enomoto [164, 165, 166] for $p = 2, 3$. While the methods in these papers are somewhat ad hoc, they are influenced by Macdonald's conjectures on how to generalize Green's treatment of $GL(n, q)$. (The ideas of Macdonald in turn helped to instigate the program of Deligne–Lusztig). The character table of $G_2(q)$ is actually much simpler to describe than the table for $SL(3, q)$ or $Sp(4, q)$, essentially due to the fact that $Z(\mathbf{G}) = 1$ here.

In our brief survey of the characters, we mainly follow the notational conventions of Chang–Ree; for $p = 2, 3$ the data look similar, but with degeneracies. (For $p = 2$ Mertens develops separate notational conventions based on Deligne–Lusztig theory, the work of Enomoto not yet being available.)

| $\lambda$ | alcove | $\dim V(\lambda)$ | $\dim L(\lambda)$ | $d_\lambda$ | $\dim Q(\lambda)/p^6$ | $\dim U(\lambda)/p^6$ |
|---|---|---|---|---|---|---|
| $(0,0)$ | 1 | 1 | 1 | 29 | 348 | 91 |
| $(1,0)$ | 1/2 | 7 | 7 | 29 | 174 | 78 |
| $(2,0)$ | 2 | 27 | 26 | 16 | 192 | 150 |
| $(0,1)$ | 2/3 | 14 | 14 | 16 | 96 | 54 |
| $(3,0)$ | 2/3 | 77 | 77 | 16 | 96 | 61 |
| $(1,1)$ | 3 | 64 | 38 | 17 | 204 | 174 |
| $(0,2)$ | 3/4 | 77 | 77 | 17 | 102 | 54 |
| $(2,1)$ | 3/4 | 189 | 189 | 17 | 102 | 96 |
| $(4,0)$ | 3/4 | 182 | 182 | 17 | 102 | 62 |
| $(1,2)$ | 4 | 286 | 248 | 18 | 216 | 186 |
| $(0,3)$ | 4/5 | 273 | 273 | 18 | 108 | 54 |
| $(3,1)$ | 4/5 | 448 | 448 | 18 | 108 | 93 |
| $(2,2)$ | 5 | 729 | 481 | 12 | 144 | 144 |
| $(1,3)$ | 5/6 | 896 | 896 | 12 | 72 | 72 |
| $(3,2)$ | 5/7 | 1547 | 1547 | 12 | 72 | 72 |
| $(4,1)$ | 5/7 | 924 | 924 | 12 | 72 | 72 |
| $(5,0)$ | 5/7 | 378 | 378 | 12 | 72 | 52 |
| $(0,4)$ | 6 | 748 | 267 | 6 | 72 | 48 |
| $(2,3)$ | 6/8 | 2079 | 532 | 6 | 36 | 36 |
| $(1,4)$ | 6/8 | 2261 | 1337 | 6 | 36 | 30 |
| $(0,5)$ | 6/8 | 1729 | 1351 | 6 | 36 | 13 |
| $(5,1)$ | 7 | 1728 | 1247 | 6 | 72 | 72 |
| $(4,2)$ | 7/8 | 2926 | 2030 | 6 | 36 | 36 |
| $(6,0)$ | 7/9 | 714 | 714 | 6 | 36 | 35 |
| $(6,1)$ | 7/9 | 3003 | 3003 | 6 | 36 | 36 |
| $(3,3)$ | 8 | 4096 | 1853 | 5 | 60 | 60 |
| $(2,4)$ | 8/11 | 4914 | 3752 | 5 | 30 | 30 |
| $(5,2)$ | 8/11 | 5103 | 1827 | 5 | 30 | 30 |
| $(4,3)$ | 11 | 7293 | 3419 | 4 | 48 | 48 |
| $(6,2)$ | 11/12 | 8372 | 5397 | 4 | 24 | 24 |
| $(1,5)$ | 11/13 | 4928 | 4368 | 4 | 24 | 21 |
| $(3,4)$ | 11/13 | 9177 | 8064 | 4 | 24 | 24 |
| $(5,3)$ | 11/13 | 12096 | 10472 | 4 | 24 | 24 |
| $(4,4)$ | 13 | 15625 | 11413 | 3 | 36 | 36 |
| $(6,3)$ | 13/14 | 19019 | 12726 | 3 | 18 | 18 |
| $(2,5)$ | 13/15 | 10206 | 9681 | 3 | 18 | 18 |
| $(5,4)$ | 13/15 | 24948 | 24661 | 3 | 18 | 18 |
| $(3,5)$ | 15 | 18304 | 6578 | 2 | 24 | 24 |
| $(4,5)$ | 15/16 | 30107 | 17374 | 2 | 12 | 12 |
| $(0,6)$ | 15/19 | 3542 | 3360 | 2 | 12 | 11 |
| $(1,6)$ | 15/19 | 9660 | 9471 | 2 | 12 | 12 |
| $(2,6)$ | 15/19 | 19278 | 19201 | 2 | 12 | 12 |
| $(5,5)$ | 16 | 46656 | 26902 | 1 | 12 | 12 |
| $(6,4)$ | 16/18 | 37961 | 11375 | 1 | 6 | 6 |
| $(6,5)$ | 16/18 | 69160 | 56266 | 1 | 6 | 6 |
| $(3,6)$ | 16/21 | 33495 | 14294 | 1 | 6 | 6 |
| $(4,6)$ | 16/21 | 53599 | 44128 | 1 | 6 | 6 |
| $(5,6)$ | 16/21 | 81081 | 77721 | 1 | 6 | 6 |

TABLE 10. Dimension data for $G_2(7)$

| | |
|---|---|
| $(0,0)$ | $(0,0)+(p-1,0)+(p-2,1)+(p-1,1)+(3,p-3)+(0,p-1)$ |
| | $+(5,p-4)+(3,p-2)+(2,p-1)+(3,p-1)+(p-3,p-1)+(p-1,p-1)$ |
| $(1,0)$ | $(1,0)+(p-1,1)+(5,p-4)+(7,p-5)+(1,p-1)+(2,p-1)$ |
| | $+(4,p-2)+(3,p-1)+(4,p-1)$ |
| $(r,0)$ | $(r,0)+(2r+3,p-r-3)+(2r+5,p-r-4)+(r,p-1)+(r+1,p-1)$ |
| | $+(r+3,p-2)+(r+2,p-1)+(r+3,p-1)$ |
| | if $2 \leq r \leq \frac{1}{2}(p-9)$ |
| $(\frac{1}{2}(p-7),0)$ | $(\frac{1}{2}(p-7),0)+(p-4,\frac{1}{2}(p+1))+(p-2,\frac{1}{2}(p-1))+(\frac{1}{2}(p-7),p-1)+(\frac{1}{2}(p-5),p-$ |
| | $+(\frac{1}{2}(p-1),p-2)+2(\frac{1}{2}(p-3),p-1)$ |
| $(\frac{1}{2}(p-5),0)$ | $(\frac{1}{2}(p-5),0)+(p-2,\frac{1}{2}(p-1))+(\frac{1}{2}(p-5),p-1)+(\frac{1}{2}(p-3),p-1)+(\frac{1}{2}(p+1),p-$ |
| $(\frac{1}{2}(p-3),0)$ | $(\frac{1}{2}(p-3),0)+(\frac{1}{2}(p-3),p-1)+(\frac{1}{2}(p+3),p-2)+(\frac{1}{2}(p+1),p-1)+(\frac{1}{2}(p+3),p-$ |
| $(r,0)$ | $(r,0)+(r,p-1)+(r+1,p-1)+(r+3,p-2)+(r+2,p-1)+(r+3,p-1)$ |
| | if $\frac{1}{2}(p-1) \leq r \leq p-6$ |
| $(p-5,0)$ | $(p-5,0)+(p-5,p-1)+(p-4,p-1)+(p-2,p-2)+2(p-3,p-1)+(p-2,p-1$ |
| $(p-4,0)$ | $(p-4,0)+(p-4,p-1)+(p-3,p-1)+2(p-1,p-2)+(p-2,p-1)+5(p-1,p-$ |
| $(p-3,0)$ | $(p-3,0)+(p-3,p-1)+4(p-2,p-1)+10(p-1,p-1)$ |
| $(p-2,0)$ | $(p-2,0)+2(p-2,p-1)+8(p-1,p-1)$ |
| $(p-1,0)$ | $(p-1,0)+(p-1,p-1)$ |
| $(0,1)$ | $(0,1)+(p-1,1)+(p-2,2)+(p-1,2)+(3,p-1)$ |
| $(0,s)$ | $(0,s)+(p-1,s)+2(p-2,s+1)$ |
| | if $p \equiv 2 \pmod 3$ and $s = \frac{1}{3}(p-5)$, or $p \equiv 1 \pmod 3$ and $s = \frac{1}{3}(2p-5)$ |
| $(0,s)$ | $(0,s)+(p-1,s)$ |
| | if $p \equiv 1 \pmod 3$ and $s = \frac{1}{3}(p-4)$, or $p \equiv 2 \pmod 3$ and $s = \frac{1}{3}(2p-4)$ |
| $(0,\frac{1}{2}(p-3))$ | $(0,\frac{1}{2}(p-3))+(p-1,\frac{1}{2}(p-3))+(p-2,\frac{1}{2}(p-1))$ |
| $(0,p-2)$ | $(0,p-2)+(p-1,p-2)+2(p-2,p-1)+5(p-1,p-1)$ |
| $(0,p-1)$ | $(0,p-1)+(p-1,p-1)$ |
| $(0,s)$ | $(0,s)+(p-1,s)+(p-2,s+1)+(p-1,s+1)$ otherwise |
| $(1,1)$ | $(1,1)+(p-1,2)+(4,p-1)$ |
| $(r,1)$ | $(r,1)+(r+3,p-1)$ if $2 \leq r \leq p-5$ |
| $(p-4,1)$ | $(p-4,1)+(p-3,p-1)+(p-2,p-1)+3(p-1,p-1)$ |
| $(1,p-2)$ | $(1,p-2)+3(p-1,p-1)$ |
| $(1,\frac{1}{2}(p-3))$ | $(1,\frac{1}{2}(p-3))+(p-1,\frac{1}{2}(p-1))+(\frac{1}{2}(p-1),p-1)+(\frac{1}{2}(p+1),p-1)$ |
| $(1,s)$ | $(1,s)+(p-1,s+1)$ otherwise, for $s \notin \{\frac{1}{3}(p-5),\frac{1}{3}(2p-5)\}$ |
| $(2r+1,\frac{1}{2}(p-3)-r)$ | $(2r+1,\frac{1}{2}(p-3)-r)+(r+\frac{1}{2}(p-1),p-1)+(r+\frac{1}{2}(p+1),p-1)$ |
| | if $1 \leq r \leq \frac{1}{2}(p-7)$ |

TABLE 11. Decomposition of $Q(\lambda)$ for $p > 7$

The entries of the Chang–Ree character table depend on a parameter

$$\varepsilon := \begin{cases} +1 \text{ if } q \equiv 1 \pmod 3 \\ -1 \text{ if } q \equiv -1 \pmod 3 \end{cases}$$

Table 12 lists the regular semisimple characters, which occur in six families attached to the conjugacy classes of $W$ as shown. For use in 18.8 we attach a label to each family. Each individual character lies in a geometric conjugacy class by itself.

At the other extreme, we list in Table 13 the 10 unipotent characters of $G$ (along with the expression of their reductions modulo $p$ in terms of Weyl characters, to

| Family | Degree | Number | $w$ | Label |
|--------|--------|--------|-----|-------|
| $X_1$ | $(q+1)^2(q^4+q^2+1)$ | $(q^2-8q+17+2\varepsilon)/12$ | $1$ | $A$ |
| $X_2$ | $(q-1)^2(q^4+q^2+1)$ | $(q^2-4q+5-2\varepsilon)/12$ | $w_\circ$ | $F$ |
| $X_a$ | $q^6-1$ | $(q-1)^2/4$ | $s_1$ | $B$ |
| $X_b$ | $q^6-1$ | $(q-1)^2/4$ | $s_2$ | $C$ |
| $X_3$ | $(q^2-1)^2(q^2-q+1)$ | $(q^2+q-1-\varepsilon)/6$ | $(s_1s_2)^2$ | $E$ |
| $X_6$ | $(q^2-1)^2(q^2+q+1)$ | $(q^2-q-1+\varepsilon)/6$ | $s_1s_2$ | $D$ |

TABLE 12. Regular semisimple characters of $G_2(q)$ [94]

| Label | Degree | Reduction modulo $p$ |
|-------|--------|----------------------|
| $X_{11}$ | $1$ | $\chi(0,0)$ |
| $X_{12}$ | $q^6$ | $\chi(q-1,q-1)$ |
| $X_{13}$ | $q(q^4+2^2+1)/3$ | $\chi(0,q-1)+\chi(0,q-2)+\chi(q-4,1)$ |
| $X_{14}$ | $q(q^4+q^2+1)/3$ | $\chi(q-1,0)+\chi(1,q-2)+\chi(q-4,0)$ |
| $X_{15}$ | $q(q+1)^2(q^2-q+1)/2$ | $\chi(q-1,0)+\chi(0,q-1)+\chi(1,q-2)+\chi(q-4,1)$ |
| $X_{16}$ | $q(q+1)^2(q^2+q+1)/6$ | $\chi(q-1,0)+\chi(0,q-1)$ |
| $X_{17}$ | $q(q-1)^2(q^2+q+1)/2$ | $\chi(q-4,0)+\chi(0,q-2)+\chi(q-4,1)+\chi(1,q-2)$ |
| $X_{18}$ | $q(q-1)^2(q^2-q+1)/2$ | $\chi(q-4,0)+\chi(0,q-2)$ |
| $X_{19}$ | $q(q^2-1)^2$ | $\chi(1,q-2)+\chi(q-4,1)$ |
| $\overline{X}_{19}$ | $q(q^2-1)^2$ | $\chi(1,q-2)+\chi(q-4,1)$ |

TABLE 13. Unipotent characters of $G_2(q)$ and reduction mod $p$

be discussed below). They comprise a single geometric conjugacy class. The first 6 of these are the constituents of the character $1_B^G$, with $X_{12}$ being the Steinberg character:

$$1_B^G = X_{11} + X_{12} + X_{13} + X_{14} + 2X_{15} + 2X_{16}.$$

The remaining 4 are cuspidal characters.

Another geometric conjugacy class consists of four characters, labelled by Chang–Ree $X_{21}, X_{22}, X_{23}, X_{24}$. Here $X_{21}$ is regular and has degree $q^2(q^4+q^2+1)$, while $X_{22}$ is semisimple and has degree $q^4+q^2+1$. The other two have degree $q(q^4+q^2+1)$.

Three characters $X_{31}, X_{32}, X_{32}$ form a geometric conjugacy class. Here $X_{31}$ is regular and has degree $q^3(q^3+\varepsilon)$, while $X_{32}$ is semisimple and has degree $q^3+\varepsilon$. The character $X_{33}$ has degree $q(q+\varepsilon)q^3+\varepsilon)$.

Finally, there are four families of geometric conjugacy classes of the type studied by Pillen (17.12). Each family contains roughly $q$ of these classes. In turn each class involves a pairs of characters, one regular and one semisimple. Degrees are constant in a given family (and repeat in two families). The Chang–Ree labels for irreducible characters are $X_{1a}$ and $X'_{1a}$, etc.

## 18.7. Reduction Modulo $p$

While the characters of $G_2(q)$ are fairly manageable, the process of reduction modulo $p$ leads to some unavoidable complexity even when $q = p$ due to the complicated composition series of some of the Weyl modules involved. But it is possible to avoid this issue at first by expressing the reductions modulo $p$ of the characters just in terms of the Brauer characters associated with Weyl modules. Then the

Weyl module decompositions can be incorporated. (This is the strategy used by Jantzen [**257**] in the case of $Sp(4, q)$.) For $p \geq 7$, Mertens works out in a uniform way the decomposition of each family of ordinary characters.

All six families of regular semisimple characters of $G_2(p)$ exhibit the same decomposition behavior: there are 119 composition factors after reduction modulo $p$, with multiplicities as high as 4. (This is the same pattern found in the case of the $\mathbf{G}_1\mathbf{T}$-modules $\widehat{Z}_1(\lambda)$ or in the case of Weyl modules with highest weight in general position in the lowest $p^2$-alcove.) For $G_2(p^r)$ the patterns would blow up, with generically $119^r$ composition factors occurring.

The initial decompositions of the 10 unipotent characters of $G_2(q)$ into Weyl characters are given in Table 13. Here again the actual number of composition factors blows up as the power of $p$ increases.

Using his methods, Mertens is able to compute explicitly the decomposition matrices for $G_2(2)$ and $G_2(3)$, which in turn recover the Cartan matrices exhibited earlier. For $p = 2$ see Table 14.

| Character | Degree | $(0,0)$ | $(1,0)$ | $(0,1)$ |
|---|---|---|---|---|
| $X_{11}$ | 1 | 1 | 0 | 0 |
| $X_{13}$ | 14 | 0 | 0 | 1 |
| $X_{14}$ | 14 | 2 | 2 | 0 |
| $X_{15}$ | 27 | 1 | 2 | 1 |
| $X_{16}$ | 21 | 1 | 1 | 1 |
| $X_{17}$ | 7 | 1 | 1 | 0 |
| $X_{18}$ | 1 | 1 | 0 | 0 |
| $X_{19}$ | 6 | 0 | 1 | 0 |
| $X'_{19}$ | 6 | 0 | 1 | 0 |
| $X_{2a}$ | 42 | 2 | 2 | 2 |
| $X'_{2a}$ | 21 | 1 | 1 | 1 |
| $X_{31}$ | 56 | 4 | 4 | 2 |
| $X_{32}$ | 7 | 1 | 1 | 0 |
| $X_{33}$ | 14 | 0 | 0 | 1 |
| $R_{w_6^2}(1,0)$ | 27 | 1 | 2 | 1 |

TABLE 14. Decomposition matrix for 1-block of $G_2(2)$ [**319**]

## 18.8. Brauer Complex of $G_2(5)$

Even though the group $G = G_2(5)$ is far too small to exhibit generic decomposition behavior, its Brauer complex (17.16) does help one to visualize better how the ordinary and modular characters fit together. Mertens works out Lie-theoretically the basic data for $p = 5$ (including Table 9 above). This can also be obtained directly in the finite group setting; in particular, the $43 \times 24$ decomposition matrix of the 1-block of $G$ can currently be found at the Modular Atlas homepage [**243**].

Without going into detail about the required bookkeeping with the parameters here, we indicate how the 25 geometric conjugacy classes containing 43 ordinary characters in the 1-block (with the Steinberg character omitted) are located in the Brauer complex. See Figure 2. (This is a top alcove version of the picture in Deriziotis [**130**, p. 268].)

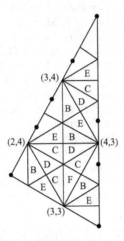

FIGURE 2. Brauer complex for $G_2(5)$

There are 17 regular semisimple characters, as in Table 12, belonging in this case to only five of the six families labelled $A, B, C, D, E, F$. These are assigned to faces of the small alcoves with corresponding labels in the Brauer complex, consistent with the orientation of the Weyl group elements to which the families correspond.

The unipotent characters (with the Steinberg character omitted) are assigned to the designated vertex of the top alcove. The geometric conjugacy class containing four characters is assigned to the vertex of the alcove which lies at the right angle corner, while the class with three characters is assigned to the vertex of the third boundary alcove.

When $p = 5$ there are five other geometric conjugacy classes, each involving one regular and one semisimple character. These are assigned to the designated walls of two alcoves sharing their long walls with the long wall of the big alcove and to the designated middle walls of three alcoves sharing these walls with the middle wall of the big alcove.

No $p$-regular weights occur here, but one can begin to see how the ordinary characters occur in (lifted) PIMs belonging to the indicated top alcove weights (see Tables 8 and 9). For example the PIM $U(\lambda)$ for $\lambda = (4, 3)$ agrees with $Q(\lambda)$ and has dimension $6 \cdot 5^6$. Its lifting to characteristic 0 involves four regular semisimple characters labelled $B, C, D, E$ (whose respective degrees are 15624, 15624, 12096, 17856) along with two regular characters (from the pairs on walls) of degrees 19530 and 13020. This data occupies one column of the decomposition matrix. Passing to lower weights in the linkage class of $\lambda$ requires inspection of larger PIMs involving more alcoves in general. In this small case, the encoding of all the data in the Brauer complex is rather complicated, but for very large $p$ the repetition of uniform patterns would become obvious.

# CHAPTER 19

# General and Special Linear Groups

In this chapter we look at some special features of representations of $GL(n, q)$ and $SL(n, q)$, most of which involve the natural action on the polynomial algebra $S := K[x_1, \ldots, x_n]$ (19.2–19.7). For example, the ring of invariants has been studied since the time of Dickson. In the case of $SL(2, q)$ there is an interesting "periodicity" phenomenon (19.8), which shows a systematic pattern in the module structure of the homogeneous summands of $S$. The polynomial ring setting lends itself to combinatorial methods involving partitions, generating functions, and the like. Since algebraic group methods often remain in the background, we provide mainly a survey of the results with occasional proofs when these are self-contained.

To conclude the chapter, we discuss in 19.11 a method of passing from modular to ordinary representations called "Brauer lifting" which in a sense reverses the earlier procedure of reduction modulo $p$.

Although these topics can be studied in purely algebraic terms, they have many connections with problems in algebraic topology, involving especially the Steenrod algebra and classifying spaces: see for example Wilkerson [432] and Henn–Schwartz [203]. Often the prime $p = 2$ is most natural in these contexts, but the algebraic framework is more general.

## 19.1. Representations of $GL(n, K)$ and $GL(n, q)$

To simplify the language of weights we have been limiting our attention to semisimple (even simple) algebraic groups, though the theory needs to be developed in the general setting of connected reductive groups: for inductive purposes, the Levi subgroups of parabolic subgroups are essential.

It is easy to adapt the conventions about roots and weights in the case of $GL(n, K)$. For background on the discussion below see [**RAGS**, II.1.21]. Taking the diagonal subgroup **T** to be a maximal torus, the coordinate functions $\varepsilon_1, \ldots, \varepsilon_n$ form a basis of the character group $X(\mathbf{T})$. (On restriction to $SL(n, K)$ one gets the extra relation $\sum \varepsilon_i = 0$.) Relative to the Borel subgroup of upper triangular matrices, the simple roots are $\varepsilon_1 - \varepsilon_2, \varepsilon_2 - \varepsilon_3, \ldots, \varepsilon_{n-1} - \varepsilon_n$. All of this can also be formulated in the language of partitions: see for example Carter–Lusztig [92].

The (rational) simple modules $L(\lambda)$ for $SL(n, K)$ all lift to simple modules for $GL(n, K)$. But $GL(n, K)$ also has an infinite family of 1-dimensional simple modules corresponding to powers of det. Tensor products of these with the $L(\lambda)$ exhaust all simple modules. In particular, there are now extra "Steinberg modules" for each power $p^r$.

On restriction to the finite group $GL(n, q)$, one gets $(q - 1)q^{n-1}$ isomorphism classes of simple modules: there are $q - 1$ distinct powers of det.

## 19.2. Action on Symmetric Powers

Our immediate concern is with the standard module $V = K^n$ for $GL(n, K)$ or $SL(n, K)$ and its dual $V^*$. Each has a natural group action, which extends canonically to the symmetric algebra. If $x_1, \ldots, x_n$ is a basis for $V^*$, the resulting algebra $S = S(V^*)$ may be identified with the polynomial algebra $K[x_1, \ldots, x_n]$. The group action preserves the $d$th homogeneous component $S_d$, which is the main object of study in the following sections. Moreover, all weight multiplicities are 1, since the basis monomials have distinct weights.

The reader has to be cautious about the conventions used in various papers, where $S(V)$ rather than $S(V^*)$ may be preferred; the group action may also occur on the right rather than the left. This has little effect on the eventual results, so we just refer here to the "natural" (left) action on $S$ or $S_d$. But in Jantzen's comparison with induced modules $H^0(\lambda)$ in [**RAGS**, II.2.16], one has to be aware of his conventions. In particular, he concludes that as a module for either group, $S_d$ is isomorphic to the induced module $H^0(d\varpi)$ for a suitable fundamental weight $\varpi$. (In the literature this may be either $\varpi_1$ or $\varpi_{n-1}$ in the usual numbering.)

In general it is almost impossible to describe the full submodule structure of $H^0(\lambda)$ as a module for $GL(n, K)$, but when $\lambda = d\varpi$ more can be said. For example, the fact that all composition factor multiplicities are 1 in this case makes it possible in principle to describe the lattice of submodules concretely in terms of diagrams. Following detailed work on the rank 1 case (by Carter–Cline, Deriziotis, Krop), Doty [**147, 148, 149**] has studied the modules $S_d$ in considerable detail. He describes their submodule lattices combinatorially, along with an algorithm for the computation of their composition factors. (See Doty–Walker [**151**] and Bardoe–Sin [**33**] for further refinements.)

The combinatorial methods used in these papers often make it possible to determine in turn how $S_d$ looks as a module for $GL(n, q)$. Usually when one restricts Weyl modules or induced modules for a reductive group such as $GL(n, K)$ to a finite subgroup $G$ such as $GL(n, q)$ or $SL(n, q)$, it is extremely difficult to work out the resulting $KG$-module structure in detail. Even in the case of the well-behaved symmetric powers $S_d$, the $G$-socle is tricky to get at. Below we consider some uniform features of the $G$-module structure which emerge only indirectly.

## 19.3. Dickson Invariants

A classical problem in invariant theory is to describe in detail the ring of invariants of a finite group acting on an associated ring. In the case of the natural action of $GL(n, q)$ or $SL(n, q)$ on the polynomial algebra $\mathbb{F}_q[x_1, \ldots, x_n]$, an elegant solution was found by Dickson [**132**]. This has since found a number of applications in algebraic topology. Modern treatments of the theorem are found in many sources, including Bourbaki [**56**, pp. 137–138], Benson [**49**, Chap. 8], Smith [**376**, 8.1], Steinberg [**397**], Wilkerson [**432**].

THEOREM. *Let $F$ be any field of characteristic $p$, and let $GL(n, q)$ act in the natural way on the polynomial ring $F[x_1, \ldots, x_n]$.*

(a) *The ring of invariants under $GL(n, q)$ is generated by $n$ algebraically independent homogeneous polynomials, of uniquely determined degrees*

$$q^n - q^{n-1}, \ldots, q^n - q, \; q^n - 1.$$

(b) *The ring of invariants under* $\mathrm{SL}(n, q)$ *is also generated by* $n$ *algebraically independent homogeneous polynomials, of uniquely determined degrees*

$$q^n - q^{n-1}, \ldots, q^n - q, \ q^{n-1} + \cdots + q^2 + q + 1.$$

In each of the two cases, invariant polynomials of the indicated degrees are unique (up to scalars) and may be called **basic invariants**.

While the statement is trivially true when $n = 1$, we always assume that $n > 1$. There are various approaches to Dickson's Theorem in the cited literature, all involving ring-theoretic techniques. As in Dickson's original proof, these can yield explicit expressions for the basic invariants in terms of determinants.

## 19.4. Complements

The study of Dickson invariants leads to some closely related questions. Write $S := K[x_1, \ldots, x_n]$.

(1) The concise proof of Dickson's Theorem by Steinberg [397] is carried out in tandem with a closely related result (which is essentially equivalent to the theorem): As an $S^G$-module, $S$ is free of rank equal to $|G|$, with a basis consisting of monomials

$$x_1^{i_1} \ldots x_n^{i_n} \text{ with } 0 \le i_k < q^n - q^{k-1}.$$

(Using more machinery from commutative algebra, Bourbaki [56, V,§5] works out in a fairly general setting such a relationship between the ring of invariants of a finite linear group and the resulting module structure over the invariants.)

(2) As an inductive tool, the invariants in $S$ of parabolic subgroups of $G$ are also worked out by Kuhn–Mitchell [292] (and later rediscovered by Hewett [205]); independent homogeneous generators of certain degrees are specified. In another direction, Tri [421] uses the Dickson invariants in order to determine the irreducible modules for parabolic subgroups.

(3) In principle, a thorough understanding of the submodule structure of dual Weyl modules would lead back to the Dickson invariants. One would need to study how each $\mathrm{GL}(n, K)$-composition factor $L(\lambda)$ of $S_d$ restricts to $G = \mathrm{GL}(n, q)$ or $\mathrm{SL}(n, q)$ and then pick out those trivial modules which actually occur in the $G$-socle of $S_d$. None of this is easy in practice. Looking for at least some uniformities in the module structure, Humphreys [236] arrives at estimates on where the basic invariants for $\mathrm{SL}(n, p)$ should be located. When $d = p^{n-1} + \cdots + p + 1$, the basic invariant in $S_d$ is shown to occur already in the socle $L(d\varpi)$ of the $\mathrm{SL}(2, K)$-module $H^0(d\varpi)$. In all other cases, it is conjectured that $L(d\varpi)$ extends another $\mathrm{SL}(2, K)$-composition factor of $H^0(d\varpi)$ whose restriction to $G$ contains a basic invariant.

(4) Elkies [160] contrasts the invariants for $\mathrm{GL}(n, q)$ with those for $\mathrm{Sp}(2n, q)$. Both groups are generated by pseudo-reflections (a necessary condition for the ring of invariants to be a polynomial ring [56, V,§5]), but for the symplectic groups the ring of invariants is in fact more complicated than a polynomial ring. He places these rings geometrically in the setting of certain Deligne–Lusztig varieties associated with the Weyl group.

## 19.5. Multiplicity of St in Symmetric Powers

Kuhn and Mitchell [292] work out explicitly the multiplicity with which the Steinberg module St for $GL(n, q)$ occurs as a composition factor (or equivalently, direct summand) of each symmetric power of the natural representation. Here one can take as the definition of St the module afforded by the top homology of the Tits building, as for example in [307, 1.13]. This can initially be defined over a more general base ring and then reduced modulo $p$.

We continue to use our previous notation $S_d$ in place of their $S(V)^d$. And we write St rather than $St_r$ for convenience, keeping fixed the power $q = p^r$. If $a_d$ is the multiplicity of St in $S_d$, define the *multiplicity series* for St to be the formal power series

$$F_{St}(t) := \sum_{d=0}^{\infty} a_d t^d.$$

THEOREM. *Let* $c_i := q^i - 1$ *and* $c := \sum_{i=1}^{n-1} c_i$. *Then*

$$F_{St}(t) = t^c \prod_{i=1}^{n} (1 - t^{c_i})^{-1}.$$

The proof involves a mixture of algebraic and topological ideas. The key ingredients are the invariants of the parabolic subgroups of $G$ in $S$ and the action of $G$ on the Tits building defined by the parabolics. It is here that St comes into play.

The motivation for developing such a formula comes originally from topology (especially in the case $p = 2$). There are a number of approaches in the literature to the special case $G = GL(n, p)$: see Mitchell–Priddy [324], Mitchell [322], Kuhn [289, 290]. In this case Carlisle–Walker [81] work out in a similar spirit the multiplicity series for certain weights adjacent to the Steinberg weight.

## 19.6. Other Simple Modules

Having examined how the trivial module and the Steinberg module occur as submodules in various symmetric powers $S_d$, we may ask more generally how to determine the multiplicities of other simple modules as either submodules or composition factors. This information is not yet accessible in most cases. But at least we see from the following theorem of Mitchell [322] that all simple modules for the finite group are involved in $S$.

THEOREM. *Let* $G = GL(n, q)$ *or* $SL(n, q)$. *Then every simple* $KG$-*module occurs as a composition factor of some* $S_d$.

Here is a quick sketch of the proof, which relies mainly on the fact that the ring of invariants $S^G$ is a polynomial algebra on $n$ homogeneous generators (19.3). As noted above in 19.4(1), $S$ is a free $S^G$-module of rank equal to $|G|$. In turn, the vector space $M := S \otimes_{S^G} K$ has dimension $|G|$. Mitchell adapts classical arguments to show that as a $G$-module, $M$ defines the same element in the Grothendieck group $R_K(G)$ as the *regular representation*. He also shows that $S$ and $S^G \otimes M$ have the same composition factors in each degree (counting multiplicity): this amounts to comparing Poincaré series. These facts combine to complete the proof.

REMARKS. (1) More complete details about composition factors of each $S_d$ have been worked out in the case of GL(3, $p$) by Doty–Walker [151], using techniques developed by Carlisle in his 1985 thesis work and later papers with various collaborators [79, 80, 81].

(2) Kuhn [291] and Krasoń–Kuhn [287] develop more powerful functorial methods to prove theorems about embeddings in direct sums of symmetric or tensor powers of the natural module over $\mathbb{F}_p$.

(3) Although the statement of the theorem involves only the action of $G$ on the various modules $S_d$, it is unclear how to prove it in this spirit without specifying somehow which $S_d$ involves a given simple $KG$-module as a composition factor.

## 19.7. Example: SL(2, $p$)

Although the submodule structure of $S_d$ for an arbitrary GL($n, q$) is combinatorially intricate, much can be said in the case $n = 2$. This and the following three sections bring out some conceptually interesting features of the situation. Here we prefer to concentrate on SL(2, $q$) rather than GL(2, $q$), to keep the parametrization of simple modules uncomplicated. We also adjust notation a little: let $S = K[x, y]$ and use names like $S_m$ and $S_n$ for typical homogeneous summands of $S$.

Consider first what can be done over $\mathbb{F}_p$. In his thesis, Glover [187] works out in great detail the actions of GL(2, $p$) and SL(2, $p$) on the space $S_n$ of homogeneous polynomials of degree $n$. Here dim $S_n = n + 1$. (His notation is different, since he indexes the modules by their dimensions.) We take $G = $ SL(2, $p$), with $p$ arbitrary. As was already well-known from the early work of Brauer and Nesbitt, the $p$ simple $KG$-modules occur immediately as $S_0, S_1, \ldots, S_{p-1}$.

As $n$ grows, the complexity of the $KG$-module structure grows. But Glover observes a striking periodicity phenomenon: If we write (uniquely) $S_n = T_n \oplus P_n$, with $P_n$ a projective summand of largest possible dimension, then the complementary modules $T_n$ satisfy

$$T_{n+p(p-1)} \cong T_n \text{ for all } n \geq 0.$$

Moreover, the modules $T_n$ are indecomposable, and each non-projective indecomposable module occurs precisely once in this list. This last feature of course reflects the special fact that SL(2, $p$) has cyclic Sylow $p$-subgroups, hence has only finitely many indecomposables. (See 8.9 and 9.10.) So this cannot generalize to SL(2, $q$). But the periodicity does generalize.

## 19.8. Periodicity for SL(2, $q$)

In [9] Alperin and Kovács extend Glover's periodicity result to SL(2, $q$), with $q = p^r$. Here $\mathbf{G} = $ SL(2, $K$).

As above, we let $S := K[x, y]$, with each $\mathbf{G}$-submodule $S_n$ isomorphic to the dual Weyl module $H^0(n\varpi)$ for the single fundamental weight $\varpi$. Thus dim $S_n = n+1$. Write $S_n = T_n \oplus P_n$, where $P_n$ is a projective $KG$-summand of largest possible dimension. Thanks to the Remak–Krull–Schmidt Theorem, this decomposition is uniquely determined within isomorphism.

THEOREM. Let $G = $ SL(2, $q$). Writing the $KG$-module $S_n = T_n \oplus P_n$ as above, we have $T_{n+q(q-1)} \cong T_n$ for all $n \geq 0$. Moreover, $S_{qn-1}$ is projective for all $n > 0$.

The proof of the theorem will be given in the following sections. It is fairly elementary, using standard general facts: projectives are injective for $KG$, while tensoring an arbitrary module with a projective module yields another projective. The Steinberg module $St_r = S_{q-1}$ also plays an essential role.

REMARKS. (1) Unpublished calculations by Kovács indicate that $T_n$ is always *indecomposable*. It then remains a problem to characterize these among the infinitely many indecomposables arising if $q \neq p$. For example, do all simple modules occur among the $T_n$?

(2) It is a challenging problem to find an appropriate analogue in higher ranks for this type of periodicity. Experimentation in rank 2 indicates that no naive generalization is likely. But a type of repetitive structure might for example occur in the family of modules $H^0(d\rho)$.

## 19.9. Key Lemma

First we prove the lemma of Glover [187] (inspired by G.E. Wall) which is fundamental to his results on $SL(2, p)$ and plays a similar role here.

LEMMA. *For all $m, n > 0$, there is a short exact sequence of* $SL(2, K)$*-modules*

$$0 \to S_{m-1} \otimes S_{n-1} \xrightarrow{\theta} S_m \otimes S_n \xrightarrow{\varphi} S_{m+n} \to 0.$$

PROOF. Observe first that the dimensions involved are compatible with the exactness of the sequence: $(m + 1)(n + 1) = mn + (m + n + 1)$.

To define $\varphi$, send $u \otimes v$ to $uv$. This is clearly a well-defined **G**-module map. Moreover, $\varphi$ is surjective: all standard monomials in $S_{m+n}$ lie in the image, since $x^i y^{m-i} \otimes x^j y^{n-j} \mapsto x^{i+j} y^{m+n-i-j}$.

Next define $\theta$ by sending $u \otimes v \mapsto ux \otimes vy - uy \otimes vx$. To see that $\theta$ is a **G**-module map, it is enough to check the effect of the standard unipotent generators of **G**. This is a routine computation.

It is clear that the image of $\theta$ lies in $\operatorname{Ker} \varphi$. Thanks to the compatibility of the dimensions involved, it is now enough to check that $\operatorname{Ker} \theta = 0$. For this we work with a normal form in $S_{m-1} \otimes S_{n-1}$: each element can be written in the form $\sum_{i=1}^m x^{m-i} y^{i-1} \otimes v_i$ for *unique* $v_i \in S_{n-1}$. If such an element lies in $\operatorname{Ker} \theta$, we get

$$0 = \sum_{i=1}^m (x^{m-i} y^{i-1} x \otimes v_i y - x^{m-i} y^{i-1} y \otimes v_i x)$$

$$= x^m \otimes v_1 y - y^m \otimes v_m x + \sum_{i=1}^{m-1} x^{m-i} y^i \otimes (v_{i+1} y - v_i x).$$

This last expression is in normal form in $S_m \otimes S_n$, forcing all summands to be 0. In turn, $0 = v_1 y = v_m x$ and $v_i x = v_{i+1} y$ for $i \leq m - 1$. Because $S$ is a domain, these equations force all $v_i = 0$, as desired. $\square$

## 19.10. Proof of Periodicity Theorem

In the following steps, we mix the original, mainly self-contained, arguments of Alperin–Kovács with some shortcuts based on general results for algebraic groups. To prove the theorem, the idea is to examine first two crucial special cases, then to combine these with Lemma 19.9.

(A) We claim that $S_{qn-1}$ is a *projective* $KG$-module for all $n > 0$.

This can be attacked inductively, as in [**9**], starting for $n = 1$ with the fact that $S_{q-1} \cong \mathrm{St}_r$ (with $q = p^r$). For the induction step, use Lemma 19.9 to get a short exact sequence of $KG$-modules:

$$0 \to S_{q-2} \otimes S_{qn-1} \to S_{q-1} \otimes S_{qn} \to S_{q(n+1)-1} \to 0.$$

By induction, the first nonzero term is projective. The next term is projective because $S_{q-1}$ is. Therefore $S_{q(n+1)-1}$ is projective as well.

As an alternative, we can appeal to a general result about dual Weyl modules with special highest weights: see Jantzen [**RAGS**, II.3.19], which implies in particular an isomorphism of **G**-modules:

$$H^0((p^r - 1)\rho + p^r\lambda) \cong H^0((p^r - 1)\rho) \otimes H^0(\lambda)^{[r]}$$

for any $\lambda \in X^+$. Here $H^0((p^r - 1)\rho) = \mathrm{St}_r$. On restriction to $G$, we get

$$H^0((q - 1)\rho + q\lambda) \cong \mathrm{St}_r \otimes H^0(\lambda).$$

In our situation, with $G = \mathrm{SL}(2,q)$, we are given a dominant weight $(qn - 1)\rho = (q - 1)\rho + q(n - 1)\rho$. So the isomorphism translates into $S_{qn-1} \cong \mathrm{St}_r \otimes S_{n-1}$. The right side being a projective $KG$-module, so is the left side.

(B) We claim (as required by the theorem) that $S_{q(q-1)}$ *is isomorphic to the direct sum of the trivial module $S_0$ and a projective module.* This is of course compatible with the dimensions involved.

How can we find a large projective submodule? Since the highest weight here is $q(q - 1)\rho$, the **G**-socle is the $r$th Frobenius twist of $\mathrm{St}_r$, which on restriction to $G$ is just $\mathrm{St}_r$. Note that the corresponding highest weight vector in $S_{q(q-1)}$ is a multiple of $x^{q(q-1)}$.

On the other hand, we can locate a projective submodule in a less direct way. It follows from (A) that $S_{q(q-2)-1}$ is a projective $KG$-module. It is easy to see (by checking the action of unipotent generators of $G$) that the polynomial $x^q y - xy^q$ is $G$-invariant (see 19.3). Multiplication by this polynomial thus defines a $KG$-homomorphism $\varphi : S_{q(q-2)-1} \to S_{q(q-1)}$. Clearly $\varphi$ is injective, since $S$ is a domain. Moreover, the image of $\varphi$ visibly does not contain $x^{q(q-1)}$, and therefore intersects the above copy of $\mathrm{St}_r$ trivially. The dimensions of our two projective submodules now add up to $q(q - 2) + q = q(q - 1)$. Since projectives are injective, we get a projective summand of this dimension in $S_{q(q-1)}$, leaving a complement of dimension 1 as required.

With (A) and (B) in hand, we can complete the proof of the theorem. Start with the exact sequence in Lemma 19.9:

$$0 \to S_{m-1} \otimes S_{n-1} \to S_m \otimes S_n \to S_{m+n} \to 0.$$

Setting $m = q(q - 1)$, this becomes:

$$0 \to S_{q(q-1)-1} \otimes S_{n-1} \to S_{q(q-1)} \otimes S_n \to S_{n+q(q-1)} \to 0.$$

Now $S_{q(q-1)-1}$ is projective, by (A), so its tensor product with $S_{n-1}$ is also projective. Since projectives are injective, the short exact sequence splits. But (B) allows us to express the middle term as the direct sum of $S_n$ and a projective module. A comparison of projective and nonprojective summands shows that $T_{n+q(q-1)}$ is isomorphic to $T_n$, as required.   □

## 19.11. Brauer Lifting

Here we look at a completely different question in which $GL(n, q)$ plays a prominent role.

Start with an arbitrary finite group $G$. As in 16.1, we consider reduction modulo $p$ of representations in the framework of a $p$-modular system $(L, R, k)$. One striking feature of the Cartan–Brauer triangle (16.4) is the *surjectivity* of the decomposition homomorphism $d : R_L(G) \to R_k(G)$. As observed by Green [193], a section of $d$ can be defined on the level of characters by applying Brauer's characterization of characters: Take any $kG$-module $V$, having Brauer character $\psi$. This is initially defined just on $G_{\text{reg}}$. Extend $\psi$ to a class function on all of $G$ by setting $\psi(g) = \psi(s)$, where $g = su$ is the decomposition of $g$ into commuting $p$-regular and $p$-singular elements. By looking closely at how $\psi$ restricts to elementary subgroups of $G$, one then deduces from Brauer's theorem that $\psi$ is a (virtual) character of $G$ over $L$. This is the **Brauer lifting** of $\psi$ (or of the $kG$-module $V$): see Curtis–Reiner [126, (18.12)]. (Quillen used this idea effectively in his proof of the Adams conjecture.)

Lusztig [307, Chap. 5] shows how to realize the Brauer lifting concretely, starting with the natural module for $GL(n, q)$. (See also Curtis [124, 4.2] and Springer [384].) He uses the $p$-modular system $(Q_k, W_k, k)$, where $k = \mathbb{F}_q$, $W_k$ is the ring of Witt vectors of $k$, and $Q_k$ is its field of fractions. In the special case $k = \mathbb{F}_p$, $W_k$ is the ring $\mathbb{Z}_p$ of $p$-adic integers while $Q_k$ is the field $\mathbb{Q}_p$ of $p$-adic numbers.

THEOREM. *If $k = \mathbb{F}_q$ and $V = k^n$ is the natural module for $G = GL(n, q)$, there are free $W_k$-modules $V^{(i)}$ on which $G$ acts, giving an exact sequence of $kG$-modules:*

$$0 \to V^{(n)} \otimes_{W_k} k \to V^{(n-1)} \otimes_{W_k} k \to \cdots \to V^{(1)} \otimes_{W_k} k \to V \to 0.$$

*Here*

$$\dim V^{(i)} = (q^{n-i+1} - 1) \cdots (q^{n-1} - 1)(q^n - 1)/(q^i - 1).$$

*The alternating sum of characters of the $V^{(i)}$ over $Q_k$ is the Brauer lifting of $V$.*

The theorem is a byproduct of Lusztig's geometric construction of an explicit discrete series representation $D(V)$ of $GL(n, q)$, realized by a free $W_k$-module of rank $(q - 1)(q^2 - 1) \cdots (q^{n-1} - 1)$. This gives the term $V^{(n)}$ in the above resolution, while the other $V^{(i)}$ are obtained by parabolic induction from the analogous discrete series representations of subgroups of type $GL(V_i)$. Lusztig shows that, except when $i = 1$ and $q = 2$, the modules $V^{(i)}$ are simple. In turn, Brauer liftings for arbitrary finite groups are obtained by considering their representations in various $GL(V)$.

Notice how the dimensions behave when $G = GL(2, q)$ or $SL(2, q)$: the discrete series module has dimension $q - 1$ and the induced (principal series) module has dimension $q + 1$, giving the alternating sum $2 = (q + 1) - (q - 1)$.

Bédard [34] goes on to show how this kind of explicit Brauer lifting, already defined by Lusztig for other classical groups, can be formulated in the setting of an arbitrary finite Chevalley group $G$. The idea is to study the formal orbit sums under the Weyl group $W$, which give a basis for $R_k(G)$. In case $G = GL(n, q)$ or $SL(n, q)$, the formal character of the natural module $V$ is just one of these orbit sums coming from a fundamental weight. For each orbit sum, Bédard describes a Brauer lifting (along lines suggested by Lusztig) as an alternating sum of irreducible representations in bijection with parts in a partition of the conjugacy class of parabolic subgroups of **G** associated with the isotropy group in $W$ of the orbit.

# CHAPTER 20

# Suzuki and Ree Groups

In this chapter we look at the simple groups found by Ree [**347**, **348**] and Suzuki [**408**], which exist only over fields of characteristic 2 or 3. The Suzuki groups are subgroups of an algebraic group of type $B_2$ $(= C_2)$ when $p = 2$, while the two families of Ree groups occur in algebraic groups of types $G_2$ when $p = 3$ and $F_4$ when $p = 2$. The idea is to combine Chevalley's special isogeny (5.3) with a standard Frobenius map, then take fixed points.

In contrast to the theory for Chevalley groups and their twisted analogues of types $A, D, E_6$, the representation theory of these groups in the defining characteristic $p = 2$ or $p = 3$ involves just one fixed prime and therefore does not raise the genericity questions considered earlier. On the other hand, there is still much to be made explicit as the power of $p$ grows.

Here we review the literature dealing with Suzuki and Ree groups, giving some examples and indicating some unsolved problems. After recalling how the groups are constructed (20.1), we discuss their simple modules (20.2). Beyond this, the treatment of projective modules, Cartan invariants, extensions, and cohomology is somewhat fragmentary. The Suzuki groups have been best studied. Similarly, while all the ordinary characters have been determined (20.7), their reduction modulo $p$ has not been worked out in detail for Ree groups.

## 20.1. Description of the Groups

Start with a simply-connected algebraic group $\mathbf{G}$ of type $B_2$ $(= C_2)$ or $F_4$ with $p = 2$, or else a group of type $G_2$ with $p = 3$. Assume that $\mathbf{G}$ is split over $\mathbb{F}_p$. In this situation there is a special isogeny (5.3) $\varphi : \mathbf{G} \to \mathbf{G}$, whose square is the standard Frobenius map $F_1$ relative to $p$. Now fix $r \geq 1$ and define an isogeny of $\mathbf{G}$ by $\sigma := F_r \circ \varphi = \varphi \circ F_r$. The group $G^\sigma$ of fixed points is called a **Suzuki group** if $\mathbf{G}$ has type $B_2$ or a **Ree group** if the type is $F_4$ or $G_2$.

Since $\varphi^2 = F_1$, we have $\sigma^2 = F_{2r+1}$. The fixed point subgroup of $\sigma^2$ is therefore a Chevalley group over the field of $p^{2r+1}$ elements, which in turn contains $G$ as a subgroup. We say that this Chevalley group is **associated** with $G$. Even though it is conventional to define Suzuki groups in terms of the $B_2$ root system, it is convenient here to regard $Sp(4, 2^{2r+1})$ as the associated Chevalley group. (But one has to be careful about which simple root is regarded as short and which as long.)

When $p = 2$, there is no distinction between the universal and adjoint Chevalley groups of type $B_2$, so the Suzuki group can equally well be constructed by starting with an adjoint group. On the other hand, there is always a single isogeny type for algebraic groups of type $F_4$ or $G_2$. Thus each root system and *odd* power of $p$ lead to a single Suzuki or Ree group over a field of that size. The resulting groups can be labelled as in Table 1, with the convention that $q^2$ in each case denotes an odd power of 2 or 3. This convention helps to make the group order formulas resemble

| Lie type | Order of group | $q^2$ |
|----------|----------------|-------|
| $^2\mathrm{B}_2(q^2)$ | $q^4(q^2-1)(q^4+1)$ | $2^{2r+1}$ |
| $^2\mathrm{G}_2(q^2)$ | $q^6(q^2-1)(q^6+1)$ | $3^{2r+1}$ |
| $^2\mathrm{F}_4(q^2)$ | $q^{24}(q^2-1)(q^6+1)(q^8-1)(q^{12}+1)$ | $2^{2r+1}$ |

TABLE 1. Suzuki and Ree groups

closely the formulas for the associated Chevalley groups, with only sign changes involved. (There are however other conventions in the literature.) We may write $\mathrm{Sz}(q^2)$ in place of $^2\mathrm{B}_2(q^2)$. We could also write $\mathrm{Ree}(q^2)$ (without ambiguity) when $q$ is an explicitly given power of 2 or 3.

The groups are all simple when $r > 0$. But $^2\mathrm{B}_2(2)$ is solvable (of order 20), while the derived group of $^2\mathrm{G}_2(3)$ has index 3 and is isomorphic to $\mathrm{SL}(2,8)$. We usually leave these groups aside in what follows. Something more interesting happens in the case of $^2\mathrm{F}_4(2)$: its derived group $^2\mathrm{F}_4(2)'$ has index 2 and is a new simple group (as shown by Tits), of order $2^{11}\,3^3\,5^2\,13$. The smallest groups are profiled in the two Atlases [**117, 243**]. Standard references for the construction of Suzuki and Ree groups are Carter [**86**, Chap. 13–14] and Steinberg [**394**, §11].

In the following sections we often use the smallest Suzuki group $\mathrm{Sz}(8)$ as an illustration; its order is $2^6 \cdot 5 \cdot 7 \cdot 13$.

REMARKS. (1) The Ree groups were found by Ree [**347, 348**] in a Lie-theoretic context similar to what is sketched above.

(2) The Suzuki groups were discovered by Suzuki [**408**], who also worked out their ordinary characters. He described them as certain doubly transitive permutation groups, but it soon became clear that (as in the case of Ree groups) they could also be characterized as the fixed points under a suitable endomorphism of a simple algebraic group.

The Suzuki and Ree groups have the structure of split $BN$-pairs, as defined in Chapter 7, which helps to unify the theory. The "Weyl group" in each case consists of the $\sigma$-fixed points $W^\sigma$ in the Weyl group $W$ of the associated Chevalley group. For types $^2\mathrm{B}_2$ and $^2\mathrm{G}_2$, one gets $|W^\sigma| = 2$, but for type $^2\mathrm{F}_4$ one gets a non-crystallographic dihedral group of order 16. The $BN$-pairs have respective ranks $1, 1, 2$.

## 20.2. Simple Modules

Let $G$ be a Suzuki or Ree group over a field of $q^2 = p^{2r+1}$ elements. According to Theorem 2.9, the number of nonisomorphic simple $KG$-modules is the number of $p$-regular classes of $G$. These are just the $\sigma$-stable classes of the associated Chevalley group. The number of such classes has been shown by Steinberg [**393**] to be $q^2$ for $^2\mathrm{B}_2(q^2)$ and $^2\mathrm{G}_2(q^2)$, and to be $q^4$ for $^2\mathrm{F}_4(q^2)$.

Just as in the case of other finite groups of Lie type, the simple $KG$-modules can all be obtained by restricting simple modules $L(\lambda)$ for the ambient algebraic group (or associated Chevalley group). This is proved by Steinberg in [**393**, §12].

To formulate the theorem, recall from 5.3 the sets of weights with short and long support, which we write briefly as $X_S$ and $X_L$.

THEOREM. *Let $G = \mathbf{G}^\sigma$ be a Suzuki or Ree group over a field of $q^2 = p^{2r+1}$ elements. As $\lambda$ ranges over those weights in $X_S$ for which $0 \le \langle \lambda, \alpha^\vee \rangle < q^2$ whenever $\alpha$ is a short simple root, the simple $\mathbf{G}$-modules $L(\lambda)$ remain simple and nonisomorphic on restriction to $G$. Thus they exhaust the isomorphism classes of simple $KG$-modules.*

The proof in Steinberg [393, §12] reworks with slight modifications his proof of the classification theorem for other groups of Lie type. We have not followed this line of argument closely in Chapter 2, so we limit ourselves here to a brief outline. The method is concrete, but requires a number of inductions as well as careful bookkeeping with weight vectors. The idea is to treat directly the case when $\lambda \in X_1$, then develop the Tensor Product Theorem to deal with other weights.

A key step is to show that when $\lambda \in X_1$ satisfies the hypothesis of the theorem, $L(\lambda)$ is simple not just for $\mathfrak{g}$ but also for the ideal $\mathfrak{g}_S$ generated by short root vectors (Remark 5.3). In turn the operators representing $G$ are built from the operators representing $\mathfrak{g}_S$, so one can keep track of how they act on weight vectors by "raising" or "lowering" and eventually conclude that any nonzero $KG$-submodule must be the whole module.

It would be interesting to approach the proof in a somewhat more conceptual way, taking full advantage of the classification of simple modules for Chevalley and twisted groups; but at the moment the only available proof seems to be the one in [393].

As a first example of the theorem, take $G = \mathrm{Sz}(8)$ with the second simple root being short. Then the 8 weights $(0, 0), (0, 1), \dots, (0, 7)$ determine the simple $KG$-modules.

REMARKS. (1) Note that all simple modules for Suzuki and Ree groups are *self-dual*, as is already true for the algebraic groups of types $B_2, G_2, F_4$: in each case $w_\circ = -1$ in the Weyl group.

(2) Since the simple $KG$-modules are restrictions of simple modules for the associated Chevalley group over a field of order $p^{2r+1}$, this field serves as a *splitting field* for $G$ (as noted in 5.2). It is actually the smallest such field.

(3) One might ask how the $L(\lambda)$ with $\lambda \in X_L$ restrict to $G$. The discussion of the special isogeny $\varphi$ in 5.3 shows that $\varphi^*$ sends $\lambda$ to $p$ times a corresponding weight with short support. Combining this with $F_r$ to get $\sigma$ (which fixes $G$ pointwise), we see that $G$ acts on $L(\lambda)$ as the $(r+1)$st Frobenius twist of a simple module whose highest weight has short support. For example, when $G = \mathrm{Sz}(8)$, with the second simple root taken as short, the $KG$-module $L(1, 0)$ is isomorphic to $L(0, 4) = L(0, 1)^{[2]}$.

(4) In [395, §13], Steinberg reformulates in a unified way the highest weight parametrization of simple modules to include all finite groups of Lie type. When $G$ is a Suzuki or Ree group as above, define $q(\alpha)$ to be $p^{r+1}$ when $\alpha$ is short and to be $p^r$ when $\alpha$ is long. Then the simple $KG$-modules are those $L(\lambda)$ for which $0 \le \langle \lambda, \alpha^\vee \rangle < q(\alpha)$ for all simple $\alpha$. (There are the expected number $q^2$ or $q^4$ of such weights.) To check this, use the reformulation indicated in (3). For example, when $G = \mathrm{Sz}(8)$, the substitution of $L(1, 0)$ for $L(0, 1)^{[2]}$ yields the list of 8 weights $(a, b)$ with $0 \le a < 2$ and $0 \le b < 4$.

## 20.3. Projective Modules and Blocks

As in the case of Chevalley groups and their twisted versions, a Suzuki or Ree group $G$ has a unique largest simple module; its dimension is the order of a Sylow $p$-subgroup: $q^m$, where $m = |\Phi^+|$ and $q^2 = p^{2r+1}$. This module is one factor in the tensor factorization (5.4) of the usual Steinberg module $L((q^2 - 1)\rho)$ for the associated Chevalley group. It is the restriction of $L((q^2 - 1)\rho_S)$ to $G$, where $\rho_S := \sum \varpi_\alpha$ (the sum taken only over the short simple roots), and may be denoted by $^2\mathrm{St}$.

This "Steinberg module" turns out to be a projective $KG$-module. One can see this either by adapting the earlier argument in 9.3 or by noting that $G$ has a (unique) ordinary character of the same degree and then applying Brauer theory as in 16.6. (See 20.7 below for references to the character data.) Table 2 summarizes the dimension data.

| Group | dim $^2\mathrm{St}$ | Case $r = 1$ |
|---|---|---|
| $^2\mathrm{B}_2(q^2)$ | $q^4$ | $2^6$ |
| $^2\mathrm{G}_2(q^2)$ | $q^6$ | $3^9$ |
| $^2\mathrm{F}_4(q^2)$ | $q^{24}$ | $2^{36}$ |

TABLE 2. Analogue of the Steinberg module

Following the reasoning in 9.4, it is easy to verify that every PIM for $KG$ can be obtained as a summand of the tensor product of $^2\mathrm{St}$ with a simple module. However, the resulting dimensions and Brauer characters do not seem to have been worked out systematically for Ree groups; Suzuki groups are discussed further in the following section. It would be interesting to imitate the theory in Chapter 10 involving higher Frobenius kernels, starting with the ideal $\mathfrak{g}_S$ in $\mathfrak{g}$ described in Remark 5.3.

What are the $p$-blocks of $KG$? As remarked in 8.6, the proof of Theorem 8.5 adapts well to the Suzuki and Ree groups. Since these groups are simple for $r > 0$, the analysis then shows that $KG$ has exactly two blocks: the one of defect 0 containing $^2\mathrm{St}$ and the principal block.

## 20.4. Cartan Invariants of Suzuki Groups

Chastkofsky–Feit [104] compute the dimensions of PIMs, as well as the first Cartan invariant, for the Suzuki groups in tandem with the associated Chevalley groups when $p = 2$. In their notation, $G_m$ is $\mathrm{Sp}(4, 2^{m/2})$ when $m$ is even and $\mathrm{Sz}(2^m)$ when $m$ is odd. Their starting point is the efficient description of simple modules as tensor products of copies of the standard 4-dimensional module (with various Frobenius twists). So the Brauer characters are manageable.

Projective modules are obtained by tensoring with the "Steinberg module" for each type of group. To keep track of multiplicities in tensor products, a graphical formalism is used together with induction on $m$. In their notation, $S := \mathbb{Z}/m\mathbb{Z}$, while the subsets of $S$ are used to label Brauer characters of PIMs. Eventually they arrive at closed formulas for degrees of PIMs and for the first Cartan invariant of each group, which involve interesting combinatorial coefficients such as Lucas numbers and Fibonacci numbers.

For Sz(8) when $p = 2$, Table 3 gives the Cartan invariants of the 1-block, with

| $\lambda$ | $\dim L(\lambda)$ | $\dim \mathrm{PIM}/2^6$ | | | | | | | |
|-----------|-------------------|-------------------------|-----|----|----|----|----|----|----|
| $(0,0)$ | 1 | 31 | 160 | 72 | 72 | 72 | 20 | 20 | 20 |
| $(1,0)$ | 4 | 14 | 72 | 34 | 32 | 32 | 10 | 8 | 9 |
| $(0,1)$ | 4 | 14 | 72 | 32 | 34 | 32 | 8 | 9 | 10 |
| $(0,2)$ | 4 | 14 | 72 | 32 | 32 | 34 | 9 | 10 | 8 |
| $(1,1)$ | 16 | 4 | 20 | 10 | 8 | 9 | 4 | 2 | 2 |
| $(1,2)$ | 16 | 4 | 20 | 8 | 9 | 10 | 2 | 4 | 2 |
| $(0,3)$ | 16 | 4 | 20 | 9 | 10 | 8 | 2 | 2 | 4 |

TABLE 3. Cartan invariants of 1-block of Sz(8) for $p = 2$

weights listed as in Remark 4 of 20.2. See Burkhardt [**71**], Liu [**304**], Zaslawsky [**460**], or the appendix to Schneider [**360**]. Labelling of the simple modules differs somewhat in these sources. Typically the Cartan invariants are found as a byproduct of the computation of decomposition numbers, using $C = D^t D$. As predicted by Brauer theory, the determinant is $2^6 = 2^3 \cdot 2 \cdot 2 \cdot 2$.

## 20.5. Cartan Invariants of the Tits Group

Much less has been done explicitly for the Ree groups.

As a byproduct of his study of the group of type $F_4$ in characteristic 2, Veldkamp [**425**] works out the 2-modular characters of the Ree group $^2F_4(2)$, whose derived group is the simple Tits group. Besides the trivial character and the Steinberg character of degree 4096, there are irreducible Brauer characters of degrees 26 and 246. It turns out that their values (at the classes of elements of orders $1, 3, 5, 13$) are in $\mathbb{Z}$.

Veldkamp's method leads to the Cartan matrix in Table 4, whose determinant

$$\begin{bmatrix} 160 & 528 & 160 \\ 528 & 1972 & 572 \\ 160 & 572 & 172 \end{bmatrix}$$

TABLE 4. Cartan invariants of 1-block of $^2F_4(2)$

is $2^{17} = 2^{12} \cdot 2^3 \cdot 2^2$. The rows and columns of the table are indexed by the simple modules of dimensions $1, 26, 246$, whose respective projective covers therefore have dimensions $13, 47, 14$ times $2^{12}$.

The Modular Atlas [**243**, p. 188] goes further in giving the five Brauer characters of $^2F_4(2)'$: on restriction to this subgroup, the character of degree 4096 splits into two characters of degree 2048 (while the class of elements of order 13 splits into two classes). Here the character values are no longer all in $\mathbb{Z}$. Each character of degree 2048 determines a block of defect 0. The characters of degree $1, 26, 246$ lie in the principal block, whose $20 \times 3$ decomposition matrix is available at the Modular Atlas homepage. With the dimensions of PIMs now cut in half, the resulting matrix $C$ is obtained from Table 4 by dividing each entry by 2.

## 20.6. Extensions and Cohomology

Here is a brief survey of some relevant work.

- As part of his study of extensions of simple modules for algebraic and finite groups in low ranks, Sin works simultaneously with Suzuki or Ree groups and the ambient Chevalley groups (as in Chastkofsky–Feit [104]).

  In [370] he writes $G_m = \mathrm{Sp}(4, 2^{m/2})$ when $m$ is even and $G_m = \mathrm{Sz}(2^m)$ when $m$ is odd. Using a more complicated version of Alperin's inductive method for $\mathrm{SL}(2, 2^n)$ [5], he is able to describe combinatorially all extensions between simple modules for $G_m$ when $m \geq 3$. Here all Ext groups turn out to be at most 1-dimensional. (For an isolated case, see Martineau [315].)

  In [373] he treats the much more complicated Chevalley and Ree groups of type $G_2$ when $p = 3$. Here the Ext groups are sometimes 2-dimensional. To handle the more difficult bookkeeping with tensor products in this case, he exploits the existence of good filtrations in tensor products of (dual) Weyl modules for the algebraic group.

  For Chevalley and Ree groups of type $F_4$ in characteristic 2, Sin gets some effective comparisons between $H^1$ for the algebraic group and $H^1$ for the finite group over a large enough field in the framework of generic cohomology [116].

- There has been little detailed study of the Loewy structure of PIMs for Suzuki and Ree groups. Relying mainly on computer methods, Schneider [360] works out explicitly the socle series of PIMs in the special case of $\mathrm{Sz}(8)$. See Table 3 for the dimension data. He relies on known character information, but has to construct simple modules explicitly with special care taken about the labelling of those which have the same dimension. Their tensor products are well-behaved, e.g., 16-dimensional simple modules are tensor products of two 4-dimensional ones (see 5.4).

- Only a few explicit higher cohomology calculations are found in the literature for Suzuki and Ree groups in the defining characteristic. Using purely group-theoretic techniques, Bell [36] studies $H^1$ and $H^2$ for a Suzuki group $G = \mathrm{Sz}(q^2)$ with coefficients in the natural 4-dimensional module $V$. Here $q^2 = 2^{2r+1}$. For example, $H^2(G, V) = 0$ when $r = 0, 1$ and is 1-dimensional when $r > 1$. The argument relies mainly on direct inspection of $H^2(B, V)$.

## 20.7. Ordinary Characters

The ordinary characters of the groups $^2B_2$ as well as the groups of type $^2G_2$ were first found using standard finite group methods. But then Deligne–Lusztig [129, §11] showed how to adapt their main results to the Suzuki and Ree groups. Lusztig went on to determine the unipotent characters for these groups. Here is a brief summary of what can be found in published sources:

($^2B_2$) In his original treatment of $G = \mathrm{Sz}(q^2)$, Suzuki [408] determines its character table. If $q^2 = 2^{2r+1}$, set $t^2 := 2q^2$. Then $G$ has $q^2 + 3$ irreducible characters, labelled as in Table 5. The indices $i, j, k$ run from 1 to (respectively) $(q^2 - 2)/2$, $(q^2 + t)/4$, $(q^2 - t)/4$. Here $X$ is the Steinberg character.

| Label | Degree |
|-------|--------|
| 1 | 1 |
| $X$ | $q^4$ |
| $W_1$ | $t(q^2 - 1)/2$ |
| $W_2$ | $t(q^2 - 1)/2$ |
| $X_i$ | $q^4 + 1$ |
| $Y_j$ | $(q^2 - 1)(q^2 - t + 1)$ |
| $Z_k$ | $(q^2 - 1)(q^2 + t + 1)$ |

TABLE 5. Characters of $Sz(q^2)$, with $q^2 = 2^{2r+1}$ and $t^2 = 2q^2$

Naturally all of this can now be recast along lines of Deligne–Lusztig theory. For example, the unipotent characters are $1, X, W_1, W_2$, with the last two being cuspidal.

($^2G_2$) Ward [431] determines the character tables of the Ree groups $^2G_2(q^2)$. He exploits various methods of Brauer, centering mainly on the behavior of the ordinary characters modulo 2. This results in a fairly dense and technical treatment. It would be interesting to have a version based on Deligne–Lusztig theory.

($^2F_4$) In [312, 14.2] Lusztig determines the 31 unipotent characters of $^2F_4(q^2)$. Then Malle [314] completes the computation of their values. Together with the decomposition of DL characters into their irreducible constituents and the Deligne–Lusztig character formula, this yields all character values.

Character tables and related information for the smallest groups can be found in the Atlas [117]. The group Sz(8) is also treated in the Appendix to Schneider [360].

## 20.8. Decomposition Numbers of Suzuki Groups

Burkhardt [71] develops combinatorial expressions for the decomposition numbers of Suzuki groups when $p = 2$ (and in the easier odd characteristic cases). As in Chastkofsky–Feit [104], he starts by writing the simple modules concretely as tensor products of algebraic conjugates of the standard 4-dimensional module for $Sp(4, K)$. Then he can study PIMs by tensoring simple modules with the Steinberg module. The Steinberg character determines a block of defect 0, while the $q^2 + 2$ other ordinary characters lie in the 1-block. Thus the decomposition matrix of this block has shape $(q^2 + 2) \times (q^2 - 1)$.

In the case of Sz(8), there are 10 irreducible characters in the 1-block, of degrees $1, 14, 14, 65, 65, 65, 35, 35, 35, 91$, along with the Steinberg character of degree 64. For this group Burkhardt obtains the $10 \times 7$ matrix in Table 6, which can be used to recover the Cartan matrix $C = D^t D$ exhibited in 20.4 above. He also computes the $34 \times 31$ decomposition matrix for the 1-block of Sz(32).

| Char | Deg | 1 | 4 | 4 | 4 | 16 | 16 | 16 |
|------|-----|---|---|---|---|----|----|----|
| 1    | 1   | 1 | 0 | 0 | 0 | 0  | 0  | 0  |
| $W_1$ | 14 | 2 | 1 | 1 | 1 | 0  | 0  | 0  |
| $W_2$ | 14 | 2 | 1 | 1 | 1 | 0  | 0  | 0  |
| $X_1$ | 65 | 5 | 3 | 2 | 2 | 1  | 0  | 1  |
| $X_2$ | 65 | 5 | 2 | 3 | 2 | 0  | 1  | 1  |
| $X_3$ | 65 | 5 | 2 | 2 | 3 | 1  | 1  | 0  |
| $Y_1$ | 35 | 3 | 1 | 2 | 1 | 0  | 0  | 1  |
| $Y_2$ | 35 | 3 | 1 | 1 | 2 | 0  | 1  | 0  |
| $Y_3$ | 35 | 3 | 2 | 1 | 1 | 1  | 0  | 0  |
| $Z_1$ | 91 | 7 | 3 | 3 | 3 | 1  | 1  | 1  |

TABLE 6. Decomposition matrix of 1-block of Sz(8) for $p = 2$

Without citing Burkhardt's earlier paper, Liu [304] develops another combinatorial algorithm for calculating decomposition numbers (and some Cartan invariants) of $Sz(q^2)$, in the framework of Chastkofsky–Feit [104]. In particular, he computes decomposition numbers explicitly for the pair of complex conjugate characters $W_i$: the nonzero ones are certain powers of 2. Like Burkhardt, he exhibits the matrix $D$ for Sz(8) and Sz(32); he then goes on to compute $D$ for Sz(128).

Recently H. Andikfar [23] has reworked the results of Liu for arbitrary Suzuki groups, in the spirit of work by Johnson [263]. He is able to make explicit many entries of both $C$ and $D$.

In earlier work, Zaslawsky [460] computes decomposition matrices for Sz(8) and Sz(32) by relying instead on the known tables of ordinary and Brauer characters. In the case of Sz(8) all of these tables are exhibited in the Appendix to Schneider [360], using labelling conventions from the two Atlases. The Modular Atlas homepage [243] currently provides the decomposition matrices for $Sz(8), Sz(32)$, and $^2F_4(2)'$.

# Bibliography

. Our primary reference for background material on representations of algebraic groups is Jantzen's book, referred to as RAGS. In addition to the sources explicitly cited, we include here some related work which may not turn up readily in a MathSciNet search.

RAGS. J.C. Jantzen, *Representations of Algebraic Groups*, Academic Press, Orlando, 1987; 2nd ed., Math. Surveys Monographs, Vol. 107, Amer. Math. Soc., Providence, RI, 2003.

1. A. Adem and R.J. Milgram, *Cohomology of Finite Groups*, Springer, Berlin, 1994; 2nd ed., 2004.

2. J.L. Alperin, *On modules for the linear fractional groups*, pp. 157–163, *International Symposium on Theory of Finite Groups*, RIMS, Kyoto, 1975.

3. _____, *Projective modules and tensor products*, J. Pure Appl. Algebra **8** (1976), 235–241.

4. _____, *Periodicity in groups*, Illinois J. Math. **21** (1977), 776–783.

5. _____, *Projective modules for* SL(2, $2^n$), J. Pure Appl. Algebra **15** (1979), 219–234.

6. _____, *Local Representation Theory*, Cambridge Univ. Press, Cambridge, 1986.

7. _____, *Weights for finite groups*, pp. 369–379, *The Arcata Conference on Representations of Finite Groups*, Proc. Sympos. Pure Math., vol. 47, Part 1, Amer. Math. Soc., Providence, RI, 1987.

8. J.L. Alperin and G.J. Janusz, *Resolutions and periodicity*, Proc. Amer. Math. Soc. **37** (1973), 403–406.

9. J. Alperin and L.G. Kovács, *Periodicity of Weyl modules for* SL(2, $q$), J. Algebra **74** (1982), 52–54.

10. J.L. Alperin and G. Mason, *On simple modules for* SL(2, $q$), Bull. London Math. Soc. **25** (1993), 17–22.

11. _____, *Partial Steinberg modules for finite groups of Lie type*, Bull. London Math. Soc. **25** (1993), 553–557.

12. _____, *Geometry, Steinberg representations and complexity*, pp. 1–5, *Groups '93 Galway / St Andrews*, London Math. Soc. Lecture Note Ser., 211, Cambridge Univ. Press, Cambridge, 1995.

13. H.H. Andersen, *Extensions of modules for algebraic groups*, Amer. J. Math. **106** (1984), 489–504.

14. _____, *Extensions of simple modules for finite Chevalley groups*, J. Algebra **111** (1987), 388–403.

15. _____, *Jantzen's filtrations of Weyl modules*, Math. Z. **194** (1987), 127–142.

16. _____ *Modular representations of algebraic groups*, pp. 23–36, *The Arcata Conference on Representations of Finite Groups*, Proc. Sympos. Pure Math., vol. 47, Part 1, Amer. Math. Soc., Providence, RI, 1987.

17. H.H. Andersen and J.C. Jantzen, *Cohomology of induced representations of algebraic groups*, Math. Ann. **269** (1984), 487–525.

18. H.H. Andersen, J.C. Jantzen, and W. Soergel, *Representations of quantum groups at a p-th root of unity and of semisimple groups in characteristic p: independence of p*, Astérisque **220** (1994).

19. H.H. Andersen, J. Jørgensen, and P. Landrock, *The projective indecomposable modules of* SL(2, $p^n$), Proc. London Math. Soc. **46** (1983), 38–52.

20. H.H. Andersen and M. Kaneda, *Loewy series for the first Frobenius kernel in a reductive algebraic group*, Proc. London Math. Soc. **59** (1989), 74–98.

21. _____, *Filtrations on* $G_1T$-modules, Proc. London Math. Soc. **82** (2001), 614–646.

22. H.H. Andersen, P. Polo, and K. Wen, *Injective modules for quantum algebras*, Amer. J. Math. **114** (1992), 571–604.

23. H. Andikfar, *Decomposition numbers of special and unitary groups in the defining characteristic*, Ph.D. thesis, Univ. Illinois at Chicago, 2005.

24. J. Archer, *Principal indecomposable modules for some three-dimensional special linear groups*, Bull. Austral. Math. Soc. **22** (1980), 439–455.

25. G. Avrunin, *A vanishing theorem for second degree cohomology*, J. Algebra **53** (1978), 382–388.

26. _____, *2-cohomology of some unitary groups*, Illinois J. Math. **24** (1980), 317–332.

27. _____, *Generic cohomology for twisted groups*, Trans. Amer. Math. Soc. **268** (1981), 247–253.

28. J.W. Ballard, *On the principal indecomposable modules of finite Chevalley groups*, Ph.D. thesis, Univ. Wisconsin, 1974.

29. _____, *Projective modules for finite Chevalley groups*, Trans. Amer. Math. Soc. **245** (1978), 221–249.

30. _____, *Injective modules for restricted enveloping algebras*, Math. Z. **163** (1978), 57–63.

31. _____, *Clifford's Theorem for algebraic groups and Lie algebras*, Pacific J. Math. **106** (1983), 1–15.

32. A.A. Baranov and I.D. Suprunenko, *Branching rules for modular fundamental representations of symplectic groups*, Bull. London Math. Soc. **32** (2000), 409–420.

33. M. Bardoe and P. Sin, *The permutation modules for* $GL(n+1, \mathbb{F}_q)$ *acting on* $\mathbb{P}^n(\mathbb{F}_q)$ *and* $\mathbb{F}_q^{n+1}$, J. London Math. Soc. (2) **61** (2000), 58–80.

34. R. Bédard, *On the Brauer liftings for modular representations*, J. Algebra **93** (1985), 332–353.

35. G.W. Bell, *On the cohomology of the finite special linear groups*, I, II, J. Algebra **54** (1978), 216–238 and 239–259.

36. _____, *Cohomology of degree 1 and 2 of the Suzuki groups*, Pacific J. Math. **75** (1978), 319–329.

37. C.P. Bendel, D.K. Nakano, and C. Pillen, *On comparing the cohomology of algebraic groups, finite Chevalley groups and Frobenius kernels*, J. Pure Appl. Algebra **163** (2001), 119–146.

38. _____, *Extensions for finite Chevalley groups*, I, Adv. Math. **183** (2004), 380–408.

39. _____, *Extensions for finite Chevalley groups*, II, Trans. Amer. Math. Soc. **354** (2002), 4421–4454.

40. _____, *Extensions for Frobenius kernels*, J. Algebra **272** (2004), 476–511.

41. _____, *Extensions for finite groups of Lie type: twisted groups*, pp. 29–46, *Finite Groups 2003*, de Gruyter, New York, 2004.

42. _____, *Second cohomology for Frobenius kernels and related structures*, preprint.

43. _____, *Extensions for finite groups of Lie type II: Filtering the truncated induction functor*, preprint.

44. D.J. Benson, *The Loewy structure of the projective indecomposable modules for* $A_8$ *in characteristic 2*, Comm. Algebra **11** (1983), 1395–1432.

45. _____, *Modular representation theory: New trends and methods*, Lecture Notes in Math., vol. 1178, Springer, Berlin, 1984.

46. _____, *Projective modules for the group of the twenty-seven lines on a cubic surface*, Comm. Algebra **17** (1989), 1017–1068.

47. _____, *Representations and Cohomology I* (Cambridge Studies in Advanced Mathematics, 30), Cambridge Univ. Press, Cambridge, 1991.

48. _____, *Representations and Cohomology II* (Cambridge Studies in Advanced Mathematics, 31), Cambridge Univ. Press, Cambridge, 1991.

49. _____, *Polynomial Invariants of Finite Groups*, London Math. Soc. Lecture Note Ser., 190, Cambridge Univ. Press., Cambridge, 1993.

50. D.J. Benson and J.F. Carlson, *Nilpotent elements in the Green ring*, J. Algebra **104** (1986), 329–350.

51. D.J. Benson and S. Martin, *Mod 2 cohomology of the unitary groups* $U_3(2^n)$, Comm. Algebra **19** (1991), 3125–3144.

52. F. M. Bleher, *Finite groups of Lie type of small rank*, Pacific J. Math. **187** (1999), 215–239.

53. B. Boe, *Geometry of the Jantzen region in Lusztig's conjecture*, Math. Comp. **70** (2001), 1265–1280.

54. A. Borel, *Properties and linear representations of Chevalley groups*, pp. 1–55, *Seminar on Algebraic Groups and Related Finite Groups*, Lecture Notes in Math., vol. 131, Springer, Berlin, 1970.

55. _____, *Linear Algebraic Groups*, 2nd enlarged ed., Springer, New York, 1991.

56. N. Bourbaki, *Groupes et algèbres de Lie*, Chapters 4–6, Hermann, Paris, 1968; 2nd ed., Masson, Paris, 1981.

57. _____, *Groupes et algèbres de Lie*, Chapters 7–8, Hermann, Paris, 1975.

58. B. Braden, *Restricted representations of classical Lie algebras of types $A_2, B_2$*, Bull. Amer. Math. Soc. **73** (1967), 482–486.

59. J. Brandt, *A lower bound for the number of irreducible characters in a block*, J. Algebra **74** (1982), 509–515.

60. R. Brauer, *Zur Darstellungstheorie der Gruppen endlicher Ordnung*, Math. Z. **63** (1956), 406–444.

61. R. Brauer and C. Nesbitt, *On the modular characters of groups*, Ann. of Math. **42** (1941), 556–590.

62. P. Brockhaus and G.O. Michler, *Finite simple groups of Lie type have non-principal p-blocks, $p \neq 2$*, J. Algebra **94** (1985), 113–125.

63. J. Brundan, *Double coset density in classical algebraic groups*, Trans. Amer. Math. Soc. **352** (2000), 1405–1436.

64. R.M. Bryant, L.G. Kovács, and R. Stöhr, *Lie powers of modules for* $GL(2, p)$, J. Algebra **260** (2003), 617–630.

65. A. Buch and N. Lauritzen, *The modular Lusztig conjecture in small rank*, http://home.imf.au.dk/abuch/dynkin/

66. N. Burgoyne, *Modular representations of some finite groups*, pp. 13–18, *Representation Theory of Finite Groups and Related Topics*, Proc. Sympos. Pure Math., vol. 21, Amer. Math. Soc., Providence, RI, 1971.

67. N. Burgoyne and C. Williamson, *Some computations involving simple Lie algebras*, pp. 162–171, *Proc. Second Sympos. Symbolic & Algebraic Manipulation*, ed. S.R. Petrick, ACM, New York, 1971.

68. _____, *On a theorem of Borel and Tits for finite Chevalley groups*, Arch. Math. **27** (1976), 489–491.

69. R. Burkhardt, *Die Zerlegungsmatrizen der Gruppen* $PSL(2, p^f)$, J. Algebra **40** (1976), 75–96.

70. _____, *Über ein kombinatorisches Problem aus der modularen Darstellungstheorie*, J. Comb. Theory A **21** (1976), 68–79.

71. _____, *Über die Zerlegungszahlen der Suzukigruppen* $Sz(q)$, J. Algebra **59** (1979), 421–433.

72. _____, *Über die Zerlegungszahlen der unitären Gruppen* $PSU(3, 2^{2f})$, J. Algebra **61** (1979), 548–581.

73. M. Cabanes, *Irreducible modules and Levi supplements*, J. Algebra **90** (1984), 84–97.

74. _____, *Brauer morphism between modular Hecke algebras*, J. Algebra **115** (1988), 1–31.

75. _____, *Extension groups for modular Hecke algebras*, J. Fac. Sci. Univ. Tokyo Sect. IA **36** (1989), 347–362.

76. _____, *A criterion for complete reducibility and some applications*, pp. 93–112, *Représentations linéaires des groupes finis* (Luminy, 1988), Astérisque **181–182** (1990).

77. M. Cabanes, ed., *Finite Reductive Groups: Related Structures and Representations*, Progr. Math., vol. 141, Birkhäuser, Boston, 1997.

78. M. Cabanes and M. Enguehard, *Representation Theory of Finite Reductive Groups*, Cambridge Univ. Press, Cambridge, 2004.

79. D. Carlisle, P. Eccles, S. Hilditch, N. Ray, L. Schwartz, G. Walker, and R. Wood, *Modular representations of* $GL(n, p)$*, splitting* $\Sigma(\mathbb{C}P^\infty \times \cdots \times \mathbb{C}P^\infty)$*, and the $\beta$-family as framed hypersurfaces*, Math. Z. **189** (1985), 239–261.

80. D. Carlisle and N. Kuhn, *Subalgebras of the Steenrod algebra and the action of matrices on truncated polynomial algebras*, J. Algebra **121** (1989), 370–387.

81. D.P. Carlisle and G. Walker, *Poincaré series for the occurrence of certain modular representations of* GL$(n,p)$ *in the symmetric algebra*, Proc. Roy. Soc. Edinburgh Sect. A **113** (1989), 27–41.

82. J.F. Carlson, *The varieties and the cohomology ring of a module*, J. Algebra **85** (1983), 104–143.

83. ———, *The cohomology of irreducible modules over* SL$(2,p^n)$, Proc. London Math. Soc. **47** (1983), 480–492.

84. J.F. Carlson, Z. Lin, and D.K. Nakano, *Support varieties for modules over Chevalley groups and classical Lie algebras*, preprint.

85. J.F. Carlson, N. Mazza, and D.K. Nakano, *Endotrivial modules for finite groups of Lie type*, preprint.

86. R.W. Carter, *Simple Groups of Lie Type*, Wiley–Interscience, New York, 1972.

87. ———, *Complex representation theory of finite groups of Lie type*, pp. 243–257, *Finite Simple Groups II*, Academic Press, London, 1980.

88. ———, *The relation between characteristic 0 representations and characteristic p representations of finite groups of Lie type*, pp. 301–311, *The Santa Cruz Conference on Finite Groups*, Proc. Sympos. Pure Math., vol. 37, Amer. Math. Soc., Providence, RI, 1980.

89. ———, *Finite Groups of Lie Type: Conjugacy Classes and Complex Characters*, Wiley–Interscience, New York, 1985.

90. ———, *Lie algebras and root systems*, pp. 1–44, *Lectures on Lie Groups and Lie Algebras*, London Math. Soc. Student Texts, No. 32, Cambridge Univ. Press, Cambridge, 1995.

91. ———, *On the representation theory of the finite groups of Lie type over an algebraically closed field of characteristic 0*, Algebra, IX, 1–120, 235–239, *Encyclopaedia Math. Sci.*, **77**, Springer, Berlin, 1995.

92. R.W. Carter and G. Lusztig, *On the modular representations of the general linear and symmetric groups*, Math. Z. **136** (1974), 193–242.

93. ———, *Modular representations of finite groups of Lie type*, Proc. London Math. Soc. **32** (1976), 347–384.

94. B. Chang and R. Ree, *The characters of* G$_2(q)$, pp. 395–413, Symposia Mathematica XIII (INDAM), Academic Press, London, 1974.

95. L. Chastkofsky, *Characters of projective indecomposable modules for finite Chevalley groups*, pp. 359–362, *The Santa Cruz Conference on Finite Groups*, Proc. Sympos. Pure Math., vol. 37, Amer. Math. Soc., Providence, RI, 1980.

96. ———, *Projective characters for finite Chevalley groups*, J. Algebra **69** (1981), 347–357.

97. ———, *Rationality of certain zeta functions associated with modular representation theory*, pp. 41–50, *Finite Groups—Coming of Age*, Contemp. Math., vol. 45, Amer. Math. Soc., Providence, RI, 1985.

98. ———, *Generic Cartan invariants for Chevalley groups*, J. Algebra **103** (1986), 466–478.

99. ———, *On the Brauer complex of a finite group of Lie type*, J. Algebra **149** (1992), 262–273.

100. ———, *On the Cartan invariants of a Chevalley group over* GF$(p^n)$, J. Algebra **150** (1992), 388–401.

101. ———, *Stability of decomposition numbers for finite Chevalley groups*, J. Algebra **202** (1998), 185–191.

102. ———, *Decomposition numbers of Chevalley groups*, J. Algebra **240** (2001), 589–607.

103. ———, *Decomposition numbers of* SL$_3(p^n)$ *and* SU$_3(p^n)$, J. Algebra **240** (2001), 608–634.

104. L. Chastkofsky and W. Feit, *On the projective characters in characteristic 2 of the groups* Suz$(2^m)$ *and* Sp$_4(2^m)$, Inst. Hautes Études Sci. Publ. Math. **51** (1980), 9–36.

105. ———, *On the projective characters in characteristic 2 of the groups* SL$_3(2^m)$ *and* SU$_3(2^m)$, J. Algebra **63** (1980), 124–142.

106. Y. Cheng, *On the projective characters of some finite groups of Lie type*, Ph.D. thesis, Yale Univ., 1981.

107. ———, *On the projective characters in characteristic 3 of the groups* Ree$(3^m)$ *and* G$_2(3^n)$, unpublished.

108. ———, *On the first Cartan invariants in characteristic 2 of the groups* SL$_3(2^m)$ *and* SU$_3(2^m)$, J. Algebra **82** (1983), 194–244.

109. ———, *On the Cartan invariants of* SL$(2,p^m)$, Comm. Algebra **14** (1986), 507–515.

110. G. Cliff and A. Stokke, *Determining irreducible* GL(n, K)*-modules*, Math. Proc. Cambridge Philos. Soc. **136** (2004), 119–137.

111. E. Cline, Ext$^1$ *for* SL$_2$, Comm. Algebra **7** (1979), 107–111.

112. E. Cline, B. Parshall, and L. Scott, *Cohomology of finite groups of Lie type* I, Inst. Hautes Études Sci. Publ. Math. **45** (1975), 169–191.

113. _____, *Cohomology of finite groups of Lie type* II, J. Algebra **45** (1977), 182–198.

114. _____, *Cohomology, hyperalgebras, and representations*, J. Algebra **63** (1980), 98–123.

115. _____, *On the tensor product theorem for algebraic groups*, J. Algebra **63** (1980), 264–267.

116. E. Cline, B. Parshall, L. Scott, and W. van der Kallen, *Rational and generic cohomology*, Invent. Math. **39** (1977), 143–163.

117. J.H. Conway, R.T. Curtis, S.P. Norton, R.A. Parker, and R.A. Wilson, *Atlas of Finite Groups*, Clarendon Press, Oxford, 1985. web.mat.bham.ac.uk/atlas/v2.0/

118. C.W. Curtis, *Representations of Lie algebras of classical type with applications to linear groups*, J. Math. Mech. **9** (1960), 307–326.

119. _____, *On the dimensions of the irreducible modules of Lie algebras of classical type*, Trans. Amer. Math. Soc. **96** (1960), 135–142.

120. _____, *On projective representations of certain finite groups*, Proc. Amer. Math. Soc. **11** (1960), 852–860.

121. _____, *Irreducible representations of finite groups of Lie type*, J. Reine Angew. Math. **219** (1965), 180–199.

122. _____, *Modular representations of finite groups with split* (B, N)*-pairs*, pp. 57–95, *Seminar on Algebraic Groups and Related Finite Groups*, Lecture Notes in Math., vol. 131, Springer, Berlin, 1970.

123. _____, *Chevalley groups and related topics*, pp. 135–189, *Finite Simple Groups (Proc. Instructional Conf., Oxford, 1969)*, Academic Press, London, 1971.

124. _____, *Representations of finite groups of Lie type*, Bull. Amer. Math. Soc. (N.S.) **1** (1979), 721–757.

125. C.W. Curtis and I. Reiner, *Representation Theory of Finite Groups and Associative Algebras*, Wiley–Interscience, New York, 1962.

126. _____, *Methods of Representation Theory*, I, Wiley–Interscience, New York, 1981.

127. _____, *Methods of Representation Theory*, II, Wiley–Interscience, New York, 1987.

128. S.W. Dagger, *On the blocks of the Chevalley groups*, J. London Math. Soc. (2) **3** (1971), 21–29.

129. P. Deligne and G. Lusztig, *Representations of reductive groups over finite fields*, Ann. of Math. **103** (1976), 103–161.

130. D.I. Deriziotis, *The Brauer complex of a Chevalley group*, J. Algebra **70** (1981), 261–269.

131. D.I. Deriziotis and G.O. Michler, *Character table and blocks of finite simple triality groups* $^3D_4(q)$, Trans. Amer. Math. Soc. **303** (1987), 39–70.

132. L.E. Dickson, *A fundamental system of invariants of the general modular linear group with a solution of the form problem*, Trans. Amer. Math. Soc. **12** (1911), 75–98.

133. F. Digne, G.I. Lehrer, and J. Michel, *The characters of the group of rational points of a reductive group with nonconnected centre*, J. Reine Angew. Math. **425** (1992), 155–192.

134. F. Digne and J. Michel, *Representations of Finite Groups of Lie Type*, London Math. Soc. Student Texts, No. 21, Cambridge Univ. Press, Cambridge, 1991.

135. R. Dipper, *Vertices of irreducible representations of finite Chevalley groups in the describing characteristic*, Math. Z. **175** (1980), 143–159.

136. _____, *On irreducible modules over finite Chevalley groups and parabolic subgroups*, Comm. Algebra **10** (1982), 1073–1088.

137. _____, *On irreducible modules of twisted groups of Lie type*, J. Algebra **81** (1983), 370–389.

138. S. Donkin, *Rationally injective modules for semisimple algebraic groups as direct limits*, Bull. London Math. Soc. **12** (1980), 99–102.

139. _____, *A note on the characters of the projective modules for the infinitesimal subgroups of a semisimple algebraic group*, Math. Scand. **51** (1982), 142–150.

140. _____, *On Ext$^1$ for semisimple groups and infinitesimal subgroups*, Math. Proc. Cambridge Philos. Soc. **92** (1982), 231–238.

141. _____, *Rational Representations of Algebraic Groups*, Lecture Notes in Math., vol. 1140, Springer, Berlin, 1985.

142. _____, *On tilting modules for algebraic groups*, Math. Z. **212** (1993), 39–60.

143. _____, *An introduction to the Lusztig Conjecture*, pp. 173–187, *Representations of Reductive Groups*, ed. R.W. Carter and M. Geck, Cambridge Univ. Press, Cambridge, 1998.

144. F. Dordowsky, *Über die Dimensionen der projektiv unzerlegbaren* $SL_3(p^n)$- *und* $SU_3(p^{2n})$- *Moduln*, Diplomarbeit, Univ. Hamburg, 1988.

145. L. Dornhoff, *Group Representation Theory*, Part A, M. Dekker, New York, 1971.

146. _____, *Group Representation Theory*, Part B, M. Dekker, New York, 1972.

147. S. Doty, *The submodule structure of certain Weyl modules for groups of type* $A_n$, J. Algebra **95** (1985), 373–383.

148. _____, *Submodules of symmetric powers of the natural module for* $GL_n$, pp. 185–191, *Invariant Theory (Denton, TX, 1986)*, Contemp. Math., 88, Amer. Math. Soc., Providence, RI, 1989.

149. _____, *The symmetric algebra and representations of general linear groups*, pp. 123–150, *Proc. Hyderabad Conference on Algebraic Groups (Hyderabad, 1989)*, Manoj Prakashan, Madras, 1991.

150. S. Doty and A. Henke, *Decomposition of tensor products of modular irreducibles for* $SL_2$, Quart. J. Math. (2004)

151. S. Doty and G. Walker, *The composition factors of* $\mathbf{F}_p[x_1, x_2, x_3]$ *as a* $GL(3, p)$-*module*, J. Algebra **147** (1992), 411–441.

152. _____, *Truncated symmetric powers and modular representations of* $GL_n$, Math. Proc. Cambridge Philos. Soc. **119** (1996), 231–242.

153. M.F. Dowd, *On the 1-cohomology of the groups* $SL_3(3^n)$ *and* $SU_3(3^n)$, Comm. Algebra **22** (1994), 4827–4839.

154. _____, *On the 1-cohomology of the groups* $SL_4(2^n)$, $SU_4(2^n)$, *and* $Spin_7(2^n)$, pp. 23–42, *Groups, Difference Sets, and the Monster* (Columbus, OH, 1993), Ohio State Univ. Math. Res. Inst. Publ., 4, de Gruyter, Berlin, 1996.

155. _____, *On the 1-cohomology of finite groups of Lie type*, J. Algebra **188** (1997), 720–732; addendum, ibid. **229** (2000), 793.

156. M.F. Dowd and P. Sin, *On representations of algebraic groups in characteristic 2*, Comm. Algebra **24** (1996), 2597–2686.

157. V. Downes, *The characters of* $Sp(6, q)$, Ph.D. thesis, Clark Univ., 1976.

158. J. Du, *The Cartan invariants of* $SL(4, 2^n)$ [Chinese], J. East China Norm. Univ. Natur. Sci. Ed. **1986**, no. 4, 17–25. MR 88m:20088

159. C.W. Eaton, *Dade's inductive conjecture for the Ree groups of type* $G_2$ *in the defining characteristic*, J. Algebra **226** (2000), 614–620.

160. N. Elkies, *Linearized algebra and finite groups of Lie type. I. Linear and symplectic groups*, pp. 77–107, *Applications of curves over finite fields (Seattle, WA, 1997)*, Contemp. Math., 245, Amer. Math. Soc., Providence, RI, 1999.

161. M.A. Elmer, *On the modular representation theory of semisimple Lie algebras*, Ph.D. thesis, Univ. Massachusetts, 1979.

162. T. Englund, *Quadratic representations for groups of Lie type over fields of characteristic two*, J. Algebra **268** (2003), 118–155.

163. V. Ennola, *Characters of finite unitary groups*, Ann. Acad. Scien. Fenn. **323** (1963), 120–155.

164. H. Enomoto, *The characters of the finite symplectic group* $Sp(4, q), q = 2^f$, Osaka J. Math. **9** (1972), 75–94.

165. _____, *The characters of the finite Chevalley group* $G_2(q)$, $q = 3^f$, Japan. J. Math. **2** (1976), 191–248.

166. _____, *The characters of* $G_2(2^n)$, Japan J. Math. **12** (1986), 325–377.

167. L. Evens, *The Cohomology of Groups*, Clarendon Press, Oxford, 1991.

168. W. Feit, *Divisibility of projective modules of finite groups*, J. Pure Appl. Algebra **8** (1976), 183–185.

169. _____, *The Representation Theory of Finite Groups*, North-Holland, Amsterdam, 1982.

170. T.J. Fisher, *A bound for the dimension of* $H_0(\mathcal{F}_V)$ *of modules* $V(\lambda_i + \lambda_j)$ *for* $A_l(q)$, J. Algebra **143** (1991), 119–129.

171. _____, *Weight operators and group geometries*, J. Algebra **158** (1993), 61–108.

172. P. Fleischmann, *Periodic simple modules for* $SU_3(q^2)$ *in the describing characteristic* $p \neq 2$, Math. Z. **198** (1988), 555–568.

173. _____, *The complexities and rank varieties of simple modules for* $(^2A_2)(q^2)$ *in the natural characteristic*, J. Algebra **121** (1989), 399–408.

174. P. Fleischmann and J.C. Jantzen, *Simple periodic modules of twisted Chevalley groups*, Pacific J. Math. **143** (1990), 229–242.

175. P. Fong and G.M. Seitz, *Groups with a* $(B, N)$-*pair of rank 2*, I, Invent. Math. **21** (1973), 1–57; II, ibid. **24** (1974), 191–239.

176. S. Foulle, *Formules de caractères pour des représentations irréductibles du groupe symplectique en caractéristique* $p$, C.R. Acad. Sci. Paris, Ser. I **335** (2002), 11–16.

177. E.M. Friedlander, *Computations of* $K$-*theories of finite fields*, Topology **15** (1976), 87–109.

178. _____, *Cohomology of irreducible modules with large weights*, pp. 187–193, *The Arcata Conference on Representations of Finite Groups*, Proc. Sympos. Pure Math., vol. 47, Part 2, Amer. Math. Soc., Providence, RI, 1987.

179. E.M. Friedlander and B.J. Parshall, *On the cohomology of Chevalley groups*, Bull. Amer. Math. Soc. (N.S.) **7** (1982), 247–250.

180. _____, *On the cohomology of algebraic and related finite groups*, Invent. Math. **74** (1983), 85–117.

181. _____, *Cohomology of Lie algebras and algebraic groups*, Amer. J. Math. **108** (1986), 235–253.

182. _____, *Cohomology of infinitesimal and discrete groups*, Math. Ann. **273** (1986), 353–374.

183. _____, *Support varieties for restricted Lie algebras*, Invent. Math. **86** (1986), 553–562.

184. M. Geck, *Finite groups of Lie type*, pp. 63–83, *Representations of Reductive Groups*, ed. R.W. Carter and M. Geck, Cambridge Univ. Press, Cambridge, 1998.

185. G. Genet, *On the commutator formula of a split BN-pair*, Pacific J. Math. **207** (2002), 177–181.

186. P.B. Gilkey and G.M. Seitz, *Some representations of exceptional Lie algebras*, Geom. Dedicata **25** (1988), 407–416.

187. D.J. Glover, *A study of certain modular representations*, J. Algebra **51** (1978), 425–475.

188. D. Gorenstein, R. Lyons, and R. Solomon, *The Classification of Finite Simple Groups*, Math. Surveys Monographs, vol. 40, no. 1, Amer. Math. Soc., Providence, RI, 1994.

189. _____, *The Classification of Finite Simple Groups, Number 2*, Math. Surveys Monographs, vol. 40, no. 2, Amer. Math. Soc., Providence, RI, 1996.

190. _____, *The Classification of Finite Simple Groups, Number 3*, Math. Surveys Monographs, vol. 40, no. 3, Amer. Math. Soc., Providence, RI, 1998.

191. R. Gow, *Construction of* $p-1$ *irreducible modules with fundamental highest weight for the symplectic group in characteristic* $p$, J. London Math. Soc. (2) **58** (1998), 619–632.

192. R. Gow and A. Kleshchev, *Connections between the representations of the symmetric group and the symplectic group in characteristic 2*, J. Algebra **221** (1999), 60–89.

193. J.A. Green, *The characters of the finite general linear groups*, Trans. Amer. Math. Soc. **80** (1955), 402–447.

194. _____, *Some remarks on defect groups*, Math. Z. **107** (1968), 133–150.

195. _____, *Walking around the Brauer tree*, J. Austral. Math. Soc. Ser. A **17** (1974), 197–213.

196. _____, *On a theorem of H. Sawada*, J. London Math. Soc. (2) **18** (1978), 247–252.

197. B.H. Gross, *Group representations and lattices*, J. Amer. Math. Soc. **3** (1990), 929–960.

198. R.M. Guralnick, *Small representations are completely reducible*, J. Algebra **220** (1999), 531–541.

199. B. Haastert and J.C. Jantzen, *Filtrations of the discrete series of* $SL_2(q)$ *via crystalline cohomology*, J. Algebra **132** (1990), 77–103.

200. _____, *Filtrations of symmetric powers via crystalline cohomology*, Geom. Dedicata **37** (1991), 45–63.

201. H.H. Hagelskjær, *Induced modules for algebraic groups*, Examensarbeit, Aarhus University, 1983.

202. P.W.A.M. van Ham, T.A. Springer, and M. van der Wel, *On the Cartan invariants of* $SL_2(\mathbf{F}_q)$, Comm. Algebra **10** (1982), 1565–1588.

203. H.-W. Henn and L. Schwartz, *Indecomposable A-module summands in $H^*V$ which are unstable algebras*, Math. Z. **205** (1990), 145–158.

204. C. Hering, W.M. Kantor, and G.M. Seitz, *Finite groups with a split BN-pair of rank 2, I*, J. Algebra **20** (1972), 435–475.

205. T.J. Hewett, *Modular invariant theory of parabolic subgroups of $GL_n(\mathbf{F}_q)$ and the associated Steenrod modules*, Duke Math. J. **82** (1996), 91–102; correction, ibid. **97** (1999), 217.

206. H.L. Hiller, *Cohomology of Chevalley groups over finite fields*, J. Pure Appl. Algebra **16** (1980), 259–263.

207. T. Holm and W. Willems, *A local conjecture on Brauer character degrees of finite groups*, Trans. Amer. Math. Soc., to appear.

208. R.R. Holmes, *Projective characters of finite Chevalley groups*, J. Algebra **124** (1989), 158–182.

209. _____, *An algorithm for the projective characters of finite Chevalley groups*, Rocky Mt. J. Math. **22** (1992), 611–630.

210. Yu-wang Hu and Jia-chen Ye, *On the first Cartan invariant for finite groups of Lie type*, Comm. Algebra **21** (1993), 935–950.

211. _____, *Cartan invariant matrix for finite group of Lie type $G_2(5)$* (Chinese), Tongji Daxue Xuebao Ziran Kexue Ban **29** (2001), 1205–1208.

212. _____, *On the first Cartan invariant for the finite group of type $G_2$*, Comm. Algebra **30** (2002), 4549–4573.

213. J.E. Humphreys, *Modular representations of classical Lie algebras and semisimple groups*, J. Algebra **19** (1971), 51–79.

214. _____, *Defect groups for finite groups of Lie type*, Math. Z. **119** (1971), 149–152.

215. _____, *Introduction to Lie Algebras and Representation Theory*, Springer, New York, 1972.

216. _____, *Projective modules for $SL(2,q)$*, J. Algebra **25** (1973), 513–518.

217. _____, *Some computations of Cartan invariants for finite groups of Lie type*, Comm. Pure Appl. Math. **26** (1973), 745–755.

218. _____, *Weyl groups, deformations of linkage classes, and character degrees for Chevalley groups*, Comm. Algebra **1** (1974), 475–490.

219. _____, *Representations of $SL(2,p)$*, Amer. Math. Monthly **82** (1975), 21–39.

220. _____, *Linear Algebraic Groups*, Springer, New York, 1975.

221. _____, *Ordinary and Modular Representations of Chevalley Groups*, Lecture Notes in Math., vol. 528, Springer, Berlin, 1976.

222. _____, *On the hyperalgebra of a semisimple algebraic group*, pp. 203–210, *Contributions to Algebra: A Collection of Papers Dedicated to Ellis Kolchin*, Academic Press, New York, 1977.

223. _____, *Modular representations of finite groups of Lie type*, pp. 259–290, *Finite Simple Groups II*, Academic Press, London, 1980.

224. _____, *Cartan invariants and decomposition numbers of Chevalley groups*, pp. 347–351, *The Santa Cruz Conference on Finite Groups*, Proc. Sympos. Pure Math., vol. 37, Amer. Math. Soc., Providence, RI, 1980.

225. _____, *Deligne–Lusztig characters and principal indecomposable modules*, J. Algebra **62** (1980), 299–303.

226. _____, *Ordinary and modular characters of $SL(3,p)$*, J. Algebra **72** (1981), 8–16.

227. _____, *On the structure of Weyl modules*, Comm. Algebra **12** (1984), 2665–2677.

228. _____, *Cartan invariants*, Bull. London Math. Soc. **17** (1985), 1–14.

229. _____, *Non-zero $Ext^1$ for Chevalley groups (via algebraic groups)*, J. London Math. Soc. (2) **31** (1985), 463–467.

230. _____, *Projective modules for $Sp(4,p)$ in characteristic $p$*, J. Algebra **104** (1986), 80–88.

231. _____, *The Steinberg representation*, Bull. Amer. Math. Soc. (N.S.) **16** (1987), 247–263.

232. _____, *Generic Cartan invariants for Frobenius kernels and Chevalley groups*, J. Algebra **122** (1989), 345–352.

233. _____, *Ordinary and modular characters of $SU(3,p)$*, J. Algebra **131** (1990), 425–431.

234. _____, *Reflection Groups and Coxeter Groups* (Cambridge Studies in Advanced Mathematics, 29), Cambridge Univ. Press, Cambridge, 1990.

235. _____, *Extremal composition factors for groups of Lie type*, pp. 303–310, *Algebraic Groups and Their Generalizations: Classical Methods*, Proc. Sympos. Pure Math., vol. 56, Part 2, Amer. Math. Soc., Providence, RI, 1994.

236. _____, *Another look at Dickson's invariants for finite linear groups*, Comm. Algebra **22** (1994), 4773–4779.

237. _____, *Conjugacy Classes in Semisimple Algebraic Groups*, Math. Surveys Monographs, vol. 43, Amer. Math. Soc., Providence, RI, 1995.

238. J.E. Humphreys and J.C. Jantzen, *Blocks and indecomposable modules for semisimple algebraic groups*, J. Algebra **54** (1978), 494–503.

239. R.S. Irving, *Projective modules in the category $\mathcal{O}_S$: Loewy series*, Trans. Amer. Math. Soc. **291** (1985), 733–754.

240. _____, *The socle filtration of a Verma module*, Ann. Sci. École Norm. Sup. (4) **21** (1988), 47–65.

241. I. Janiszczak, *Irreducible periodic modules over* $SL(m, q)$ *in the describing characteristic*, Comm. Algebra **15** (1987), 1375–1391.

242. I. Janiszczak and J.C. Jantzen, *Simple periodic modules over Chevalley groups*, J. London Math. Soc. (2) **41** (1990), 217–230.

243. C. Jansen, K. Lux, R.A. Parker, and R.A. Wilson, *An Atlas of Brauer Characters*, Clarendon Press, Oxford, 1995. www.math.rwth-aachen.de/ ~MOC/

244. J.C. Jantzen, *Darstellungen halbeinfacher algebraischer Gruppen und zugeordnete kontravariante Formen*, Bonner Math. Schriften, No. 67, 1973.

245. _____, *Zur Charakterformel gewisser Darstellungen halbeinfacher Gruppen und Lie-Algebren*, Math. Z. **140** (1974), 127–149.

246. _____, *Darstellungen halbeinfacher Gruppen und kontavariante Formen*, J. Reine Angew. Math. **290** (1977), 117–141.

247. _____, *Über das Dekompositionsverhalten gewisser modularer Darstellungen halbeinfacher Gruppen*, J. Algebra **49** (1977), 441–469.

248. _____, *Über Darstellungen höherer Frobenius-Kerne halbeinfacher algebraischer Gruppen*, Math. Z. **164** (1979), 271–292.

249. _____, *Darstellungen halbeinfacher Gruppen und ihrer Frobenius-Kerne*, J. Reine Angew. Math. **317** (1980), 157–199.

250. _____, *Weyl modules for groups of Lie type*, pp. 291–300, *Finite Simple Groups II*, Academic Press, London, 1980.

251. _____, *Zur Reduktion modulo p der Charaktere von Deligne und Lusztig*, J. Algebra **70** (1981), 452–474.

252. _____, *Zur Reduktion modulo p unipotenter Charaktere endlicher Chevalley-Gruppen*, Math. Z. **181** (1982), 97–128.

253. _____, *Filtrierungen der Darstellungen in der Hauptserie endlicher Chevalley-Gruppen*, Proc. London Math. Soc. **49** (1984), 445–482.

254. _____, *Modular representations of reductive groups*, pp. 118–154, *Group Theory, Beijing 1984*, Lecture Notes in Math., vol. 1185, Springer, Berlin, 1986.

255. _____, *Restricted Lie algebra cohomology*, pp. 91–108, *Algebraic Groups Utrecht 1986*, Lecture Notes in Math., vol. 1271, Springer, Berlin, 1987.

256. _____, *Kohomologie von p-Lie-Algebren und nilpotente Elemente*, Abhand. Math. Sem. Univ. Hamburg **56** (1986), 191–219.

257. _____, *Representations of Chevalley groups in their own characteristic*, pp. 127–146, *The Arcata Conference on Representations of Finite Groups*, Proc. Sympos. Pure Math., vol. 47, Part 1, Amer. Math. Soc., Providence, RI, 1987.

258. _____, *Low-dimensional representations of reductive groups are semisimple*, pp. 255–266, *Algebraic groups and Lie groups*, Austral. Math. Soc. Lect. Ser., 9, Cambridge Univ. Press, Cambridge, 1997.

259. G.J. Janusz, *Indecomposable representations of groups with a cyclic Sylow subgroup*, Trans. Amer. Math. Soc. **125** (1966), 288–295.

260. _____, *Indecomposable modules for finite groups*, Ann. of Math. **89** (1969), 209–241.

261. A.V. Jeyakumar, *Principal indecomposable representations for the group* $SL(2, q)$, J. Algebra **30** (1974), 444–458.

262. _____, *Periodic modules for the group* $SL(2, q)$, Comm. Algebra **8** (1980), 1721–1735.

263. B. Johnson, *Factoring Cartan matrices of group algebras*, Ph.D. thesis, Univ. Chicago, 2003.

264. W. Jones and B. Parshall, *On the 1-cohomology of finite groups of Lie type*, pp. 313–328, *Proc. Conf. Finite Groups*, Academic Press, New York, 1976.

265. M. Kaneda, *A truncation formula for certain principal indecomposable modules of finite Chevalley groups*, Comm. Algebra **14** (1986), 911–922.

266. W.M. Kantor and G.M. Seitz, *Finite groups with a split $BN$-pair of rank 1, II*, J. Algebra **20** (1972), 476–494.

267. S.-I. Kato, *On the Kazhdan–Lusztig polynomials for affine Weyl groups*, Adv. in Math. **55** (1985), 103–130.

268. S. Kawata, G.O. Michler, and K. Uno, *On Auslander–Reiten components and simple modules for finite groups of Lie type*, Osaka J. Math. **38** (2001), 21–25.

269. D. Kazhdan and G. Lusztig, *Representations of Coxeter groups and Hecke algebras*, Invent. Math. **53** (1979), 165–184.

270. G. Kempf, *Representations of algebraic groups in prime characteristic*, Ann. Sci. École Norm. Sup. **14** (1981), 61–76.

271. A.A. Khammash, *Contravariant forms on representations of finite groups with split BN-pairs*, Comm. Algebra **18** (1990), 1961–1969.

272. _____, *On the blocks of modular Hecke algebras*, J. Algebra **155** (1993), 545–554.

273. _____, *On the homological construction of the Steinberg representation*, J. Pure Appl. Algebra **87** (1993), 17–21.

274. P. Kleidman and M. Liebeck, *The Subgroup Structure of the Finite Classical Groups*, London Math. Soc. Lecture Note Ser., 129, Cambridge Univ. Press, 1990.

275. A.S. Kleshchev, *Cohomology of finite Chevalley groups with coefficients in modules with nonradical highest weights, I, II*, preprints, Inst. Mat., Acad. Nauk BSSR, Minsk, 1990.

276. _____, *1-cohomology of the special linear group with coefficients in the module of truncated polynomials*, Mat. Zametki **49** (1991), no. 6, 63–71; transl., Math. Notes **49** (1991), 597–603.

277. _____, *Extensions of modules of truncated polynomials and exterior powers over a finite special linear group*, Mat. Zametki **51** (1992), no. 4, 43–53; transl., Math. Notes **51** (1992), 356–363.

278. _____, *Cohomology of finite Chevalley groups with coefficients in modules with 1-dimensional weight spaces*, Comm. Algebra **22** (1994), 1197–1218.

279. M. Koppinen, *On the translation functors for a semisimple algebraic group*, Math. Scand. **51** (1982), 217–226.

280. _____, *Computation of simple characters of a Chevalley group*, BIT **26** (1986), 333–338.

281. S. Koshitani, *The Loewy structure of the projective indecomposable modules for SL(3, 3) and its automorphism group in characteristic 3*, Comm. Algebra **15** (1987), 1215–1253.

282. _____, *The Loewy structure of projective indecomposable modules for some modular group algebras*, pp. 447–449, *The Arcata Conference on Representations of Finite Groups*, Proc. Sympos. Pure Math., vol. 47, Part 1, Amer. Math. Soc., Providence, RI, 1987.

283. _____, *The nilpotency index of the radicals of group algebras of finite groups whose Sylow 3-subgroups are extra-special of order 27 of exponent 3*, Proc. Roy. Soc. Edinburgh Sect. A **108** (1988), 117–132.

284. F.W. Kouwenhoven, *The $\lambda$-structure of the Green ring of GL(2, $\mathbf{F}_p$) in characteristic p*, Comm. Algebra **18** (1990), 1645–1747.

285. L.G. Kovács, *Some representations of special linear groups*, pp. 207–218, *The Arcata Conference on Representations of Finite Groups*, Proc. Sympos. Pure Math., vol. 47, Part 2, Amer. Math. Soc., Providence, RI, 1987.

286. L.G. Kovács and R. Stöhr, *Lie powers of the natural module for GL(2)*, J. Algebra **229** (2000), 435–462.

287. P. Krasoń and N.J. Kuhn, *On embedding polynomial functors in symmetric powers*, J. Algebra **163** (1994), 281–294.

288. B. Krcmar, *Irreduzible, rationale Darstellungen endlicher Chevalley-Gruppen*, Ph.D. thesis, Univ. Tübingen, 1979.

289. N.J. Kuhn, *The modular Hecke algebra and Steinberg representation of finite Chevalley groups* (with an appendix by P. Landrock), J. Algebra **91** (1984), 125–141.

290. _____, *Chevalley group theory and the transfer in the homology of symmetric groups*, Topology **24** (1985), 247–264.

291. _____, *Generic representations of the finite general linear groups and the Steenrod algebra. I*, Amer. J. Math. **116** (1994), 327–360.

292. N.J. Kuhn and S.A. Mitchell, *The multiplicity of the Steinberg representation of* $GL_n$ $\mathbf{F}_q$ *in the symmetric algebra*, Proc. Amer. Math. Soc. **96** (1986), 1–6.

293. S. Kumar, *Proof of the Parthararathy–Ranga Rao–Varadarajan conjecture*, Invent. Math. **93** (1988), 117–130.

294. V.N. Landázuri, *The second degree cohomology of finite Chevalley groups*, Ph.D. thesis, Univ. Michigan, 1975.

295. P. Landrock, *A counterexample to a conjecture on the Cartan invariants of a group algebra*, Bull. London Math. Soc. **5** (1973), 223–224.

296. _____, *Finite Group Algebras and their Modules*, London Math. Soc. Lecture Note Ser., 84, Cambridge Univ. Press. Cambridge, 1983.

297. G.I. Lehrer and J. Thévenaz, *Sur la conjecture d'Alperin pour les groupes réductifs finis*, C.R. Acad. Sci. Paris **315** (1992), 1347–1351.

298. M. Liebeck, *The affine permutation groups of rank three*, Proc. London Math. Soc. **54** (1987), 477–516.

299. Z. Lin, *Extensions between simple modules for Frobenius kernels*, Math. Z. **207** (1991), 485–499.

300. _____, *Loewy series of certain indecomposable modules for Frobenius subgroups*, Trans. Amer. Math. Soc. **332** (1992), 391–409.

301. Z. Lin and D.K. Nakano, *Complexity for modules over finite Chevalley groups and classical Lie algebras*, Invent. Math. **138** (1999), 85–101.

302. Ailan Liu and Jia-chen Ye, *Cartan invariant matrix for the finite symplectic group* Sp(6,3), Algebra Colloq. **10** (2003), 33–39.

303. Jia Chun Liu and Jia-chen Ye, *Extensions of simple modules for the algebraic group of type* $G_2$, Comm. Algebra **21** (1993), 1909–1946.

304. Li-qian Liu, *The decomposition numbers of* Suz(q), J. Algebra **172** (1995), 1–31.

305. J. Looker, *The complex irreducible characters of* Sp(6,q), *q even*, Bull. Austral. Math. Soc. **17** (1977), 475–476.

306. F. Lübeck, *Small degree representations of finite Chevalley groups in defining characteristic*, LMS J. Comput. Math. **4** (2001), 135–169.

307. G. Lusztig, *The discrete series of* $GL_n$ *over a finite field*, Ann. of Math. Studies, No. 81, Princeton Univ. Press, Princeton, 1974.

308. _____, *Divisibility of projective modules of finite Chevalley groups by the Steinberg module*, Bull. London Math. Soc. **8** (1976), 130–134.

309. _____, *Representations of finite Chevalley groups*, CBMS Regional Conference Series in Mathematics, 39, Amer. Math. Soc., Providence, RI, 1978.

310. _____, *Some problems in the representation theory of finite Chevalley groups*, pp. 313–318, *The Santa Cruz Conference on Finite Groups*, Proc. Sympos. Pure Math., vol. 37, Amer. Math. Soc., Providence, RI, 1980.

311. _____, *Hecke algebras and Jantzen's generic decomposition patterns*, Adv. Math. **37** (1980), 121–164.

312. _____, *Characters of reductive groups over a finite field*, Ann. of Math. Studies, 107, Princeton Univ. Press, Princeton, NJ, 1984.

313. K. Magaard and Pham Huu Tiep, *Irreducible tensor products of representations of finite quasi-simple groups of Lie type*, pp. 239–262, *Modular Representation Theory of Finite Groups (Charlottesville, VA, 1998)*, de Gruyter, Berlin, 2001.

314. G. Malle, *Die unipotenten Charaktere von* $^2F_4(q^2)$, Comm. Algebra **18** (1990), 2361–2381.

315. R.P. Martineau, *On 2-modular representations of the Suzuki groups*, Amer. J. Math. **94** (1972), 55–72.

316. W.G. McKay, J. Patera, and D.W. Rand, *Tables of Representations of Simple Lie Algebras, Vol. I. Exceptional Simple Lie Algebras*, CRM, Université de Montreal, 1990.

317. G.J. McNinch, *Semisimple modules for finite groups of Lie type*, J. London Math. Soc. (2) **60** (1999), 771–792.

318. _____, *Filtrations and positive characteristic Howe duality*, Math. Z. **235** (2000), 651–685.

319. D. Mertens, *Zur Darstellungstheorie der endlichen Chevalley-Gruppen vom Typ* $G_2$, Diplomarbeit, Univ. Bonn., 1985.

320. W. Meyer and W. Neutsch, *Über 5-Darstellungen der Lyonsgruppe*, Math. Ann. **267** (1984), 519–535.

321. G.O. Michler, *A finite simple group of Lie type has p-blocks with different defects, p ≠ 2*, J. Algebra **104** (1986), 220–230.

322. S. Mitchell, *Finite complexes with A(n)-free cohomology*, Topology **24** (1985), 227–248.

323. \_\_\_\_\_, *On the Steinberg module, representations of the symmetric groups, and the Steenrod algebra*, J. Pure Appl. Algebra **39** (1986), 275–281.

324. S.A. Mitchell and S.B. Priddy, *Stable splittings derived from the Steinberg module*, Topology **22** (1983), 285–298.

325. D.K. Nakano, *Complexity and support varieties for finite dimensional algebras*, pp. 201–218, Lie Algebras and their Representations (Seoul, 1995), Contemp. Math., 194, Amer. Math. Soc., Providence, RI, 1996.

326. D.K. Nakano, B.J. Parshall, and D.C. Vella, *Support varieties for algebraic groups*, J. Reine Angew. Math. **547** (2002), 15–49.

327. S. Nozawa, *On the characters of the finite general unitary group* $U(4, q^2)$, J. Fac. Sci. Univ. Tokyo Sect. IA Math. **19** (1972), 257–293.

328. S. Nozawa, *Characters of the finite general unitary group* $U(5, q^2)$, J. Fac. Sci. Univ. Tokyo Sect. IA Math. **23** (1976), 23–74.

329. J.B. Olsson and K. Uno, *Dade's conjecture for general linear groups in the defining characteristic*, Proc. London Math. Soc. **72** (1996), 359–384.

330. A.A. Osinovskaya and I.D. Suprunenko, *On the Jordan block structure of images of some unipotent elements in modular irreducible representations of the classical algebraic groups*, J. Algebra **273** (2004), 586–600.

331. B.J. Parshall, *Cohomology of algebraic groups*, pp. 243–248, The Arcata Conference on Representations of Finite Groups, Proc. Sympos. Pure Math., vol. 47, Part 1, Amer. Math. Soc., Providence, RI, 1987.

332. B.J. Parshall and L.L. Scott, *Extensions, Levi subgroups and character formulas*, arXiv:math.GR/0502191.

333. C. Pillen, *Tensor products of modules with restricted highest weight*, Comm. Algebra **21** (1993), 3647–3661.

334. \_\_\_\_\_, *Reduction modulo p of some Deligne–Lusztig characters*, Arch. Math. **61** (1993), 421–433.

335. \_\_\_\_\_, *The first Cartan invariant of a finite group of Lie type for large p*, J. Algebra **174** (1995), 934–947.

336. \_\_\_\_\_, *Loewy series for principal series representations of finite Chevalley groups*, J. Algebra **189** (1997), 125–149.

337. \_\_\_\_\_, *Generic patterns for extensions of simple modules for finite Chevalley groups*, J. Algebra **212** (1999), 419–427.

338. \_\_\_\_\_, *Self-extensions for the finite symplectic groups*, preprint.

339. M. Pittaluga and E. Strickland, *On computers and modular representations of* $SL_n(K)$, pp. 120–129, Applicable Algebra, Error-correcting Codes, Combinatorics and Computer Algebra (Karlsruhe, 1986), Lecture Notes in Comput. Sci., vol. 307, Springer, Berlin, 1988.

340. \_\_\_\_\_, *A computer oriented algorithm for the determination of the dimension and character of a modular irreducible SL(n, K)-module*, J. Symbolic Comput. **7** (1989), 155–161.

341. H. Pollatsek, *First cohomology groups of some linear groups defined over fields of characteristic 2*, Illinois J. Math. **15** (1971), 393–417.

342. \_\_\_\_\_, *First cohomology groups of some orthogonal groups*, J. Algebra **28** (1974), 477–483.

343. A.A. Premet, *Weights of infinitesimally irreducible representations of Chevalley groups over a field of prime characteristic*, Mat. Sb. **133** (1987), 167–183; transl., Math. USSR-Sb. **61** (1988), 167–183.

344. A.A. Premet and I.D. Suprunenko, *Quadratic modules for Chevalley groups over fields of odd characteristics*, Math. Nachr. **110** (1983), 65–96.

345. \_\_\_\_\_, *The Weyl modules and the irreducible representations of the symplectic group with the fundamental highest weights*, Comm. Algebra **11** (1983), 1309–1342.

346. A. Przygocki, *Schur indices of symplectic groups*, Comm. Algebra **10** (1982), 279–310.

347. R. Ree, *A family of simple groups associated with the simple Lie algebra of type* ($F_4$), Amer. J. Math. **83** (1961), 401–420.

348. ———, *A family of simple groups associated with the simple Lie algebra of type* ($G_2$), Amer. J. Math. **83** (1961), 432–462.

349. F.A. Richen, *Modular representations of split BN pairs*, Trans. Amer. Math. Soc. **140** (1969), 435–460.

350. ———, *Blocks of defect zero of split* $(B, N)$ *pairs*, J. Algebra **21** (1972), 275–279.

351. M.A. Ronan and S.D. Smith, *Sheaves on buildings and modular representations of Chevalley groups*, J. Algebra **96** (1985), 319–346.

352. R. Rouquier, *The derived category of blocks with cyclic defect groups*, pp. 199–220 Derived Equivalences for Group Rings, Lecture Notes in Math., vol. 1685, Springer, Berlin, 1998.

353. N.S. Narasimha Sastry and P. Sin, *On the doubly transitive permutation representations of* $\mathrm{Sp}(2n, \mathbb{F}_2)$, J. Algebra **257** (2002), 509–527.

354. H. Sawada, *A characterization of the modular representations of finite groups with split* $(B, N)$ *pairs*, Math. Z. **155** (1977), 29–41.

355. ———, *Endomorphism rings of split* $(B, N)$-*pairs*, Tokyo J. Math. **1** (1978), 139–148.

356. ———, *On a certain covariant representation functor and irreducible modules of Chevalley groups*, Arch. Math. **52** (1989), 530–538.

357. ———, *Certain homomorphism spaces* $\mathcal{H}_K(G, U)$ *and a theorem of unification on modular representations of Chevalley groups, finite and algebraic*, J. Algebra **166** (1994), 340–355.

358. P. Schmid, *Simply connected, adjoint and universal groups of Lie type*, J. London Math. Soc. (2) **60** (1999), 817–834.

359. M. Schmidt, *Beziehungen zwischen Homologie-Darstellungen und der Hauptserie endlicher Chevalley-Gruppen*, Bonner Math. Schriften, No. 171, Bonn, 1986.

360. G.J.A. Schneider, *The structure of the principal indecomposable modules of the Suzuki group* Sz(8) *in characteristic 2*, Math. Comp. **60** (1993), 779–786, S29–S32.

361. C. Schoen, *On certain modular representations in the cohomology of algebraic curves*, J. Algebra **135** (1990), 1–18.

362. L.L. Scott, *Representations in characteristic p*, pp. 319–331, The Santa Cruz Conference on Finite Groups, Proc. Sympos. Pure Math., vol. 37, Amer. Math. Soc., Providence, RI, 1980.

363. ——— *Some new examples in 1-cohomology*, J. Algebra **260** (2003), 416–425.

364. G.M. Seitz, *The maximal subgroups of classical algebraic groups*, Mem. Amer. Math. Soc., No. 365, 1987.

365. ———, *Bounds for dimensions of weight spaces of maximal tori*, pp. 157–161, Linear Algebraic Groups and Their Representations, Contemp. Math., vol. 153, Amer. Math. Soc., Providence, RI, 1993.

366. J.-P. Serre, *Linear Representations of Finite Groups*, Springer, New York, 1977.

367. W.A. Simpson and J.A. Frame, *The character tables for* SL(3, q), SU(3, $q^2$), PSL(3, q), PSU(3, $q^2$), Canad. J. Math. **25** (1973), 486–494.

368. P. Sin, *The Green ring and modular representations of finite groups of Lie type*, J. Algebra **123** (1989), 185–192.

369. ———, *On the representation theory of modular Hecke algebras*, J. Algebra **146** (1992), 267–277.

370. ———, *Extensions of simple modules for* $\mathrm{Sp}_4(2^n)$ *and* $\mathrm{Suz}(2^m)$, Bull. London Math. Soc. **24** (1992), 159–164.

371. ———, *Extensions of simple modules for* $\mathrm{SL}_3(2^n)$ *and* $\mathrm{SU}_3(2^n)$, Proc. London Math. Soc. **65** (1992), 265–296.

372. ———, *On the 1-cohomology of the groups* $G_2(2^n)$, Comm. Algebra **20** (1992), 2653–2662.

373. ———, *Extensions of simple modules for* $G_2(3^n)$ *and* $^2G_2(3^n)$, Proc. London Math. Soc. **66** (1993), 327–357.

374. ———, *The cohomology in degree 1 of the group* $F_4$ *in characteristic 2 with coefficients in a simple module*, J. Algebra **164** (1994), 695–717.

375. ———, *Extensions of simple modules for special algebraic groups*, J. Algebra **170** (1994), 1011–1034.

376. L. Smith, *Polynomial Invariants of Finite Groups*, A K Peters, Wellesley, MA, 1995.
377. S.D. Smith, *Irreducible modules and parabolic subgroups*, J. Algebra **75** (1982), 286–289.
378. _____, *Spin modules in characteristic 2*, J. Algebra **77** (1982), 392–401.
379. _____, *Sheaf homology and complete reducibility*, J. Algebra **95** (1985), 72–80.
380. _____, *Modular representations of Chevalley groups*, pp. 277–282, *Proc. Rutgers Group Theory Year* (1983–1984), Cambridge Univ. Press., Cambridge, 1985.
381. _____, *Constructing representations from group geometries*, pp. 303–313, *The Arcata Conference on Representations of Finite Groups*, Proc. Sympos. Pure Math., vol. 47, Part 1, Amer. Math. Soc., Providence, RI, 1987.
382. T.A. Springer, *Weyl's character formula for algebraic groups*, Invent. Math. **5** (1968), 85–105.
383. _____, *Characters of special groups*, pp. 121–166, *Seminar on Algebraic Groups and Related Finite Groups*, Lecture Notes in Math., vol. 131, Springer, Berlin, 1970.
384. _____, *Relèvement de Brauer et représentations paraboliques de* $GL_n(\mathbb{F}_q)$ *[d'après G. Lusztig]*, Exposé 441, *Séminaire Bourbaki* (1973/74), Lecture Notes in Math., vol. 431, Springer, Berlin, 1975.
385. _____, *Linear Algebraic Groups*, 2nd ed., Birkhäuser, Boston, 1998.
386. T.A. Springer and R. Steinberg, *Conjugacy classes*, pp. 167–266, *Seminar on Algebraic Groups and Related Finite Groups*, Lecture Notes in Math., vol. 131, Springer, Berlin, 1970.
387. B. Srinivasan, *On the modular characters of the special linear group* $SL(2,p^n)$, Proc. London Math. Soc. (3) **14** (1964), 101–114.
388. _____, *The characters of the finite symplectic group* $Sp(4,q)$, Trans. Amer. Math. Soc. **131** (1968), 488–525.
389. _____, *Characters of finite groups of Lie type, II.*, pp. 333–339, *The Santa Cruz Conference on Finite Groups*, Proc. Sympos. Pure Math., vol. 37, Amer. Math. Soc., Providence, RI, 1980.
390. R. Steinberg, *A geometric approach to the representations of the full linear group over a Galois field*, Trans. Amer. Math. Soc. **71** (1951), 274–282.
391. _____, *The representations of* $GL(3,q), GL(4,q), PGL(3,q),$ *and* $PGL(4,q)$, Canad. J. Math. **3** (1951), 225–235.
392. _____, *Prime power representations of finite linear groups, II*, Canad. J. Math. **9** (1957), 347–351.
393. _____, *Representations of algebraic groups*, Nagoya Math. J. **22** (1963), 33–56.
394. _____, *Lectures on Chevalley Groups*, Yale Univ. Math. Dept., 1968.
395. _____, *Endomorphisms of linear algebraic groups*, Mem. Amer. Math. Soc., No. 80, 1968.
396. _____, *Conjugacy Classes in Algebraic Groups* (notes by V.V. Deodhar), Lecture Notes in Math., vol. 366, Springer, Berlin, 1974.
397. _____, *On Dickson's theorem on invariants*, J. Fac. Sci. Univ. Tokyo Sect. IA Math. **34** (1987), 699–707.
398. _____, *Collected Papers*, Amer. Math. Soc., Providence, RI, 1997.
399. E. Strickland, *An algorithm for the determination of the dimension and character of a modular irreducible* $Sp(n, K)$-*module*, Comm. Algebra **16** (1988), 2435–2445.
400. G. Stroth, *Strong quadratic modules*, Israel J. Math. **79** (1992), 257–279.
401. I.A.I. Suleiman, *The modular characters of the twisted Chevalley group* ${}^2D_4(2)$ *and* ${}^2D_4(2).2$, Math. Japon. **39** (1994), 107–117.
402. I.D. Suprunenko, *Minimal polynomials of elements of order p in irreducible representations of Chevalley groups over fields of characteristic p*, pp. 389–400, *Proc. Intern. Conf. on Algebra* (Novosibirsk, 1989), Contemp. Math., vol. 131, Part 1, Amer. Math. Soc., Providence, RI, 1992.
403. _____, *Irreducible representations of simple algebraic groups containing matrices with big Jordan blocks*, Proc. London Math. Soc. **71** (1995), 281–332.
404. _____, *p-large representations and asymptotics: a survey and conjectures*, J. Math. Sci. (New York) **100** (2000), no. 1, 1861–1870.
405. I.D. Suprunenko and A.E. Zalesskii, *Irreducible representations of finite classical groups containing matrices with simple spectra*, Comm. Algebra **26** (1998), 863–888.

406. _____, *Representations of finite exceptional groups of Lie type containing matrices with simple spectra*, Comm. Algebra **28** (2000), 1789–1833.

407. M. Suzuki, *A new type of simple groups of finite order*, Proc. Nat. Acad. Sci. U.S.A. **46** (1960), 868–870.

408. _____, *On a class of doubly transitive groups*, Ann. Math. **75** (1962), 105–145.

409. _____, *Finite groups with a split BN-pair of rank one*, pp. 139–147, *The Santa Cruz Conference on Finite Groups*, Proc. Sympos. Pure Math., vol. 37, Amer. Math. Soc., Providence, RI, 1980.

410. J.G. Thackray, *Modular representations of some finite groups* Ph.D. thesis, Cambridge Univ., 1980.

411. C.B. Thomas, *Modular representations and the cohomology of finite Chevalley groups*, Bol. Soc. Mat. Mexicana **37** (1992), 535–543.

412. Pham Huu Tiep and A.E. Zalesskii, *Mod p reducibility of unramified representations of finite groups of Lie type*, Proc. London Math. Soc. **84** (2002), 439–472.

413. _____, *Reducibility modulo p of complex representations of finite groups of Lie type: Asymptotical result and small characteristic cases*, Proc. Amer. Math. Soc. **130** (2002), 3177–3184.

414. _____, *Unipotent elements of finite groups of Lie type and realization fields of their complex representations*, J. Algebra **271** (2004), 327–390.

415. F.G. Timmesfeld, *A remark on irreducible modules for finite Lie-type groups*, Arch. Math. **46** (1986), 499–500.

416. N.B. Tinberg, *Some indecomposable modules of groups with split (B, N)-pairs*, J. Algebra **61** (1979), 508–526.

417. _____, *Modular representations of finite groups with unsaturated split (B, N)-pairs*, Canad. J. Math. **32** (1980), 714–733.

418. _____, *The Levi decomposition of a split (B, N)-pair*, Pacific J. Math. **91** (1980), 233–238.

419. _____, *Weights and admissible pairs*, Comm. Algebra **12** (1984), 1257–1263.

420. J. Tits, *Buildings of Spherical Type and Finite BN-Pairs*, Lecture Notes in Math., vol. 386, Berlin, Springer, 1974.

421. Ton That Tri, *The irreducible modular representations of parabolic subgroups of general linear groups*, Comm. Algebra **26** (1998), 41–47.

422. Y. Tsushima, *On certain projective modules for finite groups of Lie type*, Osaka J. Math. **27** (1990), 947–962.

423. B.S. Upadhyaya, *Composition factors of the principal indecomposable modules for the special linear group* SL(2, q), J. London Math. Soc. (2) **17** (1978), 437–445.

424. _____, *Filtrations with Weyl module quotients of the principal indecomposable modules for the group* SL(2, q), Comm. Algebra **7** (1979), 1469–1488.

425. F.D. Veldkamp, *Representations of algebraic groups of type $F_4$ in characteristic 2*, J. Algebra **16** (1970), 326–339.

426. D.-N. Verma, *Role of affine Weyl groups in the representation theory of algebraic Chevalley groups and their Lie algebras*, pp. 653–705, *Lie Groups and Their Representations* (Proc. Summer School on Group Representations of the Bolyai János Math. Soc., Budapest, 1971), Halsted, New York, 1975.

427. H. Völklein, *The 1-cohomology of the adjoint module of a Chevalley group*, Forum Math. **1** (1989), 1–13.

428. _____, *1-cohomology of Chevalley groups*, J. Algebra **127** (1989), 353–372.

429. _____, *On extensions between simple modules of a Chevalley group*, Proc. Second Intern. Group Theory Conf. (Bressanone, 1989), Rend. Circ. Mat. Palermo (2) Suppl. No. 23 (1990), 337–346.

430. G. Walker, *Modular Schur functions*, Trans. Amer. Math. Soc. **346** (1994), 569–604.

431. H.N. Ward, *On Ree's series of simple groups*, Trans. Amer. Math. Soc. **121** (1966), 62–89.

432. C. Wilkerson, *A primer on the Dickson invariants*, pp. 421–434, *Proc. Northwestern Homotopy Theory Conf.* (Evanston, Ill., 1982), Contemp. Math., vol. 19, Amer. Math. Soc., Providence, RI, 1983.

433. W. Willems, *Blocks of defect zero in finite simple groups of Lie type*, J. Algebra **113** (1988), 511–522.

434. _____, *On degrees of irreducible Brauer characters*, Trans. Amer. Math. Soc., to appear.

435. P.W. Winter, *On the modular representations of the two dimensional special linear group over an algebraically closed field*, J. London Math. Soc. (2) **16** (1977), 237–252.

436. W.J. Wong, *Representations of Chevalley groups in characteristic p*, Nagoya Math. J. **45** (1972), 39–78.

437. _____, *Irreducible modular representations of finite Chevalley groups*, J. Algebra **20** (1972), 355–367.

438. Bao-xing Xu and Jia-chen Ye, *Irreducible characters for algebraic groups in characteristic two. I*, Algebra Colloq. **4** (1997), 281–290.

439. Jia-chen Ye, *The Cartan invariants of* $SL(3, p^n)$ [Chinese], J. Math. Res. Exposition **2** (1982), no. 4, 9–19. MR 84g:20087

440. _____, *The Cartan invariants of* $Sp(4, p^n)$ [Chinese], J. East China Norm. Univ. Natur. Sci. Ed. (1985), no. 2, 18–25. MR 87k:20084

441. _____, *On the first Cartan invariant of the groups* $SL(3, p^n)$ *and* $SU(3, p^n)$, pp. 388–400, *Group Theory, Beijing 1984*, Lecture Notes in Math., vol. 1185, Springer, Berlin, 1986.

442. _____, *On the first Cartan invariant of the group* $Sp(4, p^n)$, Acta Math. Sinica (N.S.) **4** (1988), 18–27.

443. _____, *Cartan invariants of finite groups of Lie type*, pp. 235–241, *Classical Groups and Related Topics (Beijing, 1987)*, Contemp. Math., 82, Amer. Math. Soc., Providence, RI, 1989.

444. _____, *Cartan invariants of finite groups of Lie type (I)*, Dongbei Shuxue **5** (1989), 93–101.

445. _____, *Extensions of simple modules for the group* $Sp(4, K)$, J. London Math. Soc. (2) **41** (1990), 51–62.

446. _____, *Extensions of simple modules for the group* $Sp(4, K)$. *II*, Chinese Sci. Bull. **35** (1990), 450–454.

447. _____, *Extensions of simple modules for* $G_2(p)$, Comm. Algebra **22** (1994), 2771–2802.

448. _____, *The Cartan invariant matrix for the finite group* $SL(5, 2)$, Algebra Colloq. **4** (1997), 203–211.

449. Jia-chen Ye and Zhong-guo Zhou, *Irreducible characters for algebraic groups in characteristic two. II*, Algebra Colloq. **6** (1999), 401–411.

450. _____, *Irreducible characters of algebraic groups in characteristic two. III*, Comm. Algebra **28** (2000), 4227–4247.

451. _____, *Irreducible characters for algebraic groups in characteristic three*, Comm. Algebra **29** (2001), 201–223.

452. _____, *Irreducible characters for algebraic groups in characteristic three. II*, Comm. Algebra **30** (2002), 273–306.

453. S. El B. Yehia, *Extensions of simple modules for the universal Chevalley group and its parabolic subgroups*, Ph.D. thesis, Warwick Univ., 1982.

454. A.E. Zalesskii, *A fragment of the decomposition matrix of the special unitary group over a finite field*, Izv. Akad. Nauk SSSR Ser. Math. **54** (1990), 26–41; transl., Math. USSR–Izv. **36** (1991), 23–39.

455. _____, *Decomposition numbers modulo p of certain representations of the groups* $SL_n(p^k), SU_n(p^k), Sp_{2n}(p^k)$, pp. 491–500, *Topics in Algebra, Part 2* (Warsaw, 1988), Banach Center Publ., 26, PWN, Warsaw, 1990.

456. _____, *Spectra of p-elements in representations of the group* $SL_n(p^\alpha)$, Uspekhi Mat. Nauk **45** (1990), no. 4, 155–156; transl., Russian Math. Surveys **45** (1990), 194–195.

457. A.E. Zalesskii and I.D. Suprunenko, *Representations of dimension* $(p^n \pm 1)/2$ *of the symplectic group of degree 2n over a field of characteristic p* [Russian], Vestsi Akad. Navuk BSSR Ser. Fiz.-Mat. Navuk **1987**, no. 6, 9–15.

458. _____, *Truncated symmetric powers of the natural realizations of the groups* $SL_m(P)$ *and* $Sp_m(P)$ *and their restrictions to subgroups*, Sibirsk. Mat. Zh. **31** (1990), no. 4, 33–46; transl., Siberian Math. J. **31** (1991), 555–566.

459. _____, *Permutation representations and a fragment of the decomposition matrix of symplectic and special linear groups over a finite field*, Sibirsk. Mat. Zh. **31** (1990), no. 5, 46–60; transl., Siberian Math. J. **31** (1991), 744–755.

460. E. Zaslawsky, *Computational methods applied to ordinary and modular characters of some finite simple groups*, Ph.D. thesis, Univ. California at Santa Cruz, 1974.

# Frequently Used Symbols

| Symbol | Description | Section |
|---|---|---|
| $K$ | algebraically closed field of characteristic $p > 0$ | 1.1 |
| $\mathbf{G}$ | algebraic group over $K$ | 1.1 |
| $\mathbf{T}$ | maximal torus of $\mathbf{G}$ | 1.1 |
| $\mathbf{B}$ | Borel subgroup of $\mathbf{G}$ | 1.1 |
| $\mathbf{U}$ | unipotent radical of $\mathbf{B}$ | 1.1 |
| $\ell$ | rank of $\mathbf{G}$   $(= \dim \mathbf{T})$ | 1.1 |
| $\Phi$ | root system of $\mathbf{G}$ | 1.1 |
| $X$ | weight lattice $(= X(\mathbf{T}))$ of $\mathbf{G}$ | 1.1 |
| $\mathbb{Z}\Phi$ | root lattice of $\mathbf{G}$ | 1.1 |
| $F$ | Frobenius morphism $\mathbf{G} \to \mathbf{G}$ | 1.3 |
| $\pi$ | nontrivial graph automorphism | 1.3 |
| $G$ | finite group (usually of Lie type $\mathbf{G}^F$) | 1.3 |
| $\Delta$ | simple roots in $\Phi$ | 2.1 |
| $\Phi^+$ | positive roots relative to $\Delta$ | 2.1 |
| $m$ | $|\Phi^+|$ | 2.1 |
| $\Phi^\vee$ | dual root system | 2.1 |
| $\alpha^\vee$ | coroot of $\alpha$ | 2.1 |
| $\langle x, \alpha^\vee \rangle$ | pairing for $x \in \mathbb{R} \otimes_\mathbb{Z} X$, $\alpha \in \Phi$ | 2.1 |
| $X^+$ | dominant weights | 2.1 |
| $W$ | Weyl group | 2.1 |
| $w_\circ$ | longest element of $W$ | 2.1 |
| $\mathbb{Z}[X]$ | group ring of $X$ | 2.1 |
| $e(\lambda)$ | basis element of additive group $\mathbb{Z}[X]$ | 2.1 |
| $\mathcal{X}$ | ring of invariants $\mathbb{Z}[X]^W$ | 2.1 |
| $\operatorname{ch} M$ | formal character of $\mathbf{G}$-module $M$ | 2.1 |
| $L(\lambda)$ | simple $\mathbf{G}$-module, $\lambda \in X^+$ | 2.2 |
| $v^+$ | maximal vector | 2.2 |
| $\lambda^*$ | $-w_\circ \lambda$, for $\lambda \in X^+$ | 2.2 |
| $H^0(\lambda)$ | induced module | 2.3 |
| $\mathfrak{g}$ | Lie algebra of $\mathbf{G}$ | 2.5 |
| $\mathbf{G}_r$ | $r$th Frobenius kernel | 2.5 |
| $X_r$ | $\{\lambda \in X \mid 0 \leq \langle \lambda, \alpha^\vee \rangle < p^r \text{ for all } \alpha \in \Delta\}$ | 2.5 |

| | | |
|---|---|---|
| $\widetilde{\nu}$ | $\pi(\nu)$, for $\nu \in X$ | 2.6 |
| $M^{[r]}$ | $r$th Frobenius twist of **G**-module $M$ | 2.7 |
| $V(\lambda)$ | Weyl module | 3.1 |
| $\chi(\lambda)$ | ch $V(\lambda)$ | 3.1 |
| $\chi_p(\lambda)$ | ch $L(\lambda)$ | 3.1 |
| $\rho$ | sum of fundamental dominant weights | 3.3 |
| $W_p$ | affine Weyl group relative to $p$ | 3.4 |
| $w \cdot x$ | dot action $w(x + \rho) - \rho$ of $W$ or $W_p$ | 3.5 |
| $\alpha_0$ | highest short root in $\Phi$ | 3.5 |
| $C_\circ$ | lowest $p$-alcove | 3.5 |
| $h$ | Coxeter number of $W$ | 3.5 |
| $\mathrm{St}_r$ | Steinberg module $L((p^r - 1)\rho)$ | 3.7 |
| $G_{\mathrm{reg}}$ | set of $p$-regular elements in $G$ | 5.5 |
| $R_K(G)$ | Grothendieck group of $KG$-modules | 5.5 |
| $U_r(\lambda)$ | PIM for $KG$, $\lambda \in X_r$ | 9.2 |
| $L_r(\lambda)$ | simple **G**$_r$-module | 10.1 |
| $Z_r(\lambda)$ | (co)induced **G**$_r$-module | 10.1 |
| $Q_r(\lambda)$ | PIM for **G**$_r$ | 10.1 |
| $\widetilde{W}_p$ | extended affine Weyl group | 10.1 |
| $\widehat{L}_r(\lambda)$ | simple **G**$_r$**T**-module | 10.2 |
| $\widehat{Z}_r(\lambda)$ | induced **G**$_r$**T**-module | 10.2 |
| $\widehat{Q}_r(\lambda)$ | projective **G**$_r$**T**-module | 10.2 |

Printed in the United States
By Bookmasters